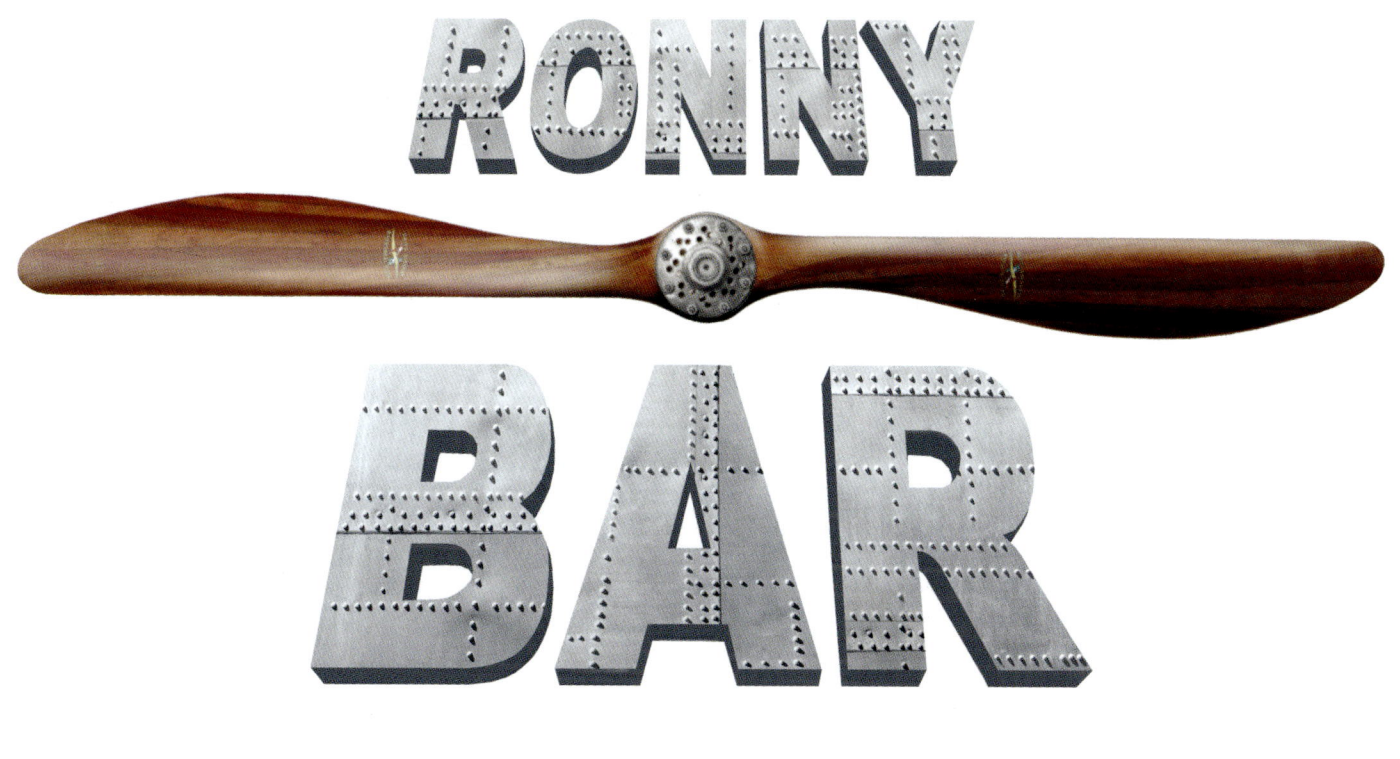

Ronny Bar
Profiles

German Fighters
of the Great War
Volume I

Dedicated to:
Layla and Mónica, my daughters.

CONTENTS

Foreword	7
Preface	9
Acknowledgements	11
Fokker E-I	13
Fokker E-II	23
Fokker E-III	31
Fokker E-IV	45
Pfalz E-I	59
Pfalz E-II	63
Pfalz E-IV	67
Fokker D-I	71
Fokker D-II	75
Fokker D-III	81
Halberstadt D-II	85
Halberstadt D-II (Han) & (Av)	89
Halberstadt D-III	95
Halberstadt D-V	99
Albatros D-I	103
Albatros D-II	113
Albatros D-II (L.V.G.)	127
Albatros D-II (O.A.W.)	133
Albatros D-III	137
Albatros D-III (O.A.W.)	165
Roland D-II & D-IIa	183
Roland D-II & D-IIa (Pfal)	191
Siemens-Schuckert D-I	193
Albatros D-V	197

FOREWORD

World War I aviation history enthusiasts around the world recognise Ronny Bar as a very fine visual artist. His meticulously researched and detailed aircraft colour profiles seem 'real' enough to take off at any moment. Ronny's artworks have long graced the pages of international journals and magazines, as well as book dust jackets and other media.

His previous book, devoted to World War I British two-seaters, was a masterful creation that showcased some of his best work.

Now, Ronny's new book, featuring a fine array of German fighter aircraft from 1915 through 1917, is another tour de force. Indeed, viewing this large-format volume of carefully done work is like visiting a fine art exhibition, where each creation can be studied and enjoyed; it also gives the viewer the impression of lurking about several German airfields well stocked with often colourful – yet fully-armed – combat aircraft. The final reality is that the airplanes depicted have more than aesthetic value; they show working machines of war, some as fresh as the day they arrived from the manufacturers, others bearing oil stains and other signs of war earned during aerial combat.

This collection features scrupulously done profiles of Fokker and Pfalz monoplanes, as well as Fokker, Halberstadt, Albatros, Roland and Siemens-Schuckert biplanes. They represent aircraft flown by famous highly-decorated aces and men whose achievements were more modest, but every bit as dedicated as the famous air heroes.

This book remembers a broad span of air combatants during the early stages of World War I.

The companion volume covering the remainder of that conflict is much anticipated.

Peter Kilduff

PREFACE

Arguably the Germans invented the single seat fighter as we know it today… Despite numerous attempts by other countries to outfit an airplane with a forward-firing machine gun, including Roland Garros fitting his Morane-Saulnier Parasol with steel plates on the propeller blades, it was when Anthony Fokker introduced his famous *Eindecker* with a practical synchronization system that would allow him to fire a machine gun on the line of flight between the propeller rotating blades that the true fighter airplane was born, dedicated exclusively to fighting enemy machines in the air.

Throughout most of World War I, German fighter pilots fought against a numerically superior enemy, and yet, thanks to the technical superiority of many of their aircraft, to clever tactics, an immense courage, and an *esprit de corps* that bonded them in a band of brothers, they fought their enemy until the very end of the war. In the aftermath, many of them would bitterly but proudly declare that they did not lose *their* war.

Probably, due to a government-backed advertising campaign, with constant newspaper articles chronicling their exploits, collectible cards with their portraits (the famous Sanke Cards), and promotion and awards granted for their aerial victories, plus the flamboyant and colourful finishes of their mounts, we could say today that German fighter pilots were the *rock star*s of the Great War!

After the kind reception of my first book, *British Two-Seaters of the Great War*, many readers and enthusiasts kept asking me for a book dedicated to German fighters, and I must confess that, even before I've been asked to, I couldn't wait to get started on this project!

I decided not to elaborate much on a text, since everything I can say about these aircraft has already been written by authors much more authorised than me, and since my forte is the profile artwork, and that's what the readers will be looking for when buying this book, I had come to the conclusion that what I should do is focus on my aircraft profiles, and try to display them in a size that they can be observed to the smallest details, with a few captions that provide the basic data and some comments when I deemed necessary.

As I began to gather material, I realised that due to the large number of potential profiles I had at my disposal, I would need to split the work into two volumes. Volume I covers most of the German-built single-seat fighter aircraft that saw action in the Great War, from the first Fokker *Eindeckern* of the summer of 1915 to the Albatros D-V of the summer of 1917.

All the profiles are illustrated in a constant style, so that it is possible to see the comparative size between each type of aircraft, but the top and bottom views are in a slightly smaller scale, for obvious space reasons, but still keeping a constant scale between them.

Each subject was based on photographic evidence and thoroughly researched, and although in some cases I have had no choice but to resort to a somehow educated guess, I tried to be as accurate as is possible to the facts of what happened more than 100 years ago.

Some of these profiles have been published previously by Albatros Publications, Wingnut Wings etc. But most were done especially for this book.

Ronny Bar

ACKNOWLEDGEMENTS

I'll be forever grateful...

To Peter Kilduff: For honouring me by writing the Foreword to this work, and for choosing me to illustrate several of his wonderful books.

To Greg Van Wynngarden: For taken the time to review each profile in this book and advising and helping me make this work as accurate as possible.

To Josef Scott: For generously sharing with me his expertise on the Fokker Eindeckern.

To Sir Peter Jackson: Because working with him on his amazing Wingnut Wings project for more than 15 years was an unforgettable experience which also boosted my career in an incredible way.

To Richard Alexander, for the support and patience during all the time we worked together in the already legendary Wingnut Wings and currently in the new Kotare Models, and for the friendship we built during my times in New Zealand.

To Ray Rimell: For his enduring and generous support throughout all these years publishing my work in the iconic Windsock Datafiles.

To Ray again, and also to my good friend Colin Huston from Cross & Cockade, Jack Herris from Aeronaut Books, Aaron Weaver from Over The Front, Ruth Sheppard from Casemate Publishers, Michael Spilling from Amber Books Ltd, Nigel Dingley from Air Britain Ltd, Steve O'Hara from Mortons Books, and to many other publishers: For publishing my work and making it known by thousands of readers and enthusiasts all around the world.

To authors and writers, Greg VanWynngarden again, to my friend Colin Owers, Peter Kilduff, Alan Toelle, Josef Scott, Jim Wilberg, Bruno Schmaelling, Gregory Alegi, Paelo Varriale, Mark Wilkins, Paul Hare, JS Alcorn, Xavier Chevallier, Phillip Jarret, Edward Ward: For trusting me to illustrate their magnificent books and articles and for generously sharing their huge knowledge, expertise and advice to help me improve my work.

To my colleague artists: Juanita Franzi, Russell Smith, Robert Karr, Bob Pearson, Jim Miller, Dave Douglas, Steve Anderson, Mark Miller, Tomasz Gronczewski, Piotr Mrozoswski, Jerry Boucher: For their exquisite and always inspiring artwork, and for the good vibes I've always received from them.

To Eugene Ushakow, who whenever I asked provided me with good information.

To my good old friends: Claudio Meunier, Diego Fernetti and Hector Martin Afflitto Echagile, for encouraging and supporting me to enter this fascinating world of professional aircraft profiling.

Ronny Bar

FOKKER E.I

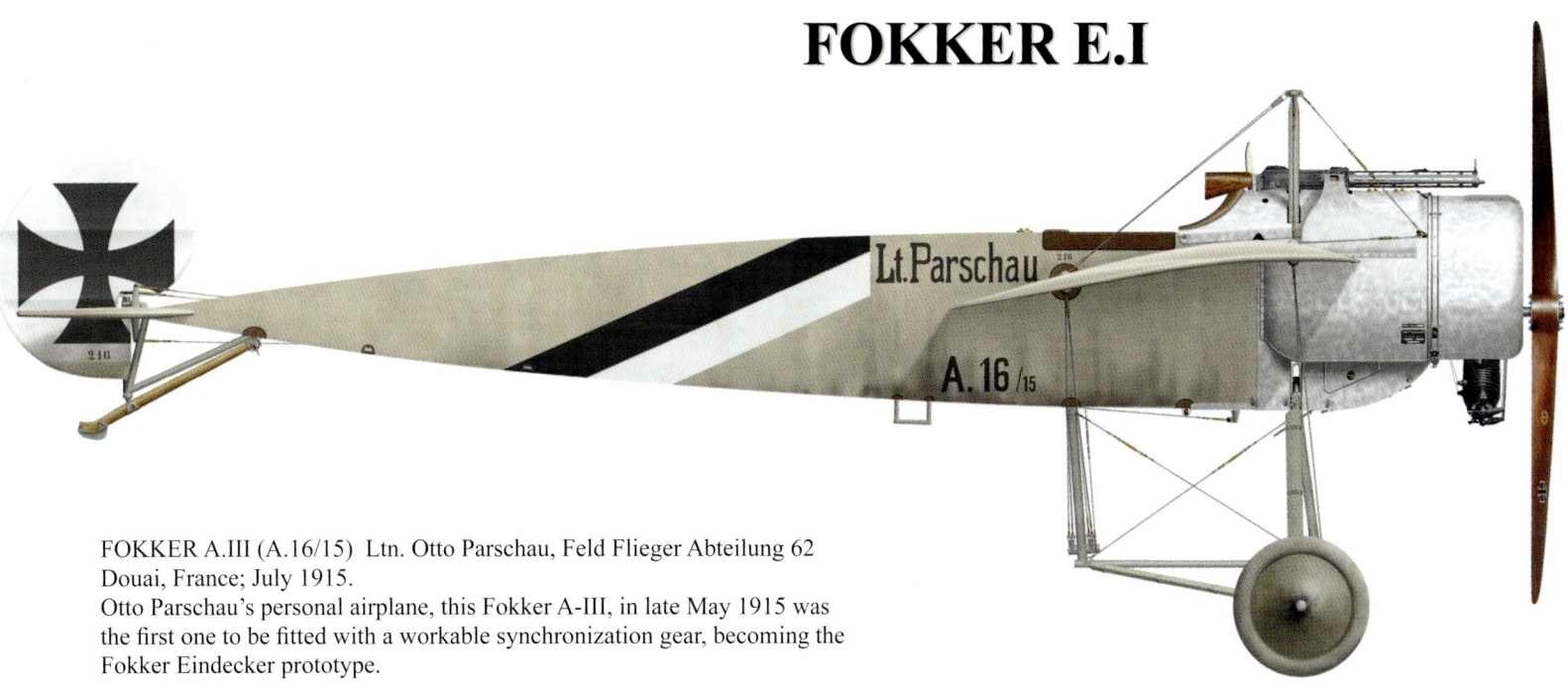

FOKKER A.III (A.16/15) Ltn. Otto Parschau, Feld Flieger Abteilung 62
Douai, France; July 1915.
Otto Parschau's personal airplane, this Fokker A-III, in late May 1915 was
the first one to be fitted with a workable synchronization gear, becoming the
Fokker Eindecker prototype.

FOKKER E.I (E.5/15) Ltn. Kurt Wintgens, Feld Flieger Abteilung 6b
Bühl-Saarburg, Germany; August 1915.
Kurt Wintgens claimed the first ever victory of a Fokker Eindecker with
a synchronized machine gun on this airplane on July 1, 1915.

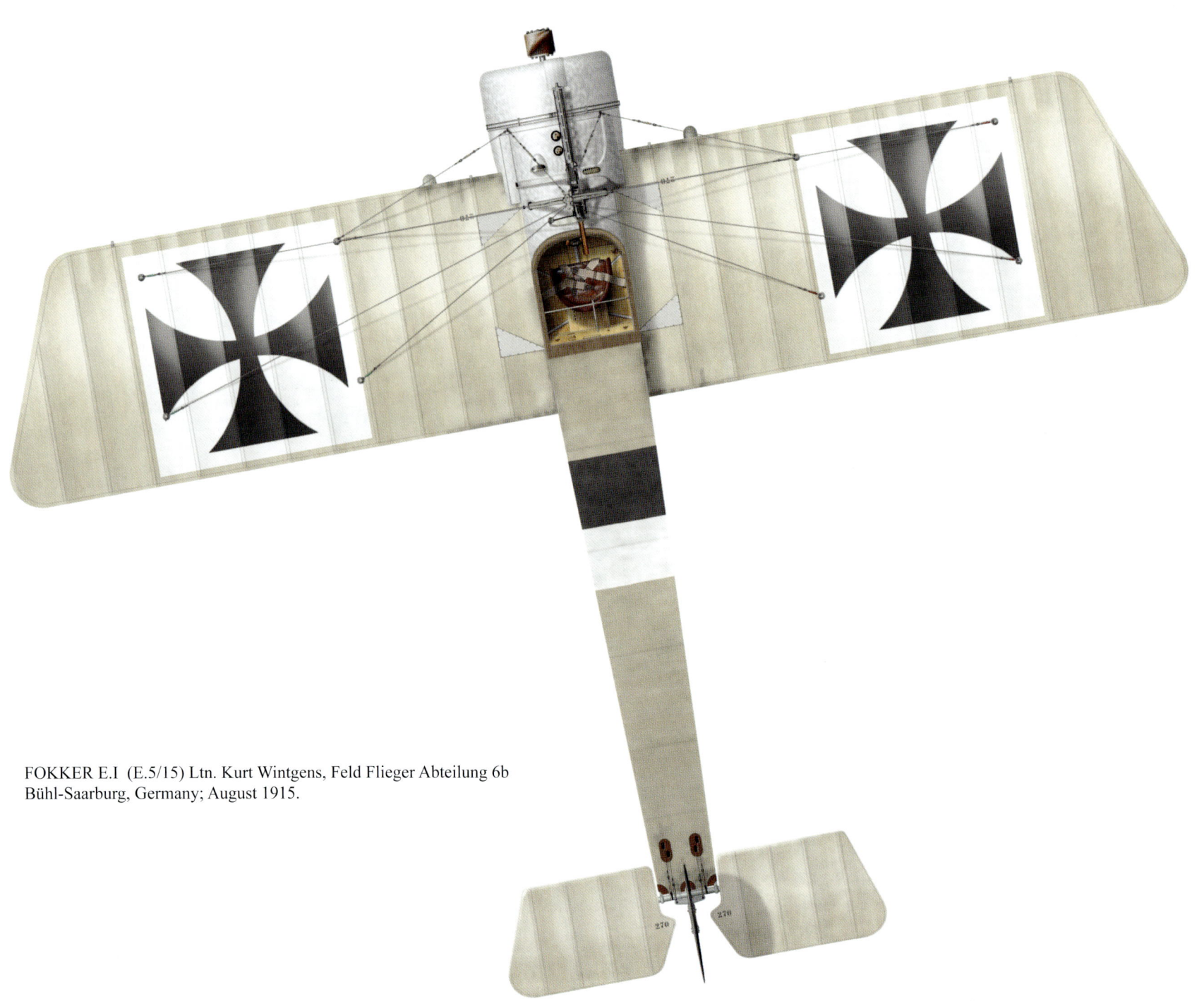

FOKKER E.I (E.5/15) Ltn. Kurt Wintgens, Feld Flieger Abteilung 6b
Bühl-Saarburg, Germany; August 1915.

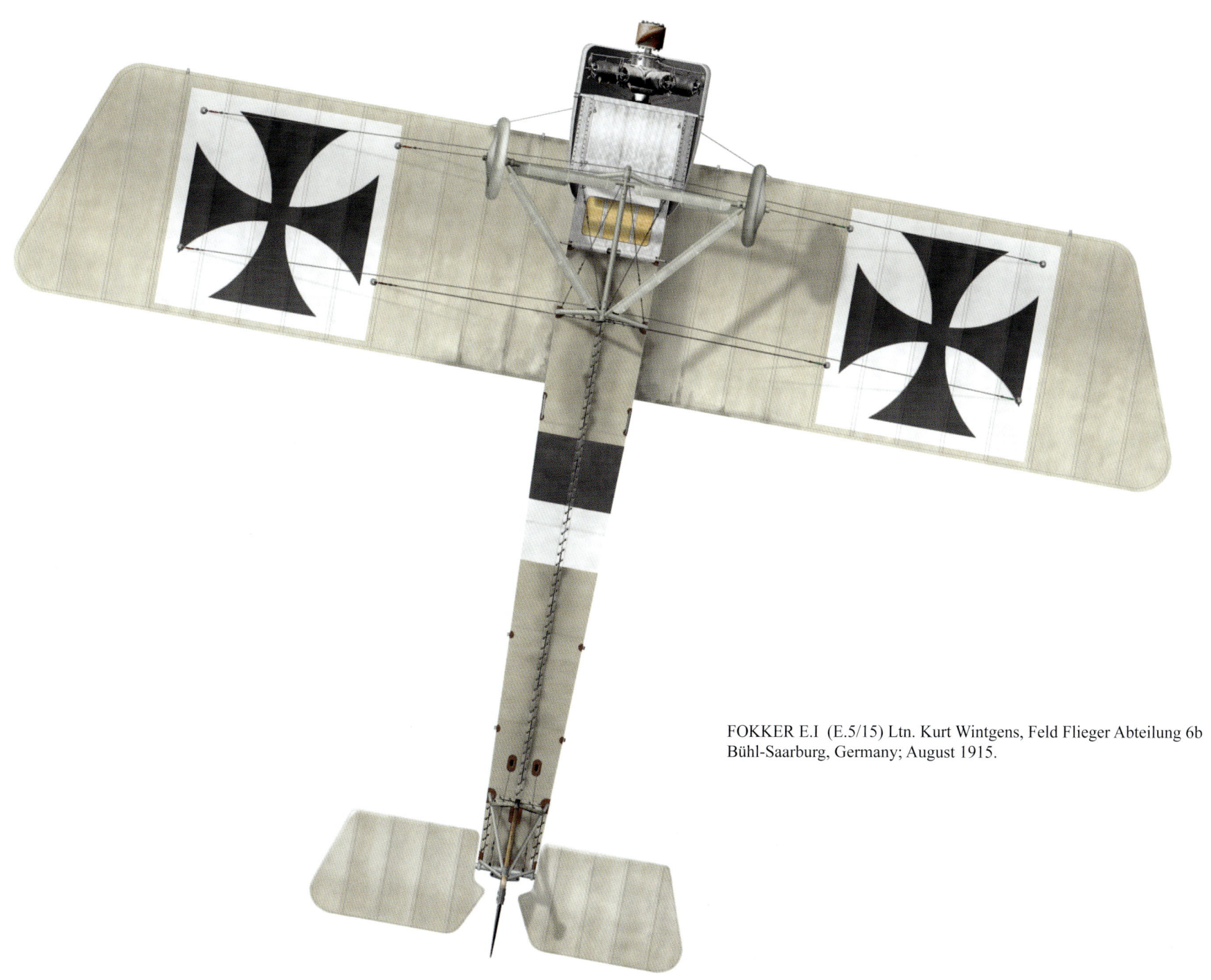

FOKKER E.I (E.5/15) Ltn. Kurt Wintgens, Feld Flieger Abteilung 6b
Bühl-Saarburg, Germany; August 1915.

FOKKER E-1 (6/15) Uffz. Richard Dietrich, Feld Flieger Abteilung 24
Lille, France; August 1915.

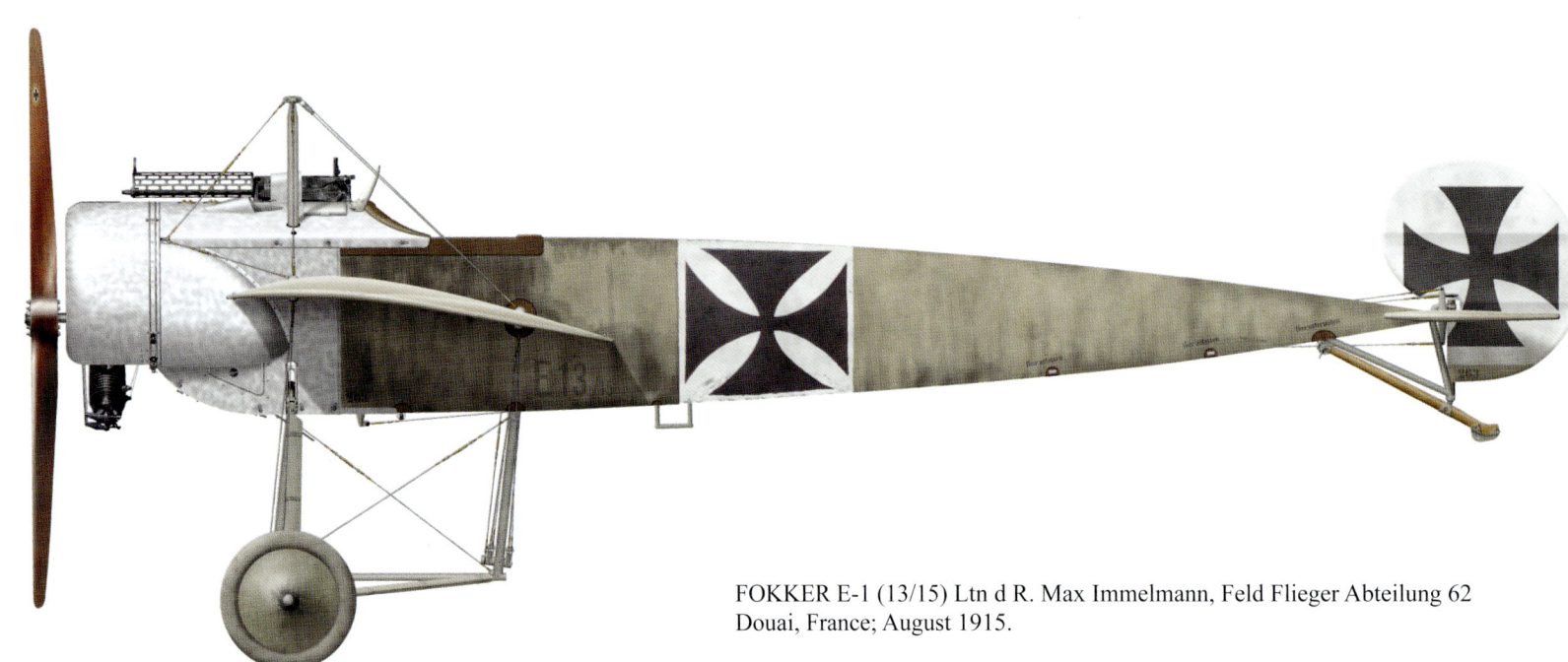

FOKKER E-1 (13/15) Ltn d R. Max Immelmann, Feld Flieger Abteilung 62
Douai, France; August 1915.

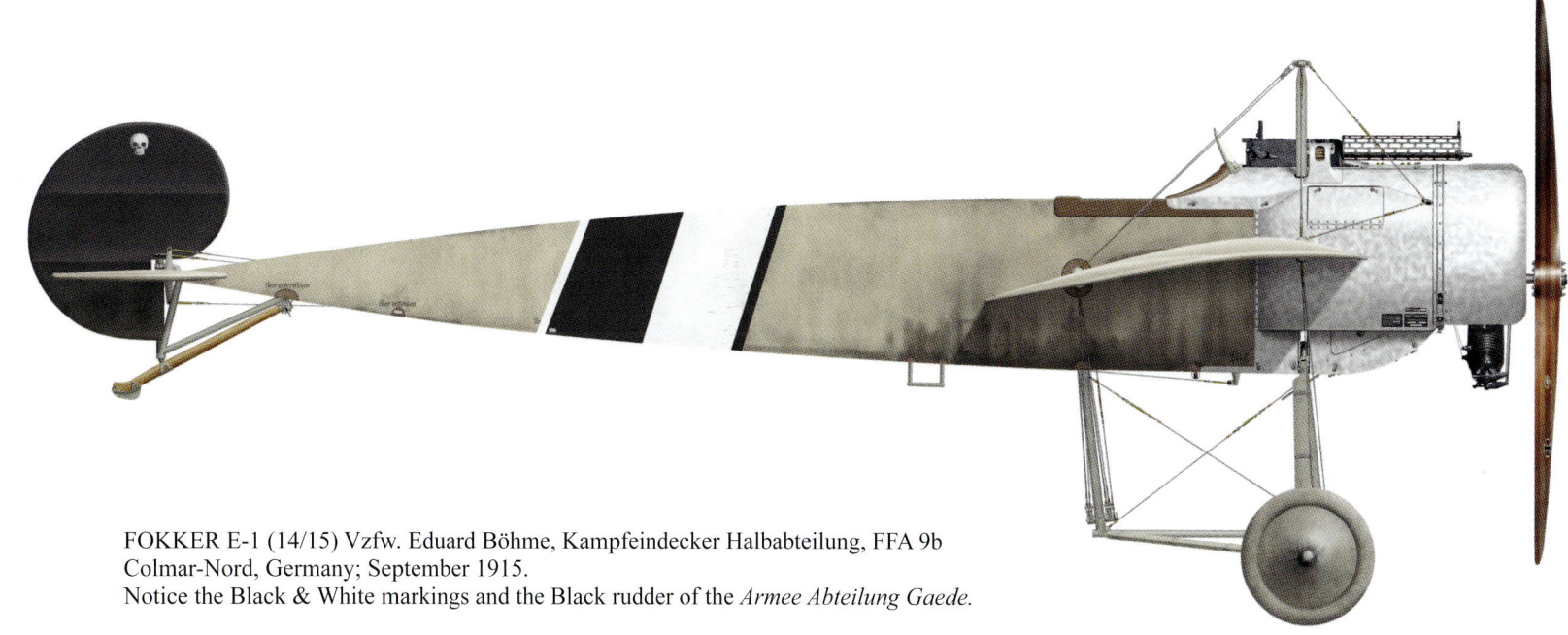

FOKKER E-1 (14/15) Vzfw. Eduard Böhme, Kampfeindecker Halbabteilung, FFA 9b
Colmar-Nord, Germany; September 1915.
Notice the Black & White markings and the Black rudder of the *Armee Abteilung Gaede*.

FOKKER E-1 (15/15) Ltn. Hans-Joachim Buddecke, Feldflieger Abteilung 23
Roupy, France; September 1915.

FOKKER E-I (43/15) Offz Stv. Willy Rosenstein, Feld Flieger Abteilung 19
Porcher, Germany; September 1915.

FOKKER E-I (29/15) Ltn. Walter von Bülow-Bothkamp, Feldflieger Abteilung 22
St Vith, Belgium; October 1915.
Photos suggest that there was a light and a dark stripe in front of the fuselage Cross.
Being the von Bülow family colours Dark Blue and Yellow, speculatively those
colours were used for this illustration.

FOKKER E-I (54/15) Vzfw. Ernst Udet, Kampfeinsitzer Kommando Habsheim
Habsheim, Germany; January 1916.
Note the *Armee Abteilung Gaede* Black & White markings were overpainted with
Field Grey paint, leaving a White field for the *Eiserne Kreuze*.

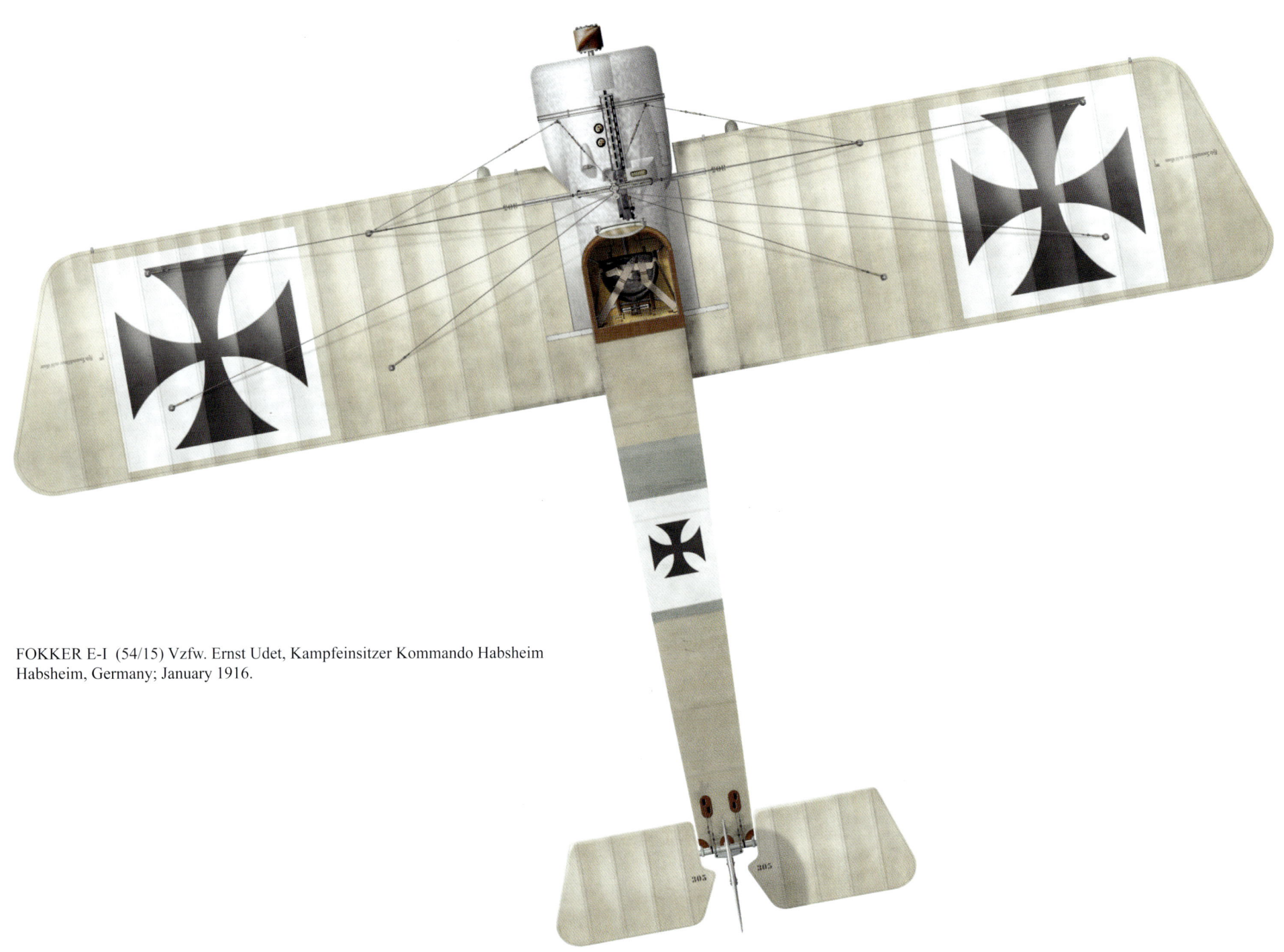

FOKKER E-I (54/15) Vzfw. Ernst Udet, Kampfeinsitzer Kommando Habsheim
Habsheim, Germany; January 1916.

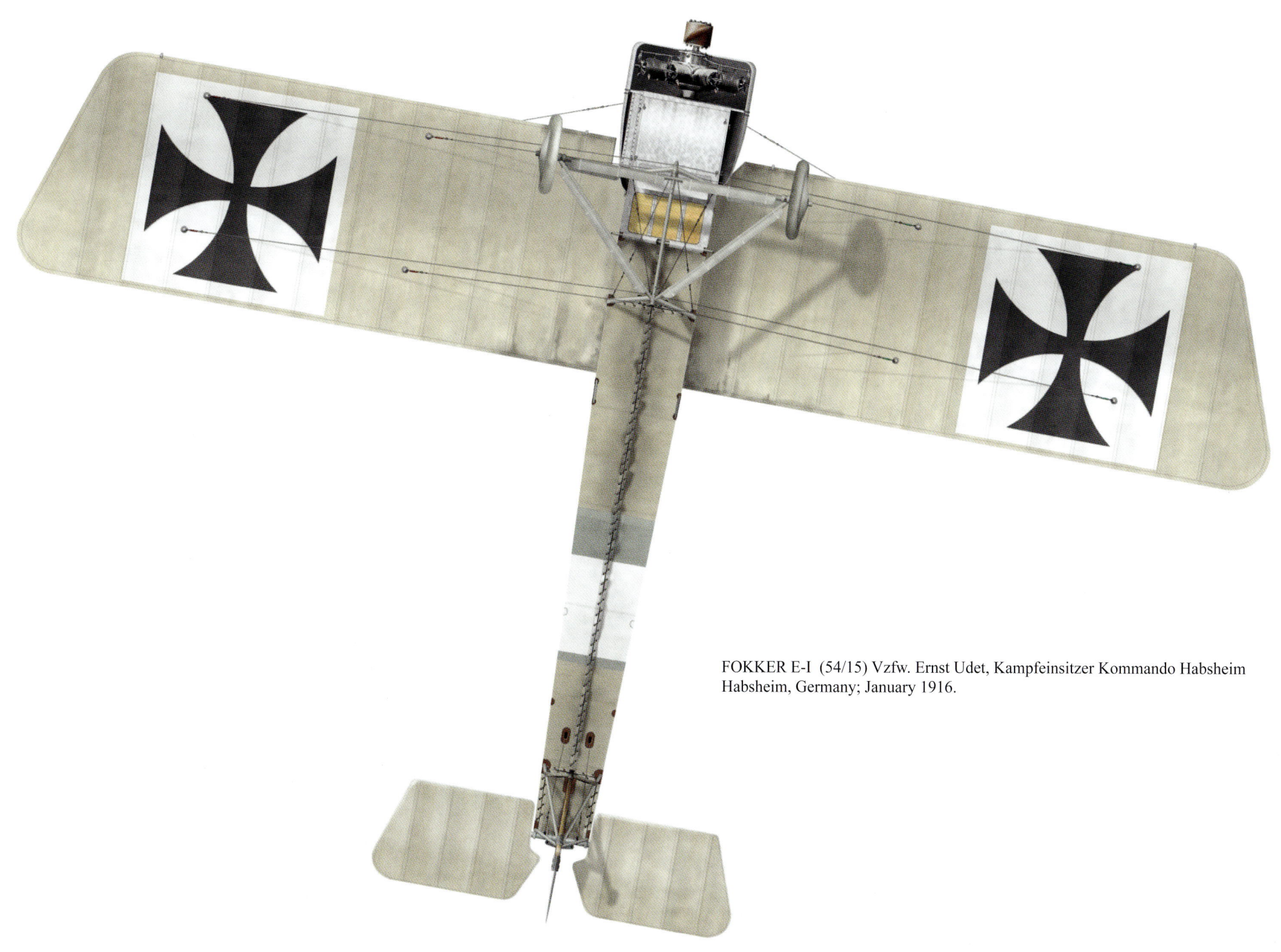

FOKKER E-I (54/15) Vzfw. Ernst Udet, Kampfeinsitzer Kommando Habsheim
Habsheim, Germany; January 1916.

FOKKER E-I (45/15) Feld Flieger Abteilung 42
Metz-Frescaty, Germany; Early 1916.
Note Teddy Bear attached to the top cabane struts
and compass fixed outside the cockpit.

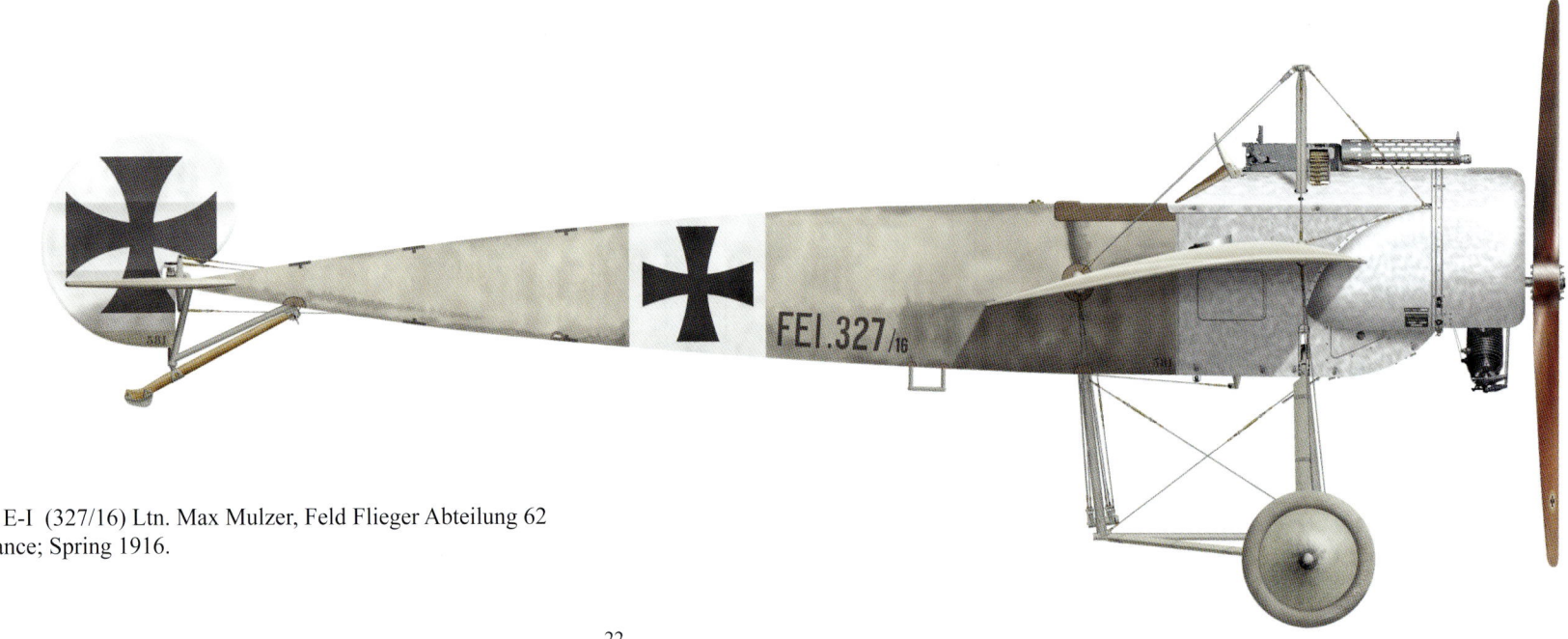

FOKKER E-I (327/16) Ltn. Max Mulzer, Feld Flieger Abteilung 62
Douai, France; Spring 1916.

FOKKER E-II

FOKKER E-II (7/15) Feldflieger Abteilung 62
Douai, France; June 1915.
This machine was flown by Tony Fokker as a
demonstration aircraft for the frontline units.

FOKKER E-II (69/15) Ltn. Kurt von Crailsheim, Feldflieger Abteilung 53
Monthois, France; September 1915.

FOKKER E-II (serial unknown) Ltn. Kurt von Crailsheim, Feldflieger Abteilung 53
Monthois, France; October 1915.

FOKKER E-II (36/15) Obit. Ernst von Althaus, Kampfeinsitzer Kommando Vaux
Vaux, France; October 1915.

FOKKER E-II (37/15) Ltn. Max Immelmann, Feldflieger Abteilung 62
Douai, France; October 1915.

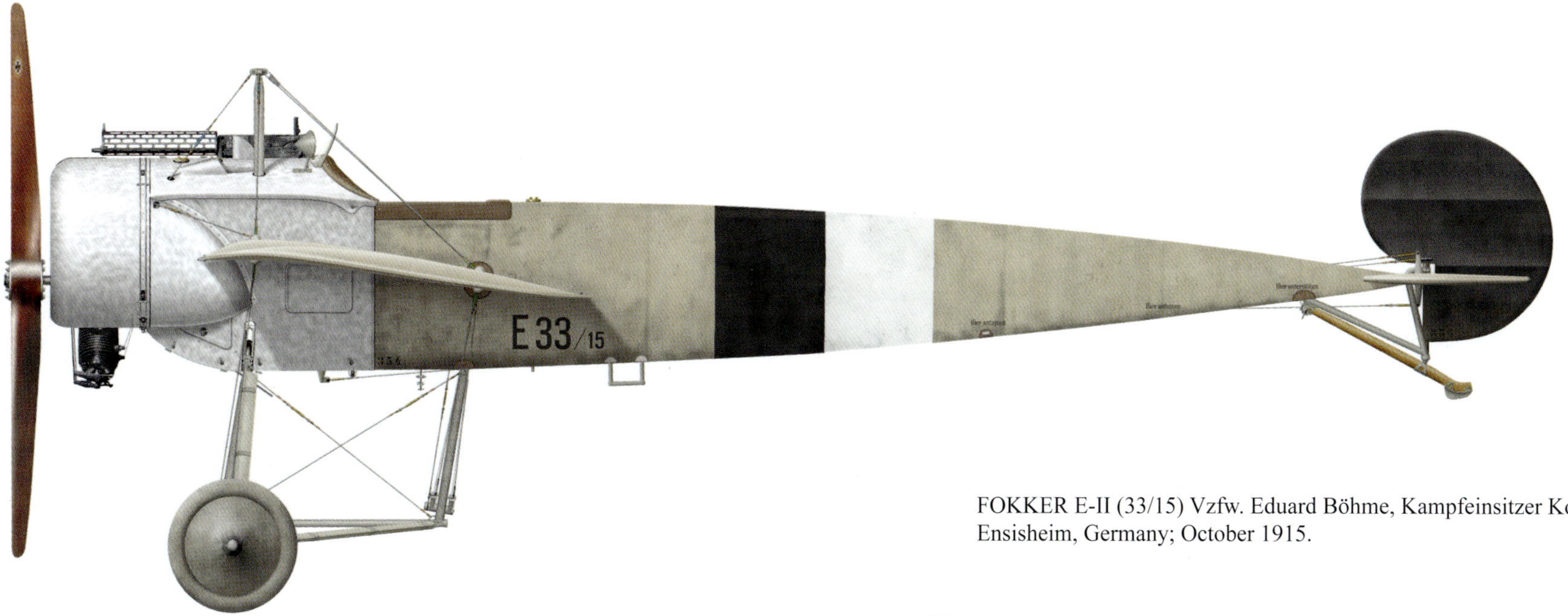

FOKKER E-II (33/15) Vzfw. Eduard Böhme, Kampfeinsitzer Kommando Ensisheim
Ensisheim, Germany; October 1915.

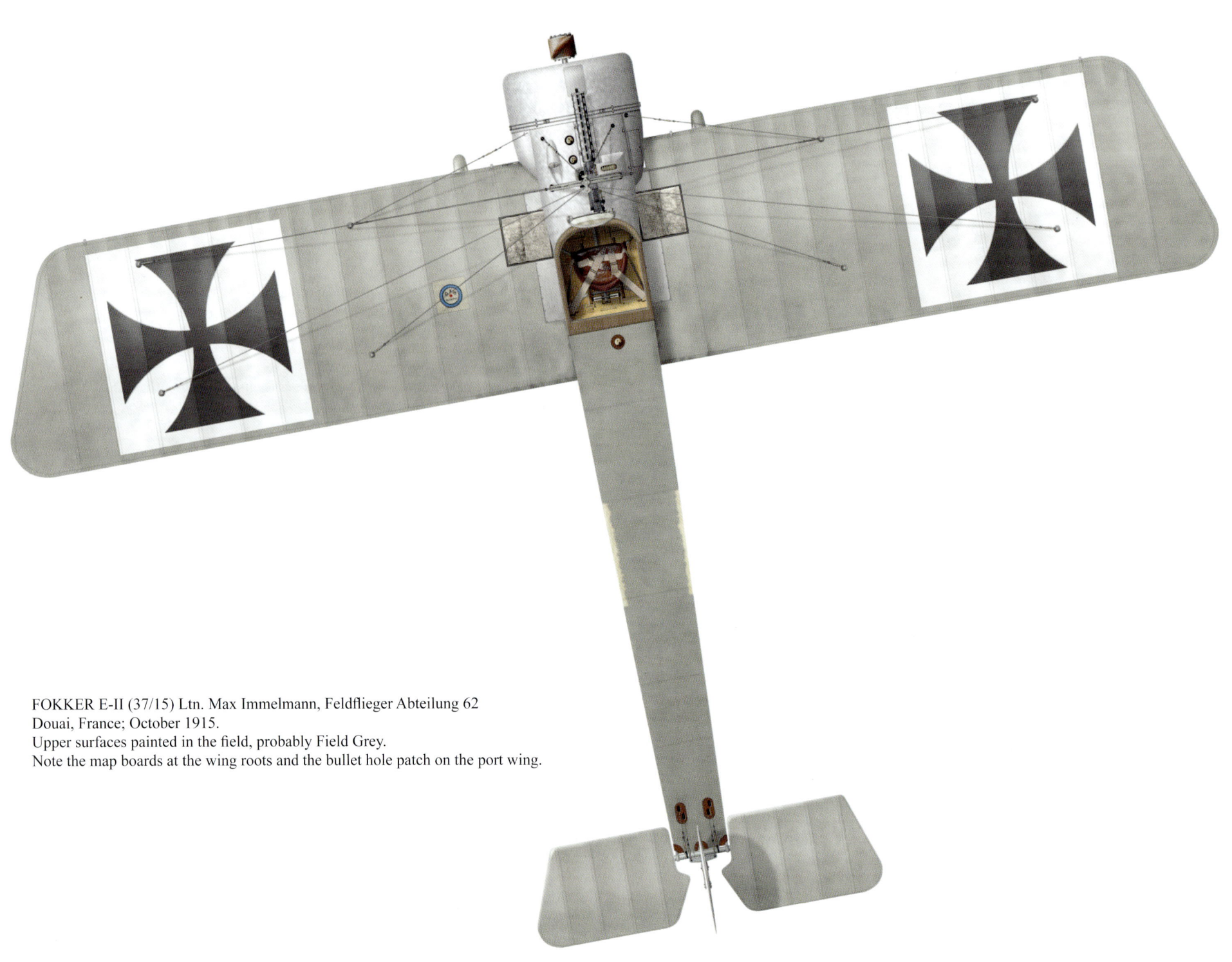

FOKKER E-II (37/15) Ltn. Max Immelmann, Feldflieger Abteilung 62
Douai, France; October 1915.
Upper surfaces painted in the field, probably Field Grey.
Note the map boards at the wing roots and the bullet hole patch on the port wing.

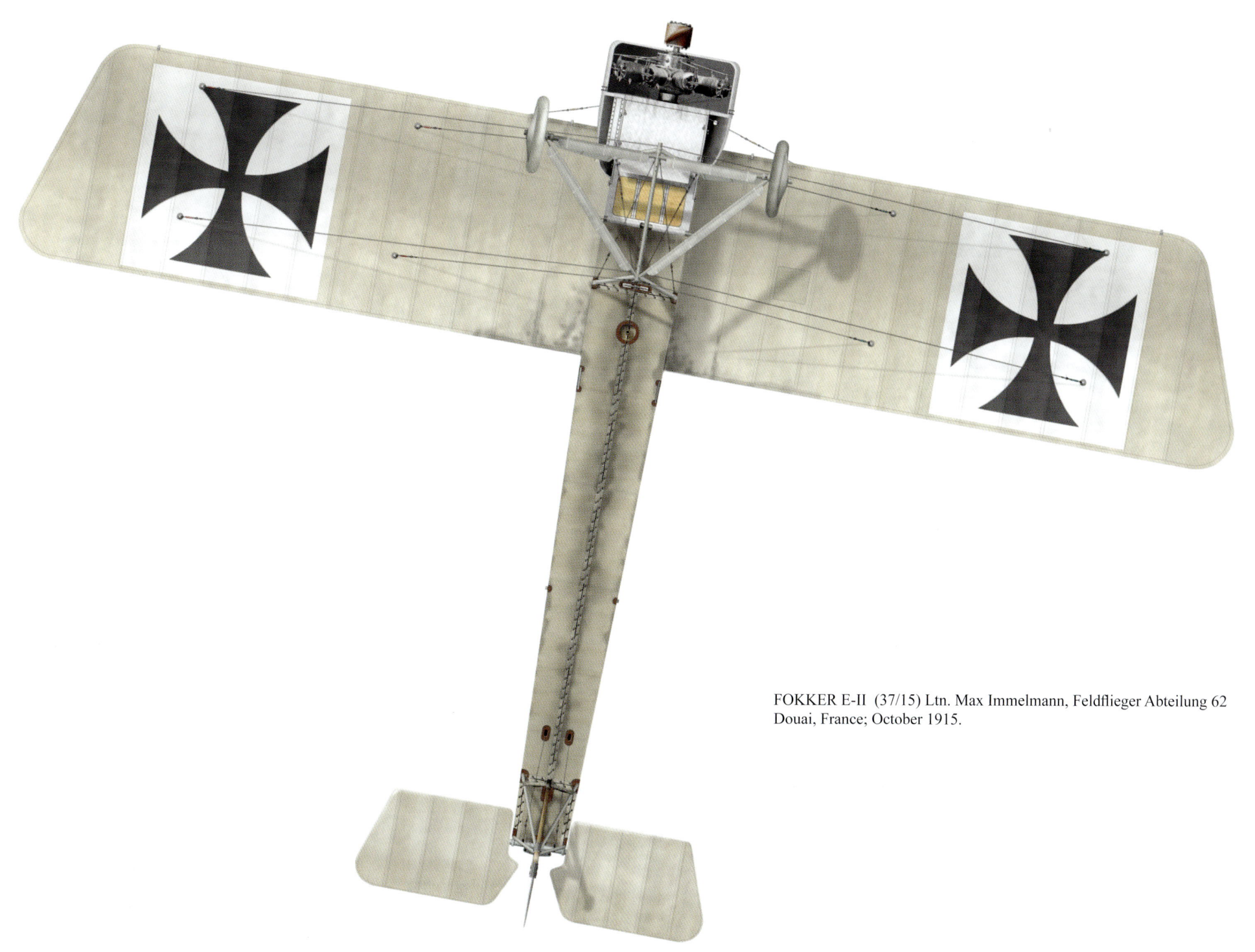

FOKKER E-II (37/15) Ltn. Max Immelmann, Feldflieger Abteilung 62
Douai, France; October 1915.

FOKKER E-II (22/15) Vzfw. Hugo Stöber, Kampfeinsitzer Kommando Ensisheim
Ensisheim, Germany; October 1915.

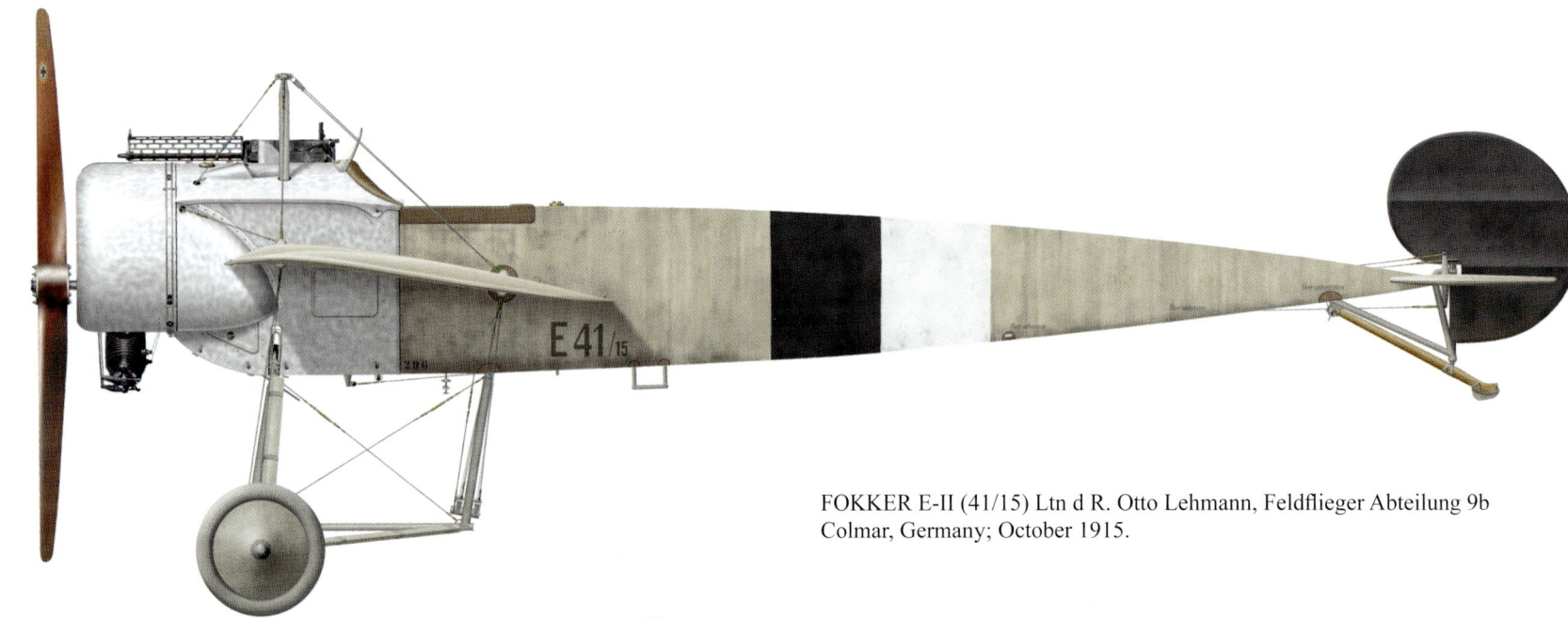

FOKKER E-II (41/15) Ltn d R. Otto Lehmann, Feldflieger Abteilung 9b
Colmar, Germany; October 1915.

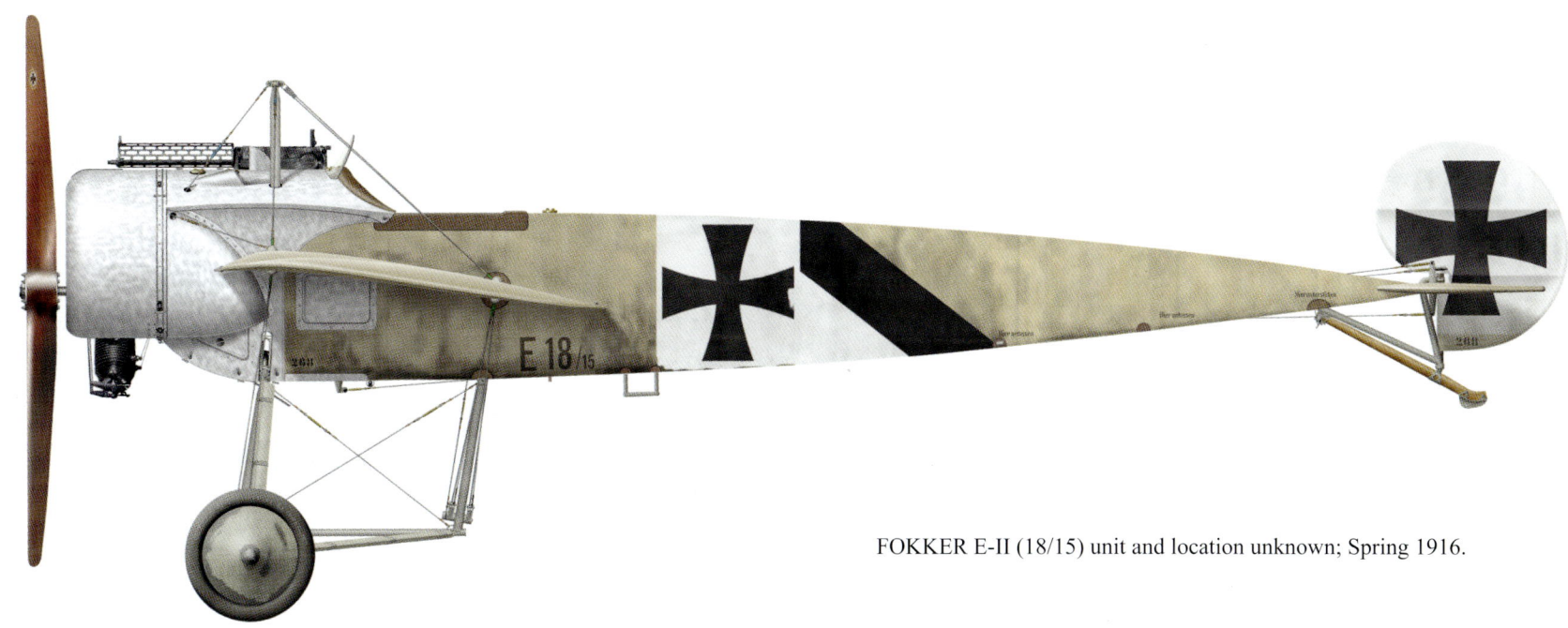

FOKKER E-II (18/15) unit and location unknown; Spring 1916.

FOKKER E-II (20/15) Ltn. Bruno Loerzer, Kampfeinsitzer Kommando Jametz
Jametz, France; March 1916.

FOKKER E-II (90/15) Feldflieger Abteilung 34
Cunel, France; Spring 1916.

FOKKER E-II (89/15) Feldflieger Abteilung 34
Cunel, France; Spring 1916.

FOKKER E-III

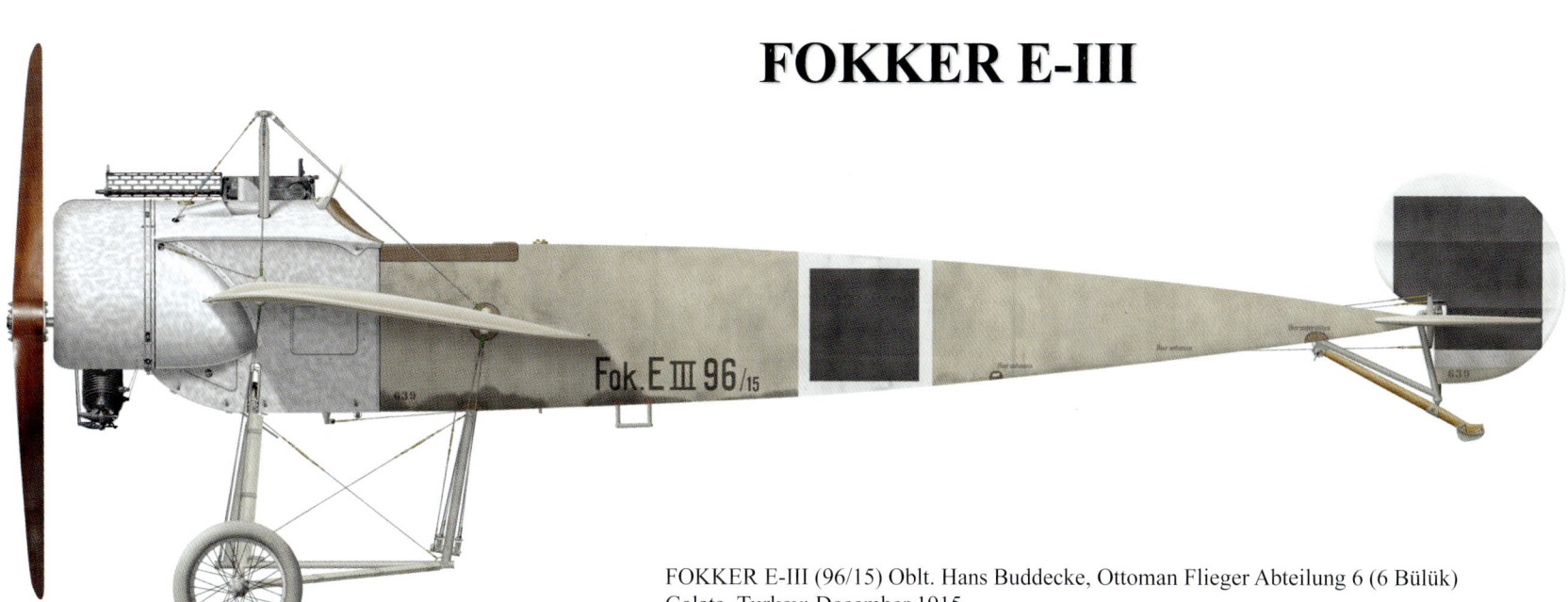

FOKKER E-III (96/15) Oblt. Hans Buddecke, Ottoman Flieger Abteilung 6 (6 Bülük)
Galata, Turkey; December 1915.

FOKKER E-III (105/15) Vzfw. Ernst Udet, Kampfeinsitzer Kommando Habsheim
Habsheim, Germany; December 1915.

FOKKER E-III (105/15) Vzfw. Ernst Udet, Kampfeinsitzer Kommando Habsheim
Habsheim, Germany; Early 1916.
Same machine as the previous view with the Black and White markings overpainted
with Field Gray paint.

FOKKER E-III (105/15) Vzfw. Ernst Udet, Kampfeinsitzer Kommando Habsheim
Habsheim, Germany; March 1916.
Same machine painted overall in Field Gray, probably for camouflage purposes.
Note Luger P-08 pistol with shoulder stock attached to the cockpit side.

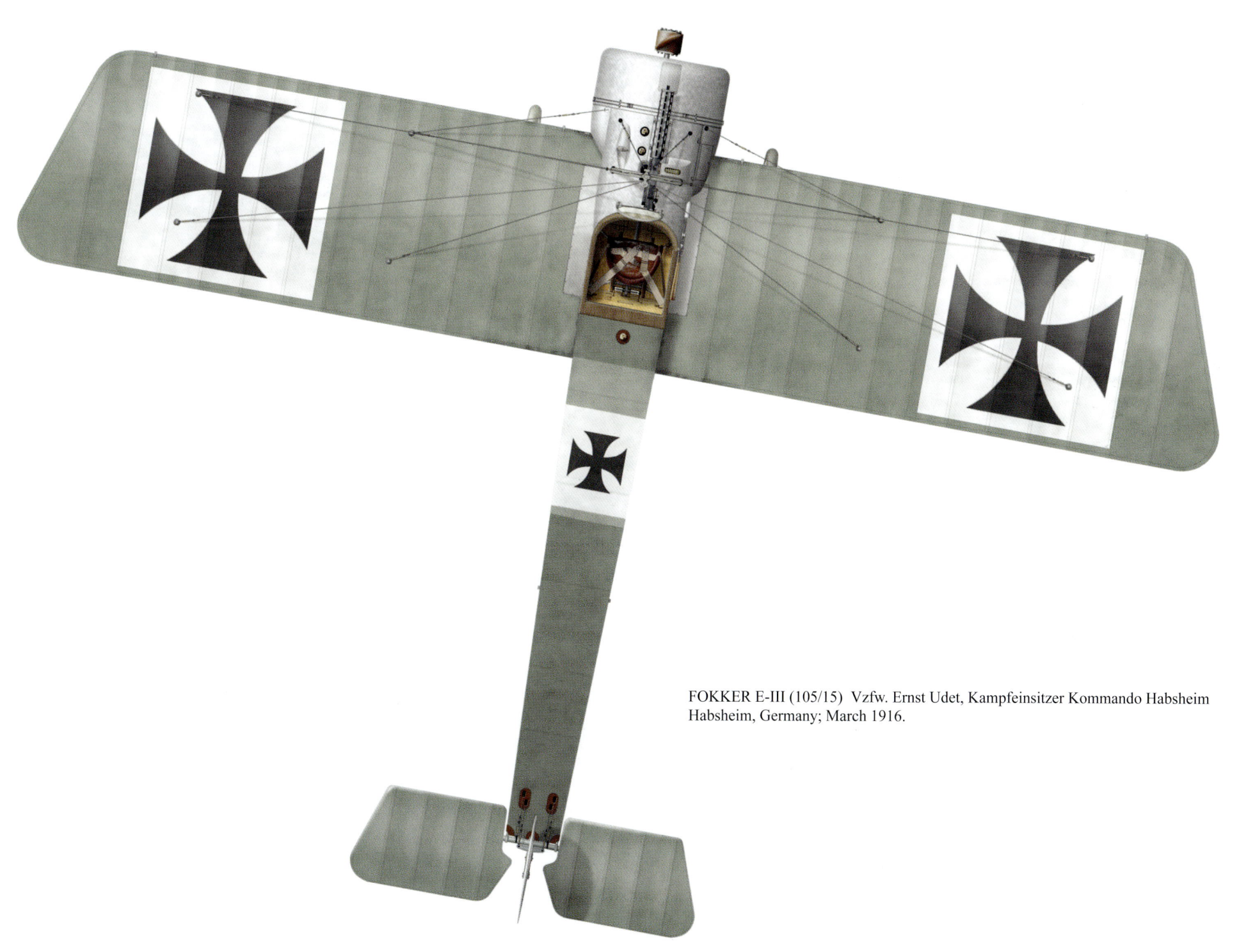

FOKKER E-III (105/15) Vzfw. Ernst Udet, Kampfeinsitzer Kommando Habsheim
Habsheim, Germany; March 1916.

FOKKER E-III (107/15) Fokkerstaffel AOK4
Rumbeke, Belgium; Early 1916.

FOKKER E-III (404/15) Vzfw. Ernst Udet, Kampfeinsitzer Kommando Habsheim
Habsheim, Germany; Spring 1916.
Note anemometer attached over the port wing leading edge.

FOKKER E-III (401/15) Ltn. Helmuth von Zastrow, Kampfeinsitzer Kommando Jametz
Jametz, France; Spring 1916.
Note compass outside the cockpit.

FOKKER E-II1 (422/15) Feldflieger Abteilung 34
Cunel, France; Spring 1916.

FOKKER E-III (406/15) Ltn d R. Hans Gutermuth, Feldflieger Abteilung 44
Metz, Germany; Spring 1916.
Note nonstandard propeller spinner and E-IV style rear top deck.

FOKKER E-III (LFl 96) Ltn z See. Gotthard Sachsemberg, Marine Feldflieger Abteilung I
Mariakerke, Belgium; April 1916.

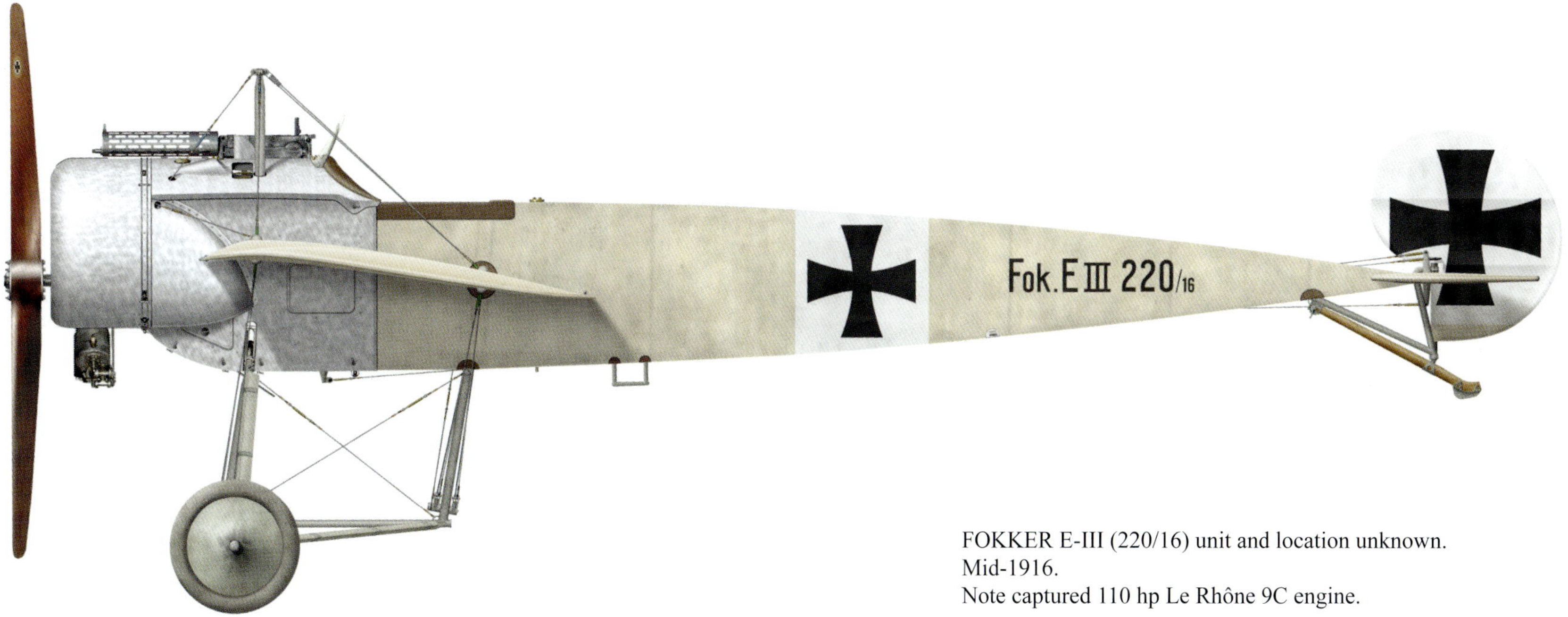

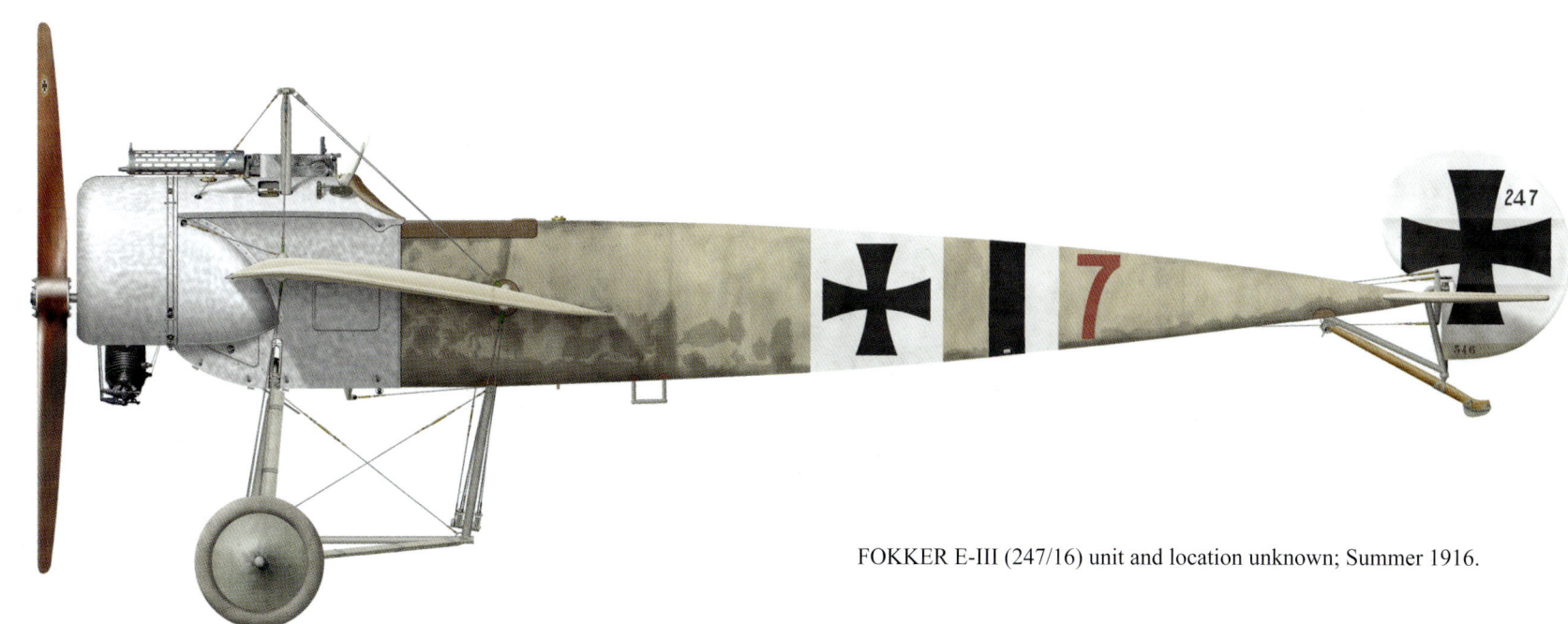

FOKKER E-III (220/16) unit and location unknown.
Mid-1916.
Note captured 110 hp Le Rhône 9C engine.

FOKKER E-III (247/16) unit and location unknown; Summer 1916.

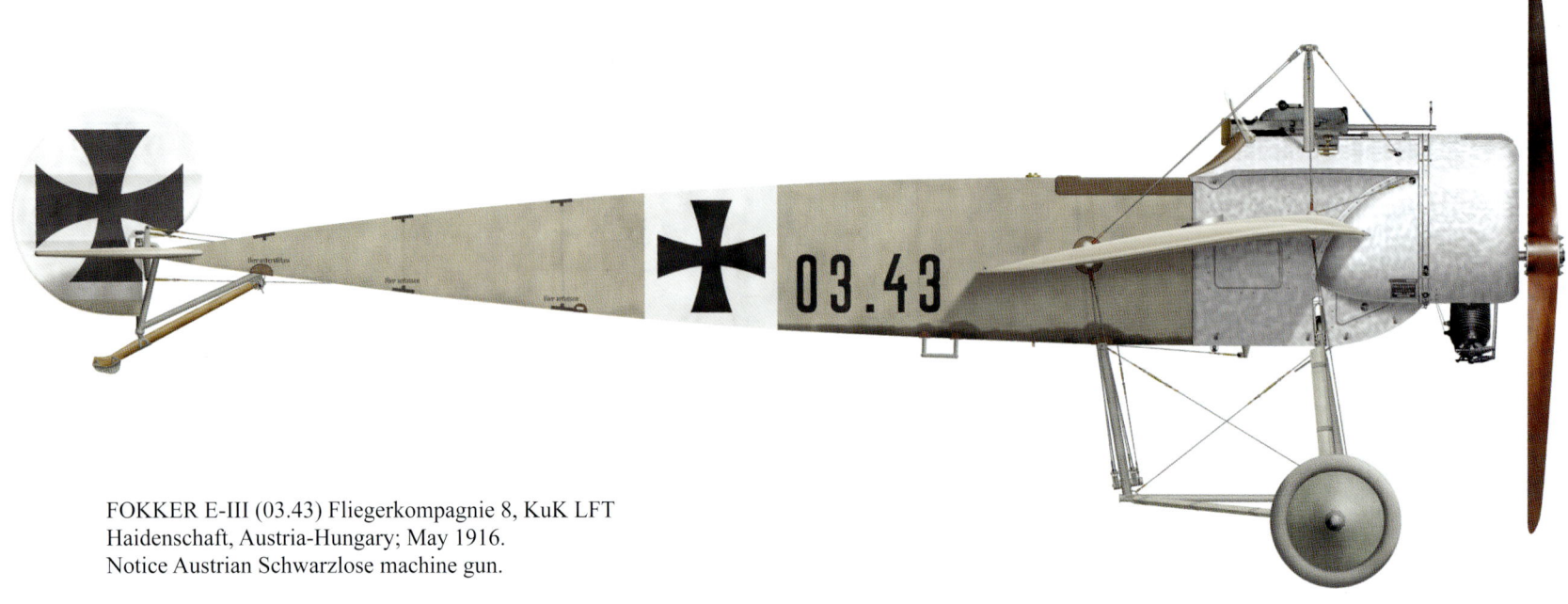

FOKKER E-III (03.43) Fliegerkompagnie 8, KuK LFT
Haidenschaft, Austria-Hungary; May 1916.
Notice Austrian Schwarzlose machine gun.

FOKKER E-III (608/15) Ltn d R. Josef Jacobs, Fokkerstaffel West
Le Faux Ferme, France; May 1916.

FOKKER E-III (246/16) Ltn. Max Immelmann, Kampfeinsitzer Kommando 3
Douai, France; June 1916.
Immelmann was shot down and killed in this machine on June 18, 1916.

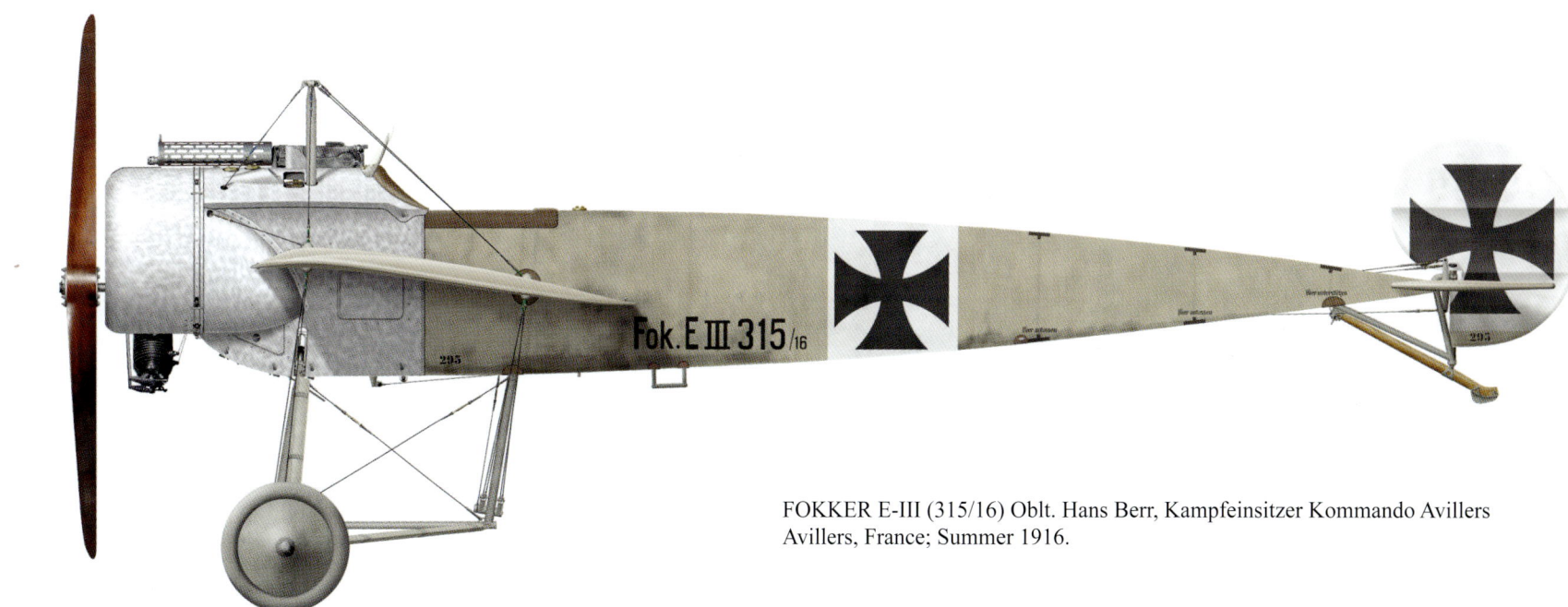

FOKKER E-III (315/16) Oblt. Hans Berr, Kampfeinsitzer Kommando Avillers
Avillers, France; Summer 1916.

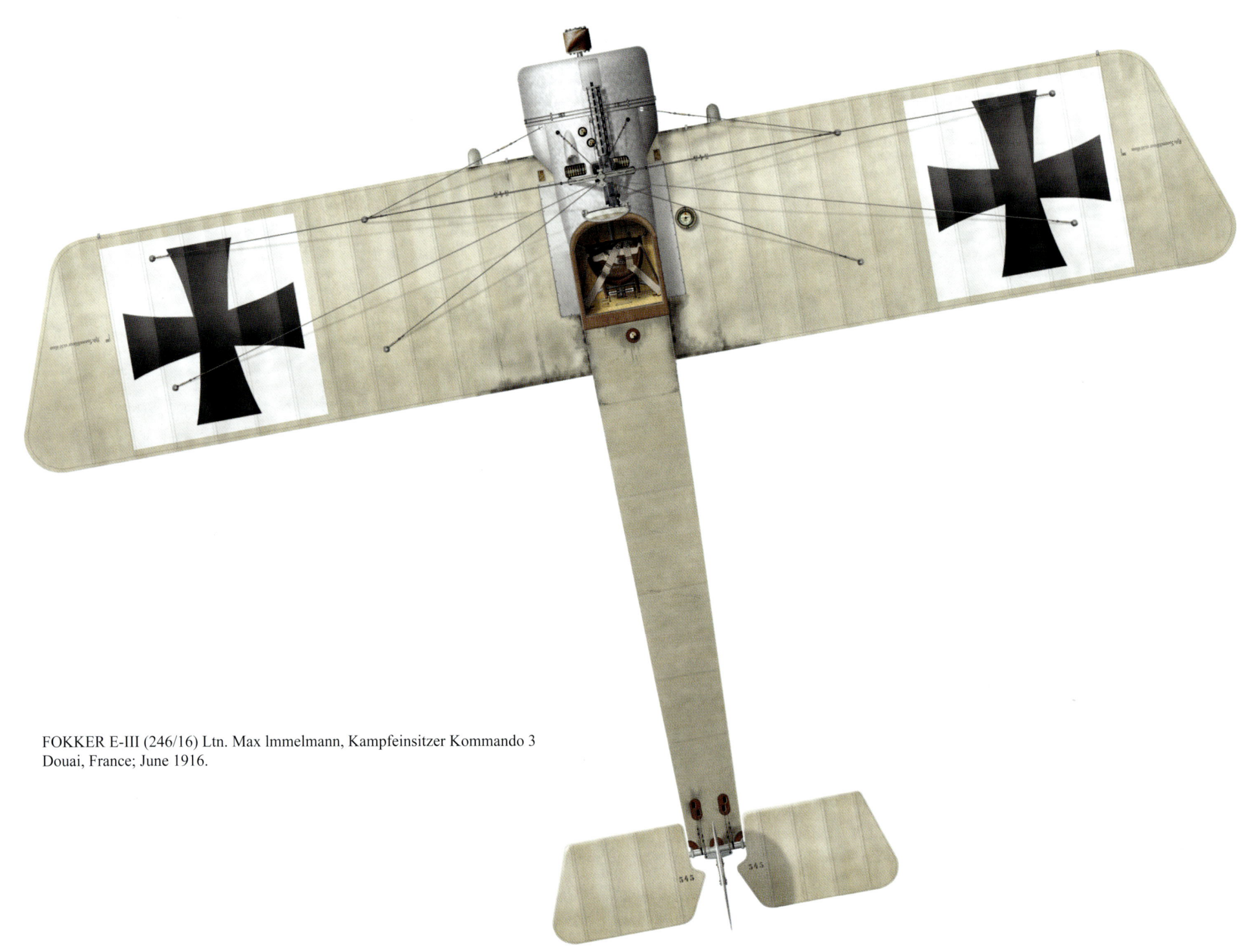

FOKKER E-III (246/16) Ltn. Max Immelmann, Kampfeinsitzer Kommando 3
Douai, France; June 1916.

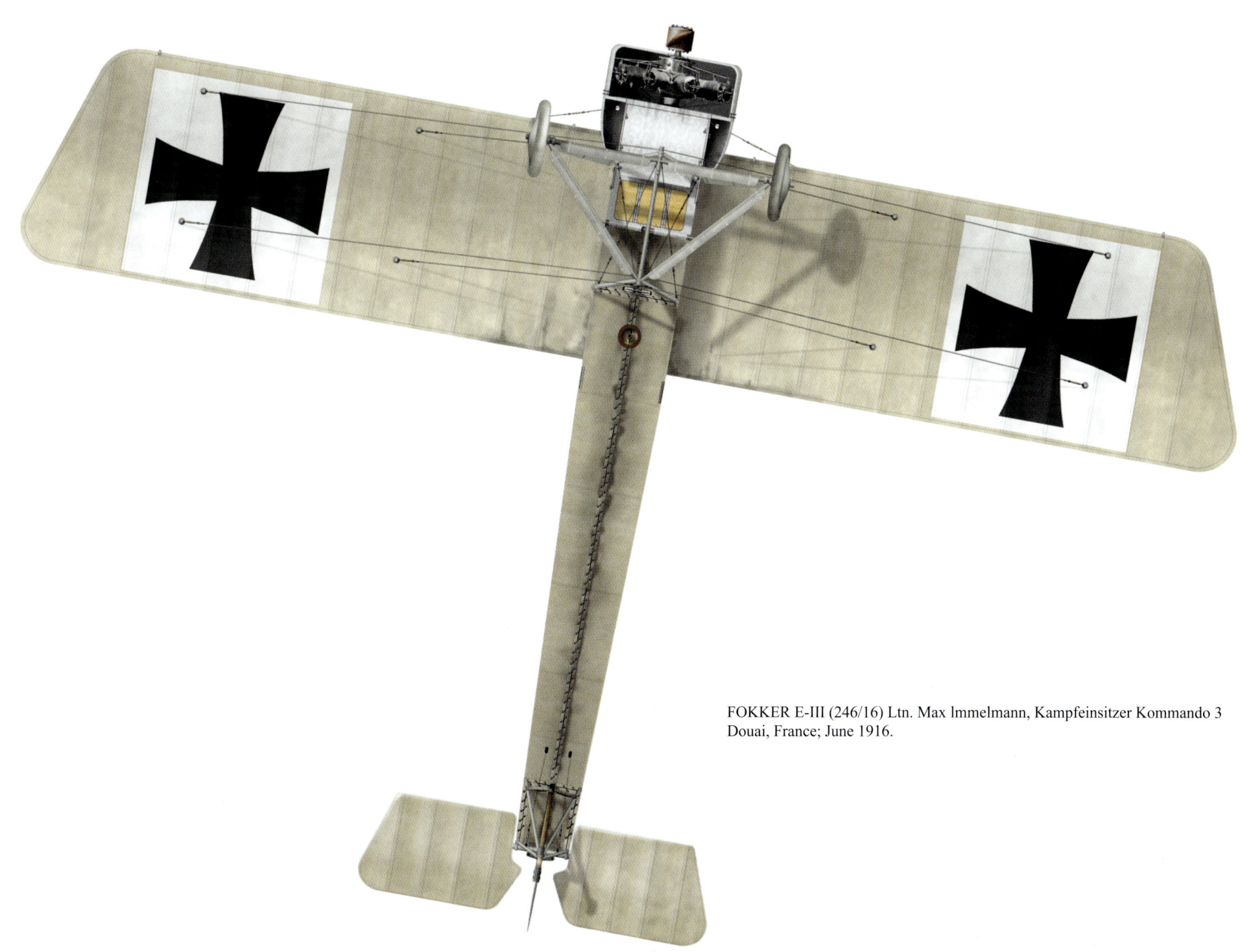

FOKKER E-III (246/16) Ltn. Max Immelmann, Kampfeinsitzer Kommando 3
Douai, France; June 1916.

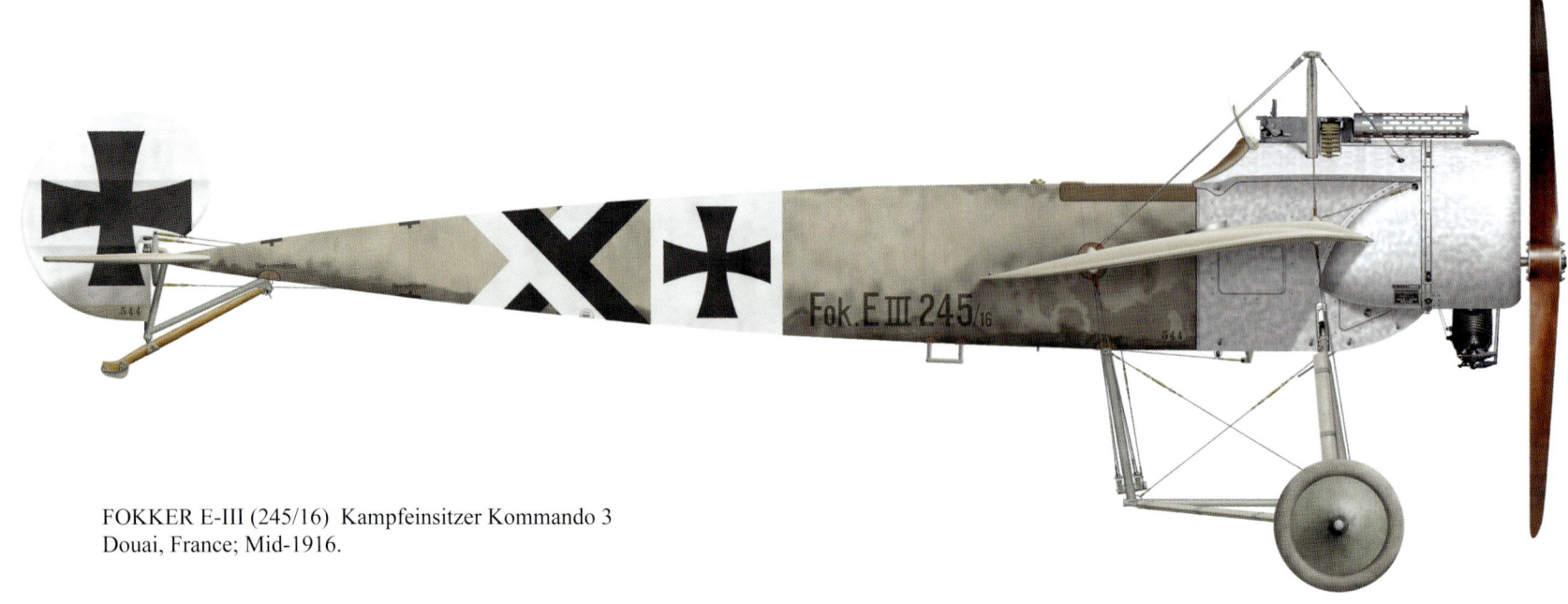

FOKKER E-III (245/16) Kampfeinsitzer Kommando 3
Douai, France; Mid-1916.

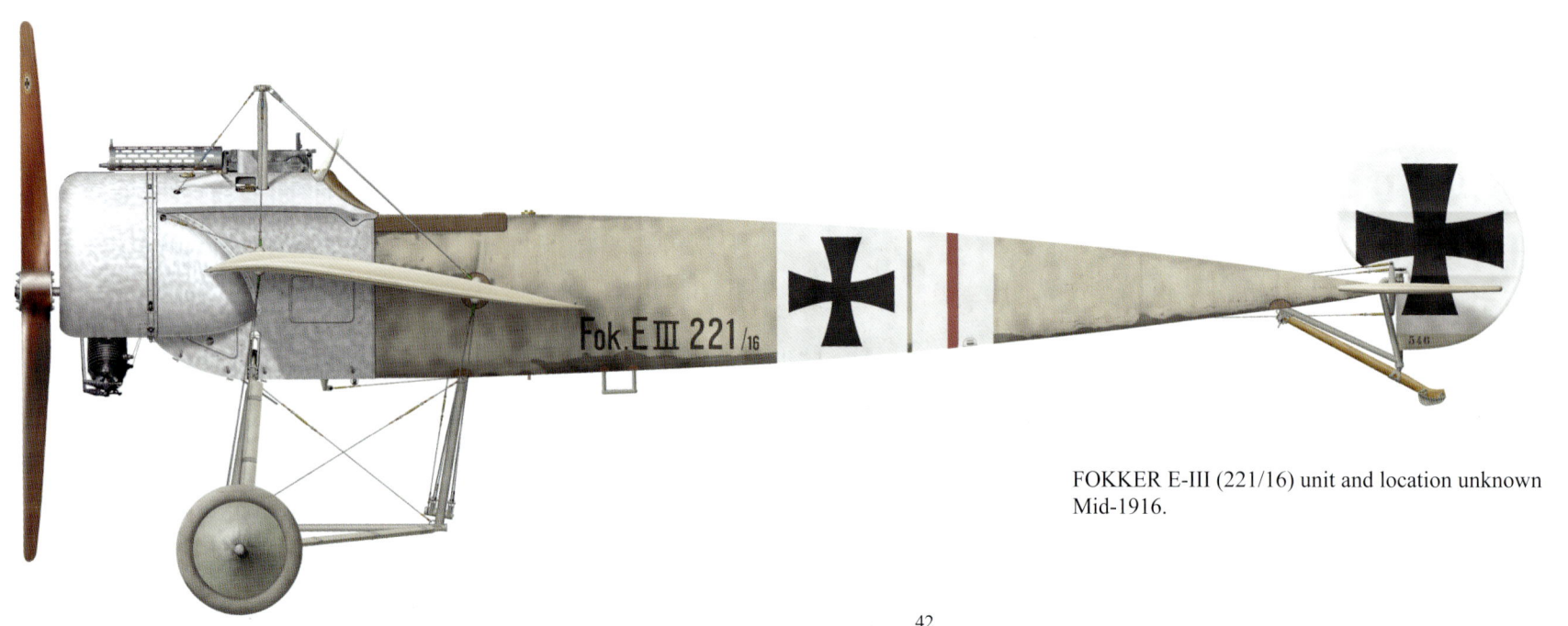

FOKKER E-III (221/16) unit and location unknown
Mid-1916.

FOKKER E-III (LF222) unknown naval unit and location
Mid-1916.

FOKKER E-II1 (604/15) unit and location unknown
Mid-1916.

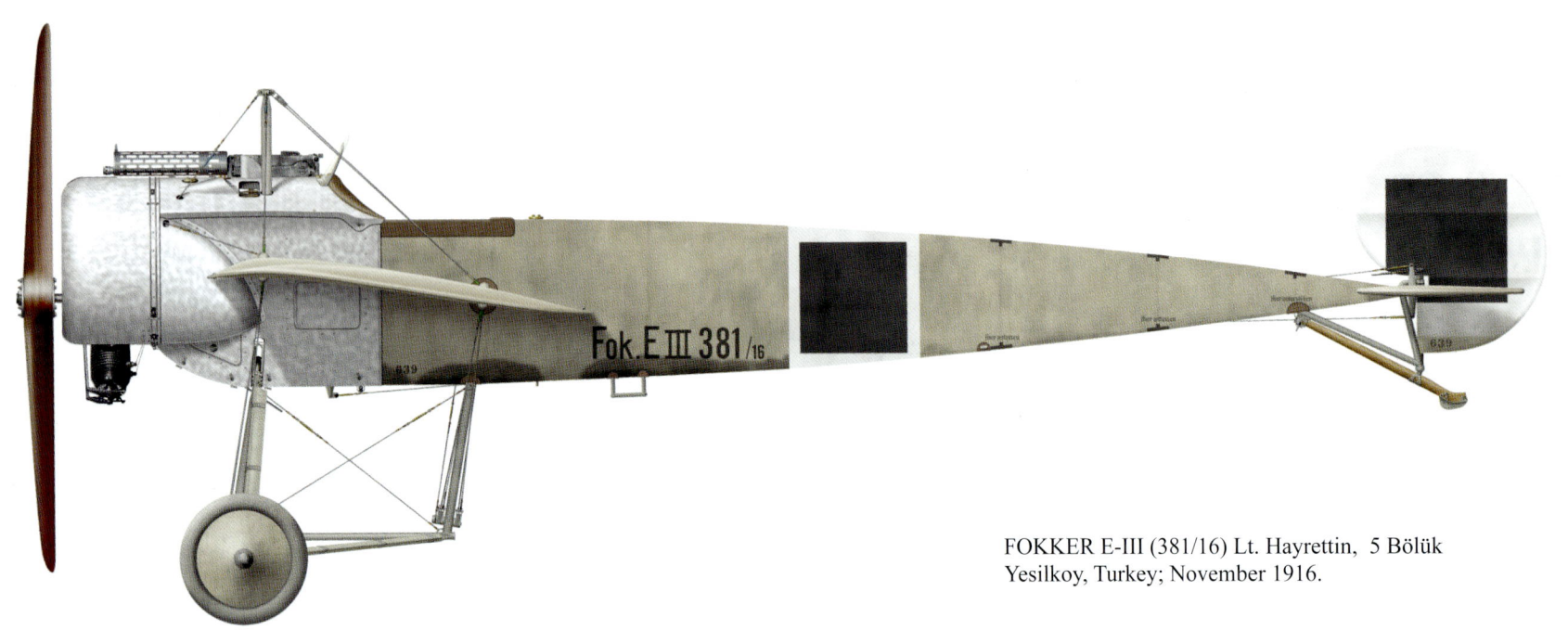

FOKKER E-III (381/16) Lt. Hayrettin, 5 Bölük
Yesilkoy, Turkey; November 1916.

FOKKER A-III (A6) Fliegerkompagnie 42J, KuK LFT
Prosecco, Austria-Hungary; Summer 1917.
Notice nonstandard cooling jacket for the Spandau machine gun.

FOKKER E-IV

FOKKER E-IV (122/15) Ltn. Otto Parschau, Feldflieger Abteilung 62
Douai, France; November 1915.

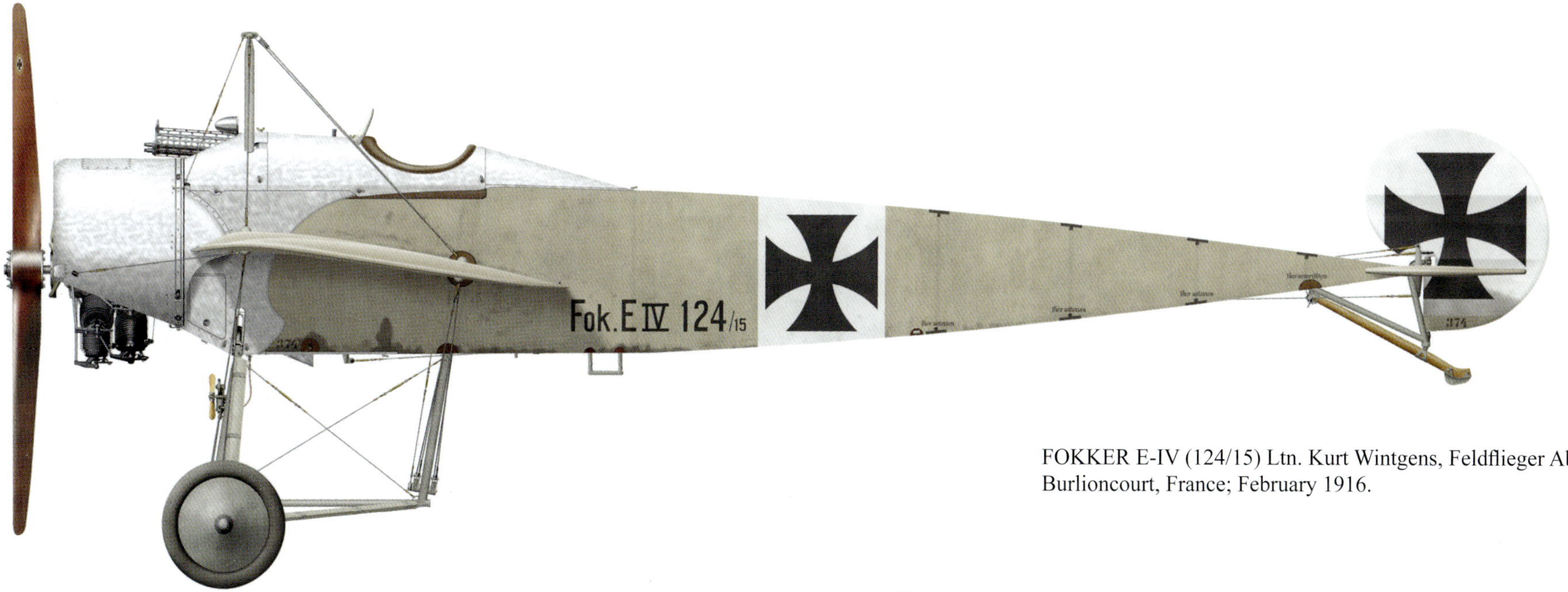

FOKKER E-IV (124/15) Ltn. Kurt Wintgens, Feldflieger Abteilung 12
Burlioncourt, France; February 1916.

FOKKER E-IV (639/15) Ltn. Hartmuth Baldamus, Feldflieger Abteilung 20
Aussonce, France; Spring 1916.

FOKKER E-IV (127/15) Oblt. Max Immelmann, Kampfeinsitzer Kommando 3
Douai, France; Spring 1916.
Notice nonstandard louvers on the fuselage side.

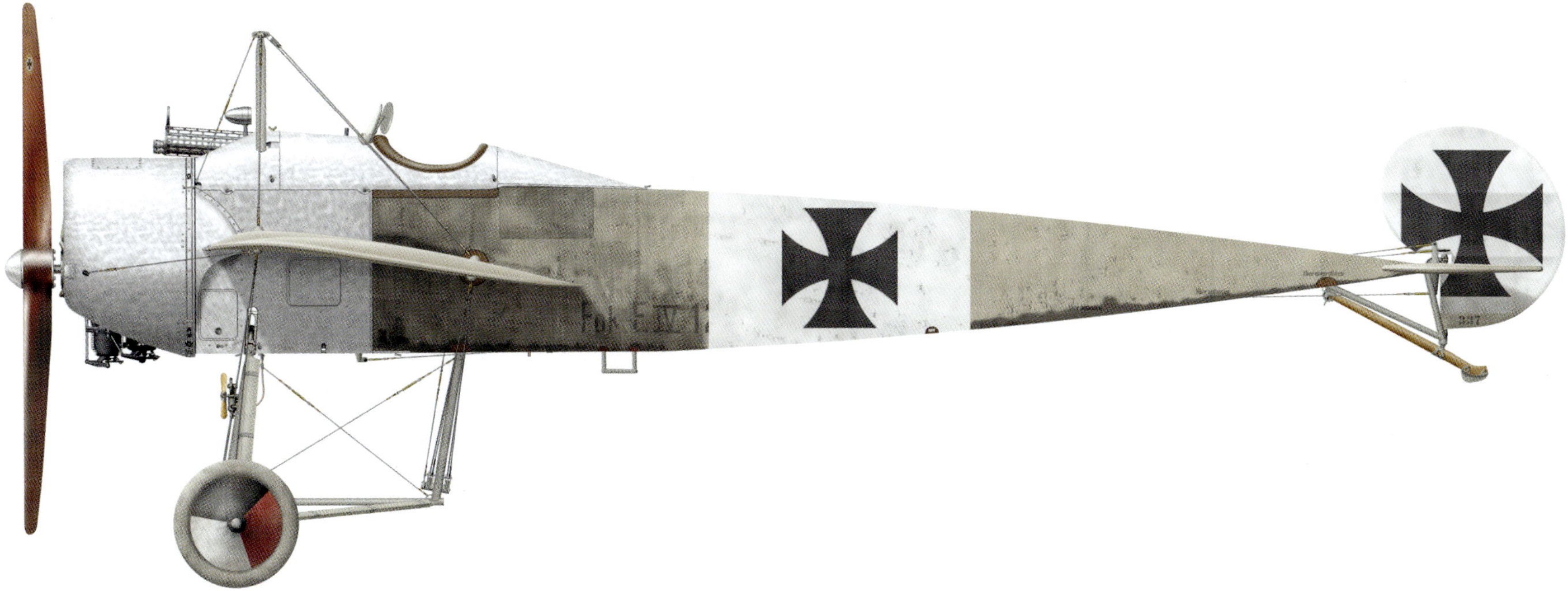

FOKKER E-IV (123/15) Oblt. Oswald Boelcke, Kampfeinsitzer Kommando Sivry
Sivry, France; April 1916.

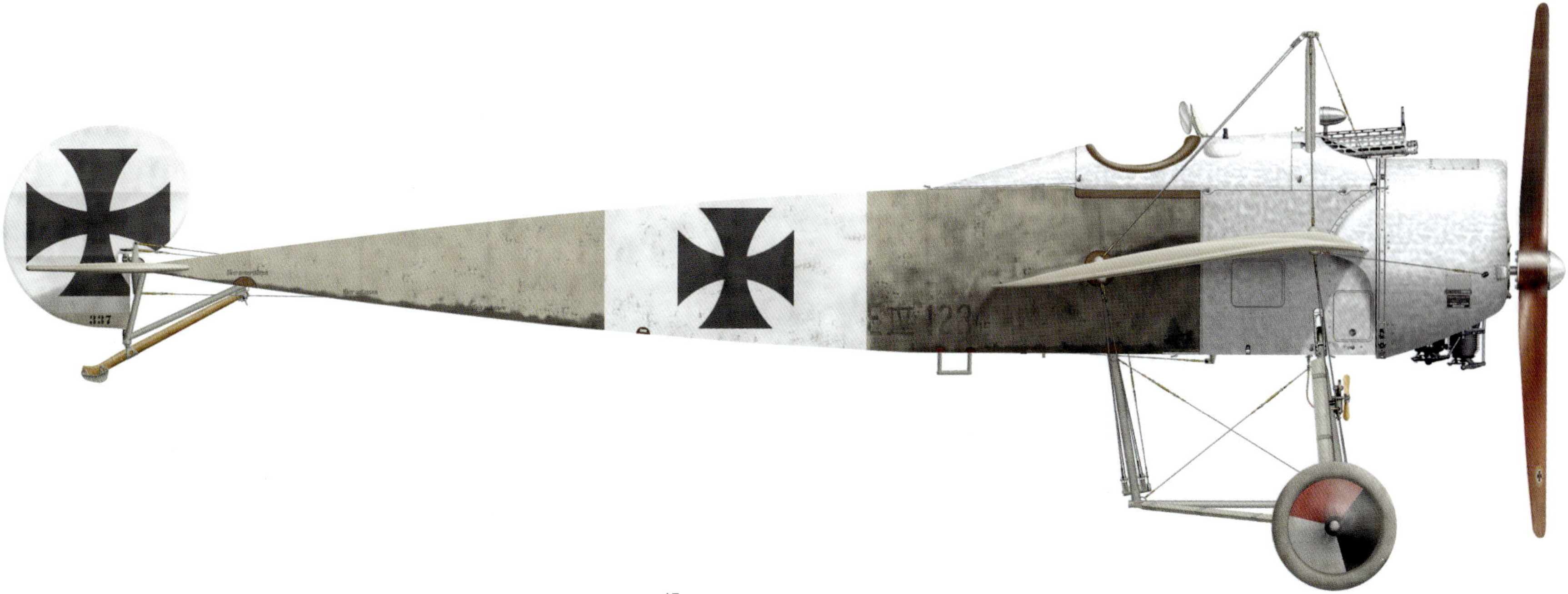

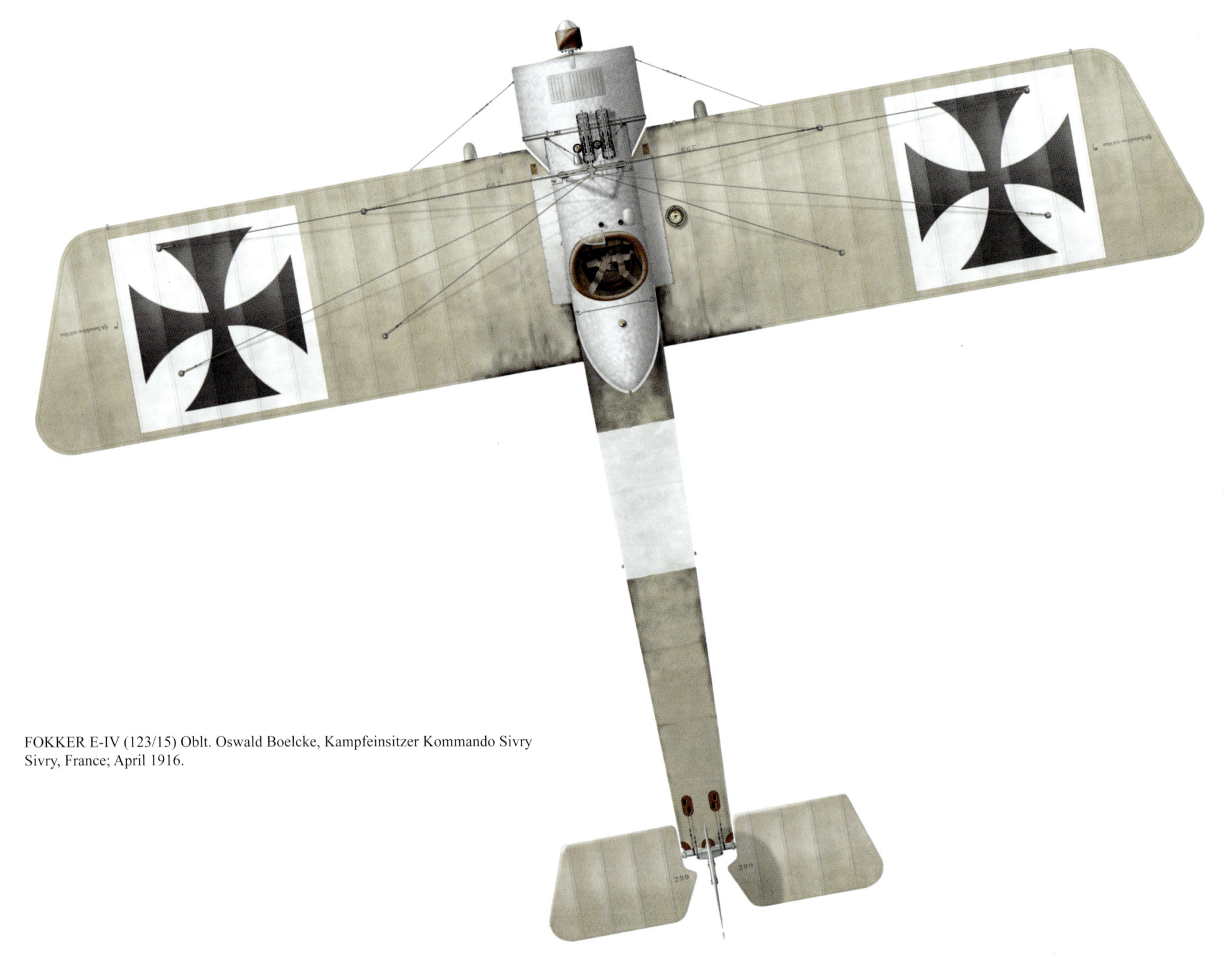

FOKKER E-IV (123/15) Oblt. Oswald Boelcke, Kampfeinsitzer Kommando Sivry
Sivry, France; April 1916.

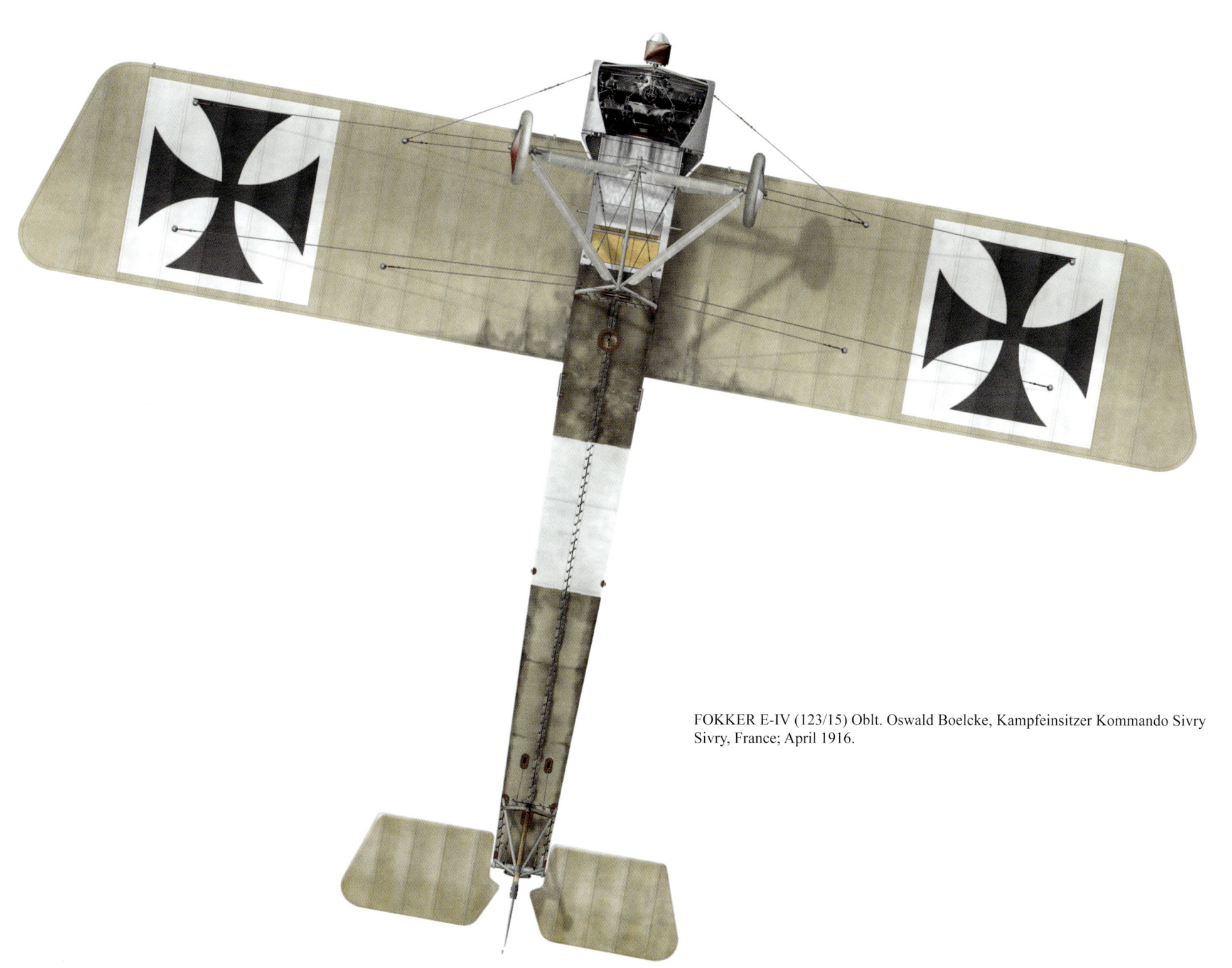

FOKKER E-IV (123/15) Oblt. Oswald Boelcke, Kampfeinsitzer Kommando Sivry
Sivry, France; April 1916.

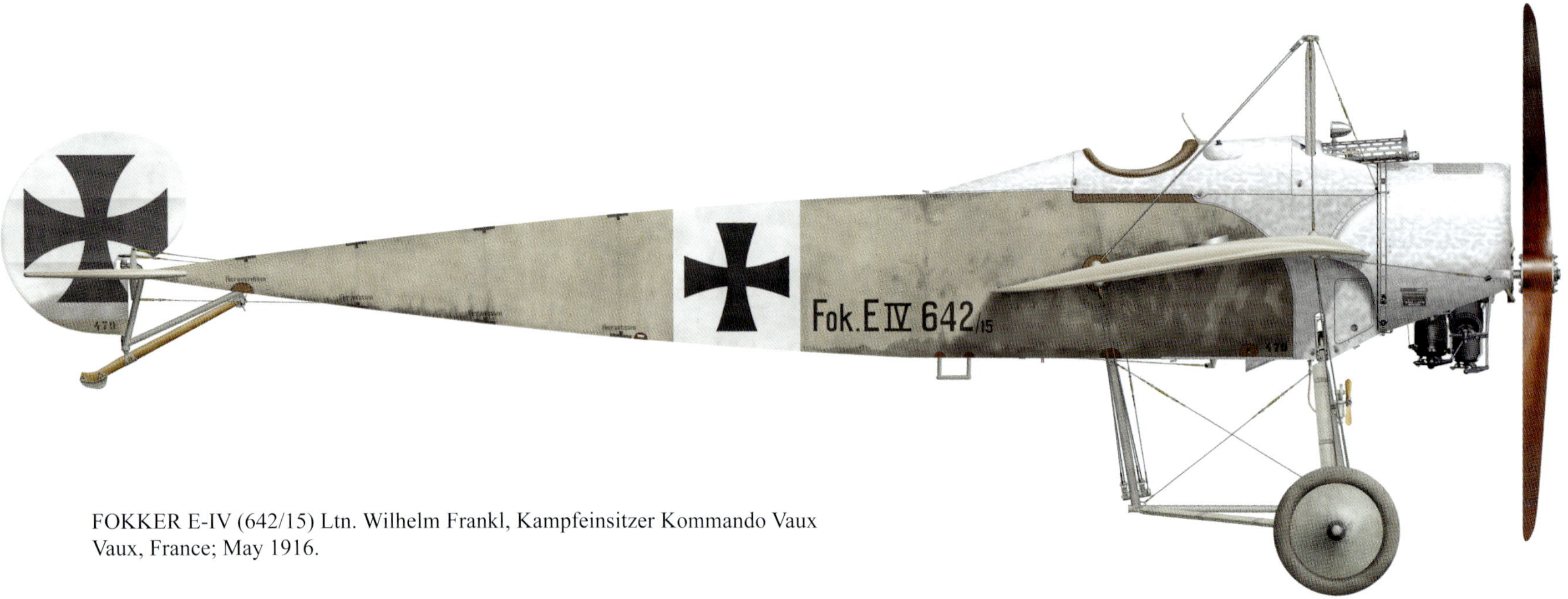

FOKKER E-IV (642/15) Ltn. Wilhelm Frankl, Kampfeinsitzer Kommando Vaux
Vaux, France; May 1916.

FOKKER E-IV (437/15) Ltn. Walter Höhndorf, Kampfeinsitzer Kommando Vaux
Vaux, France; Summer 1916.

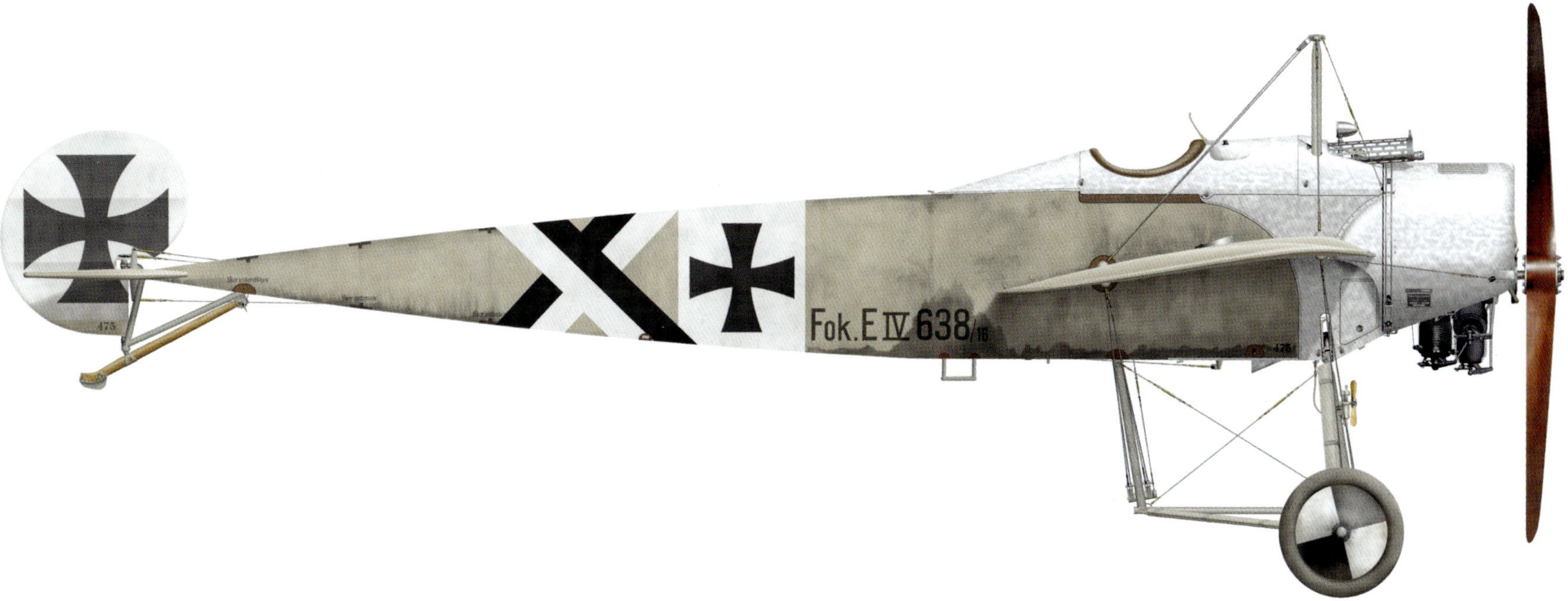

FOKKER E-IV (638/15) Kampfeinsitzer Kommando 3
Douai, France; Summer 1916.

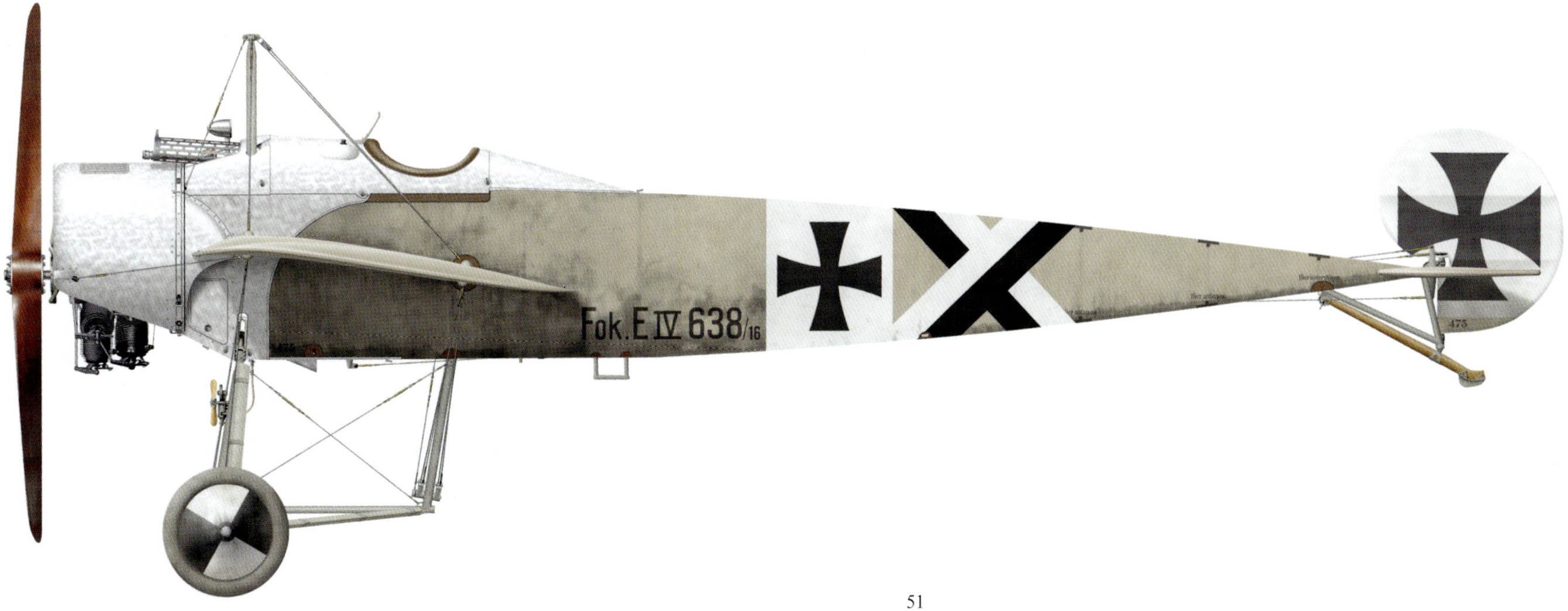

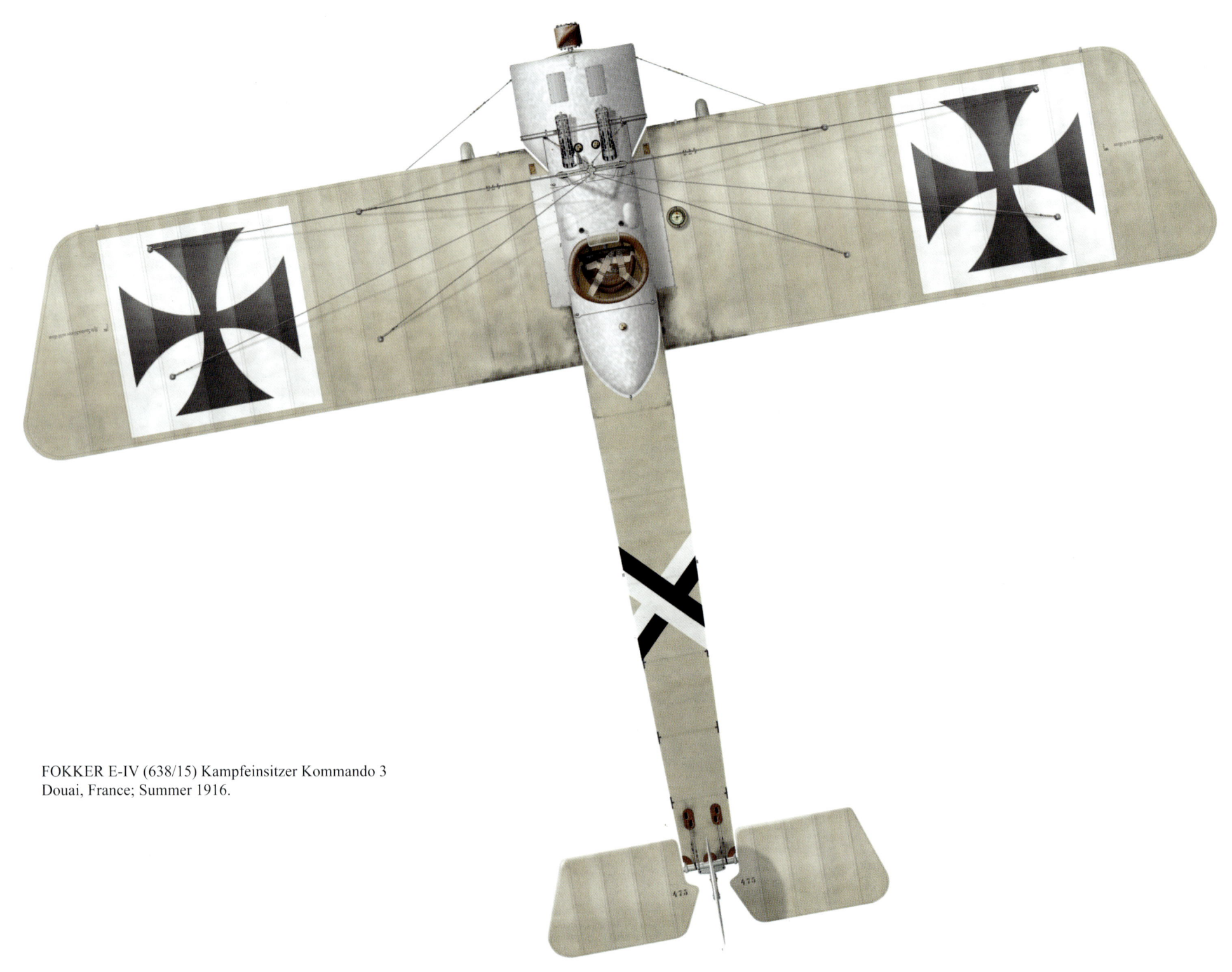

FOKKER E-IV (638/15) Kampfeinsitzer Kommando 3
Douai, France; Summer 1916.

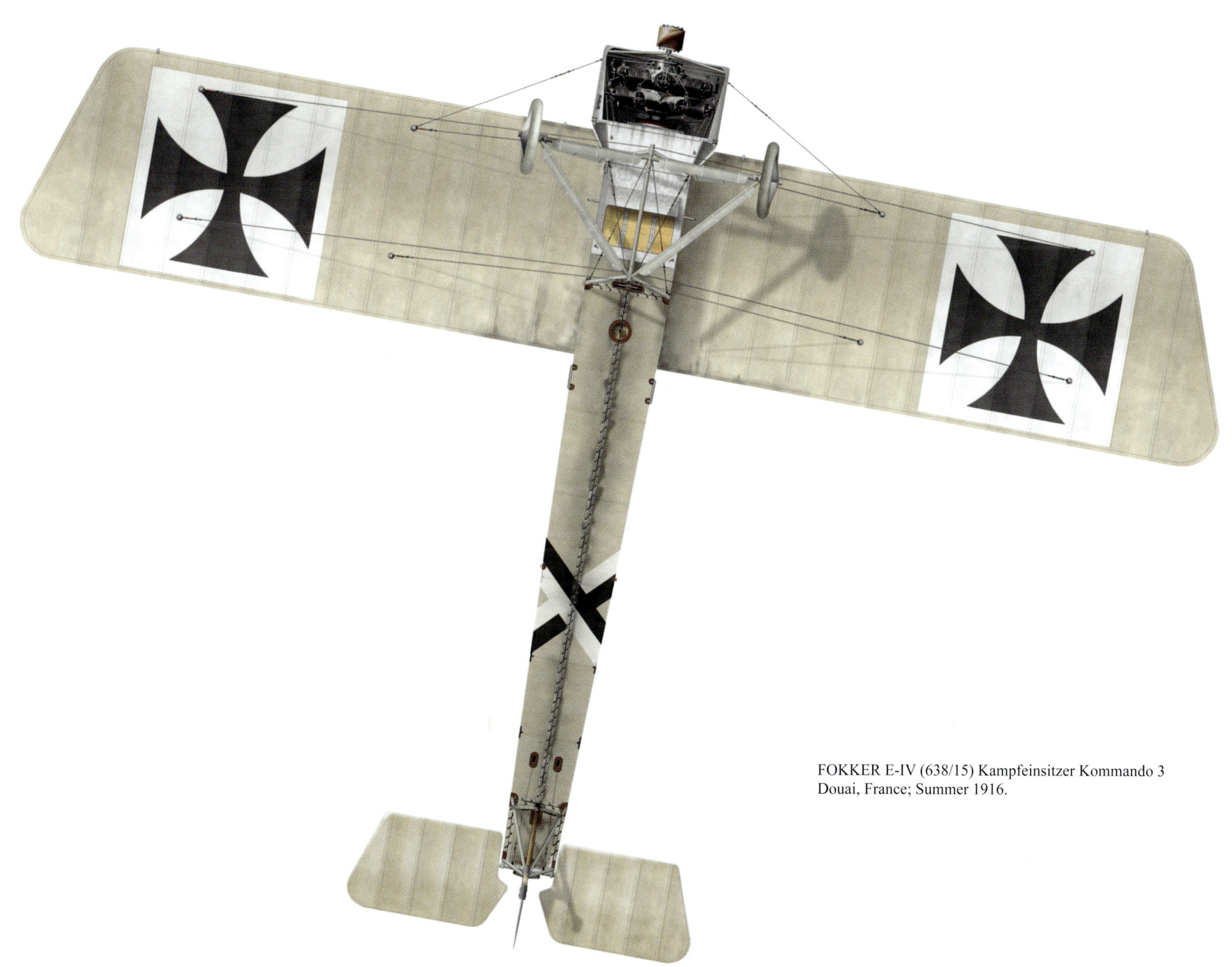

FOKKER E-IV (638/15) Kampfeinsitzer Kommando 3
Douai, France; Summer 1916.

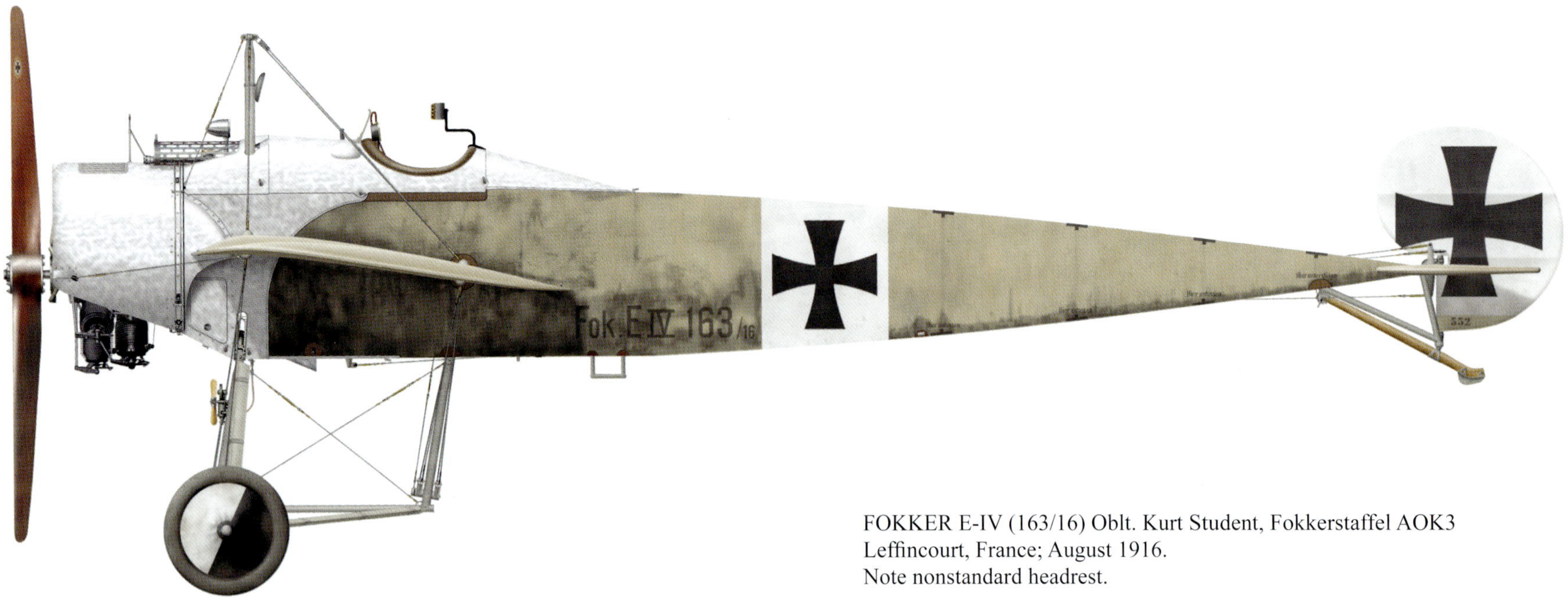

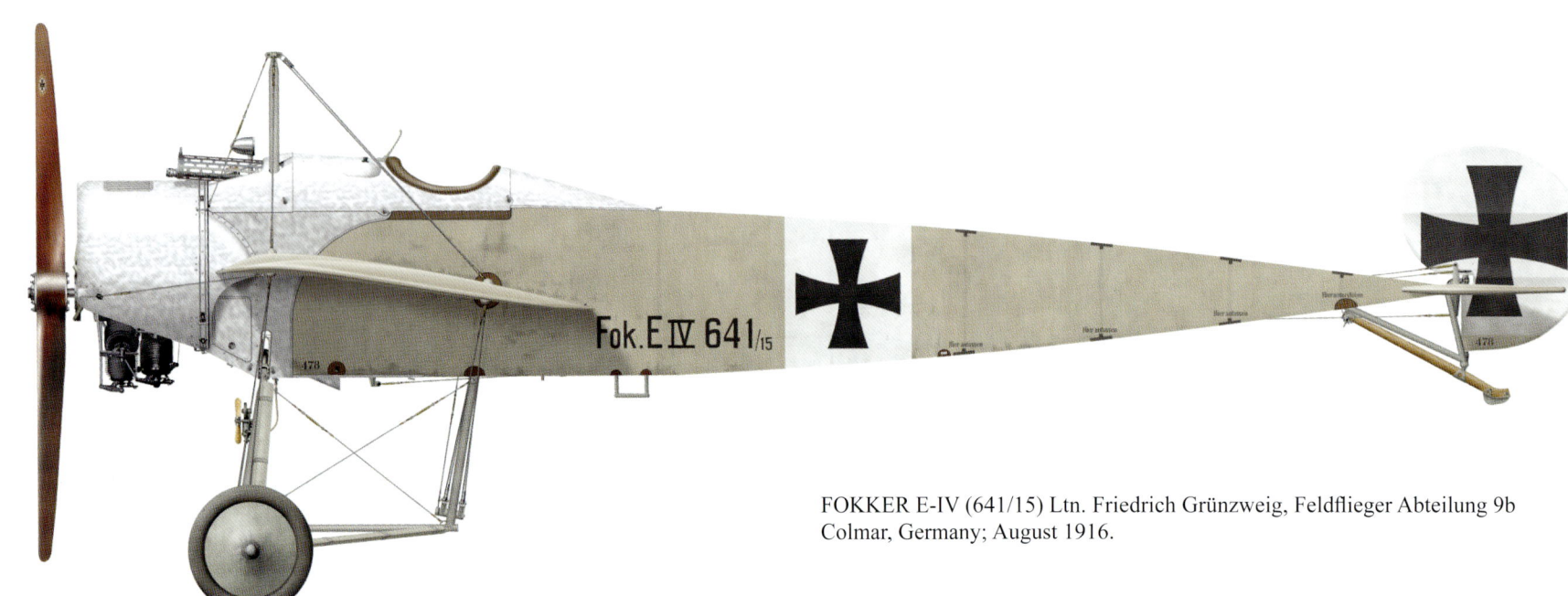

FOKKER E-IV (163/16) Oblt. Kurt Student, Fokkerstaffel AOK3
Leffincourt, France; August 1916.
Note nonstandard headrest.

FOKKER E-IV (641/15) Ltn. Friedrich Grünzweig, Feldflieger Abteilung 9b
Colmar, Germany; August 1916.

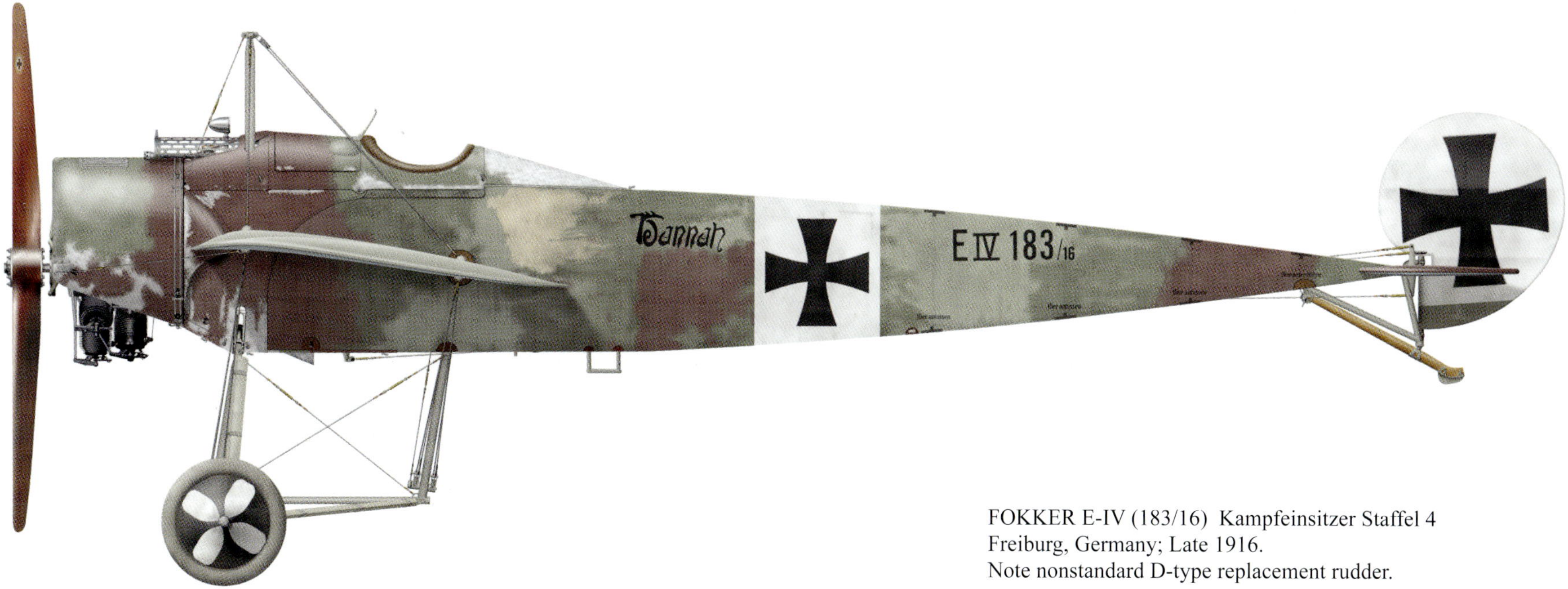

FOKKER E-IV (183/16) Kampfeinsitzer Staffel 4
Freiburg, Germany; Late 1916.
Note nonstandard D-type replacement rudder.

FOKKER E-IV (186/16) unit and location unknown; Late 1916.

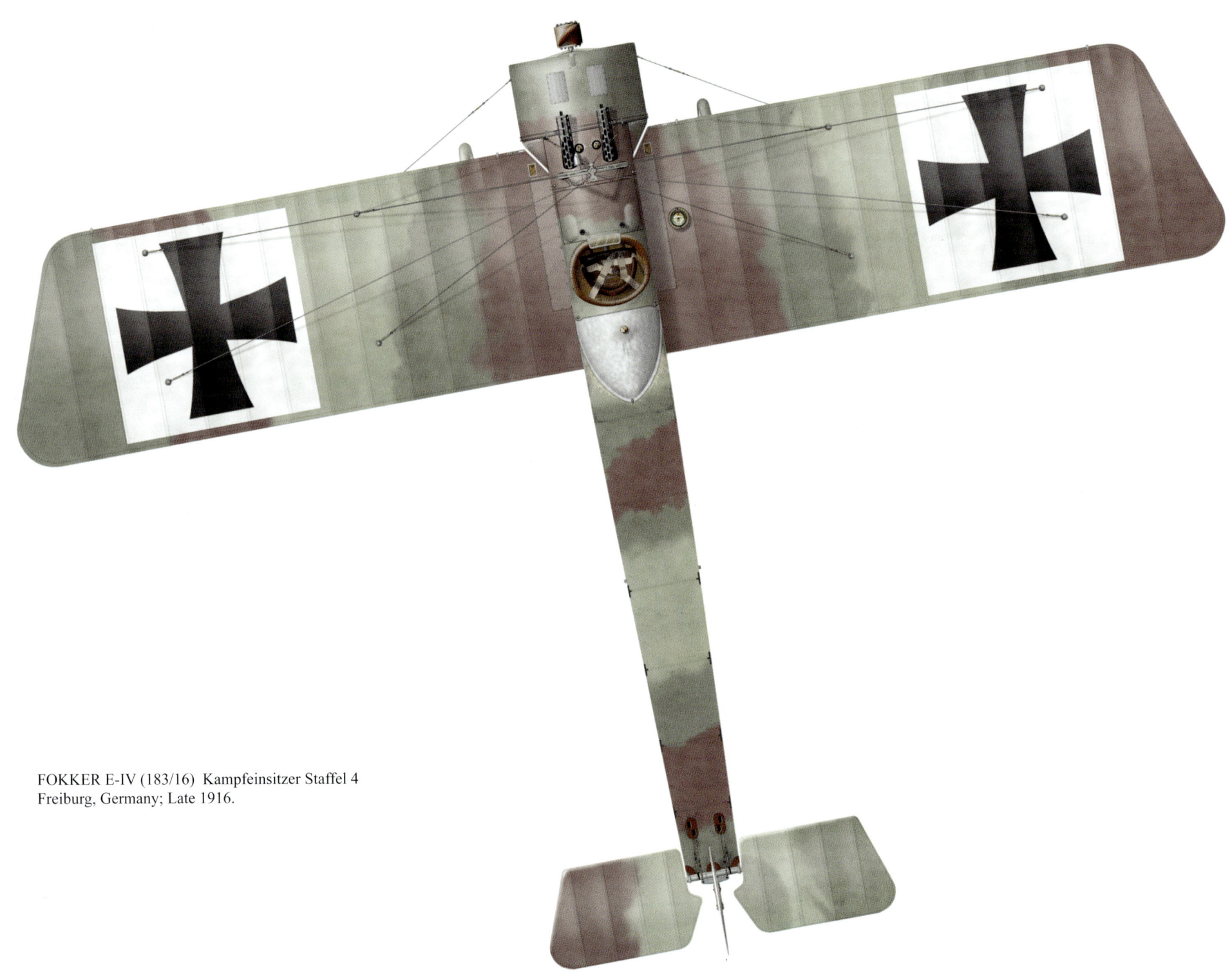

FOKKER E-IV (183/16) Kampfeinsitzer Staffel 4
Freiburg, Germany; Late 1916.

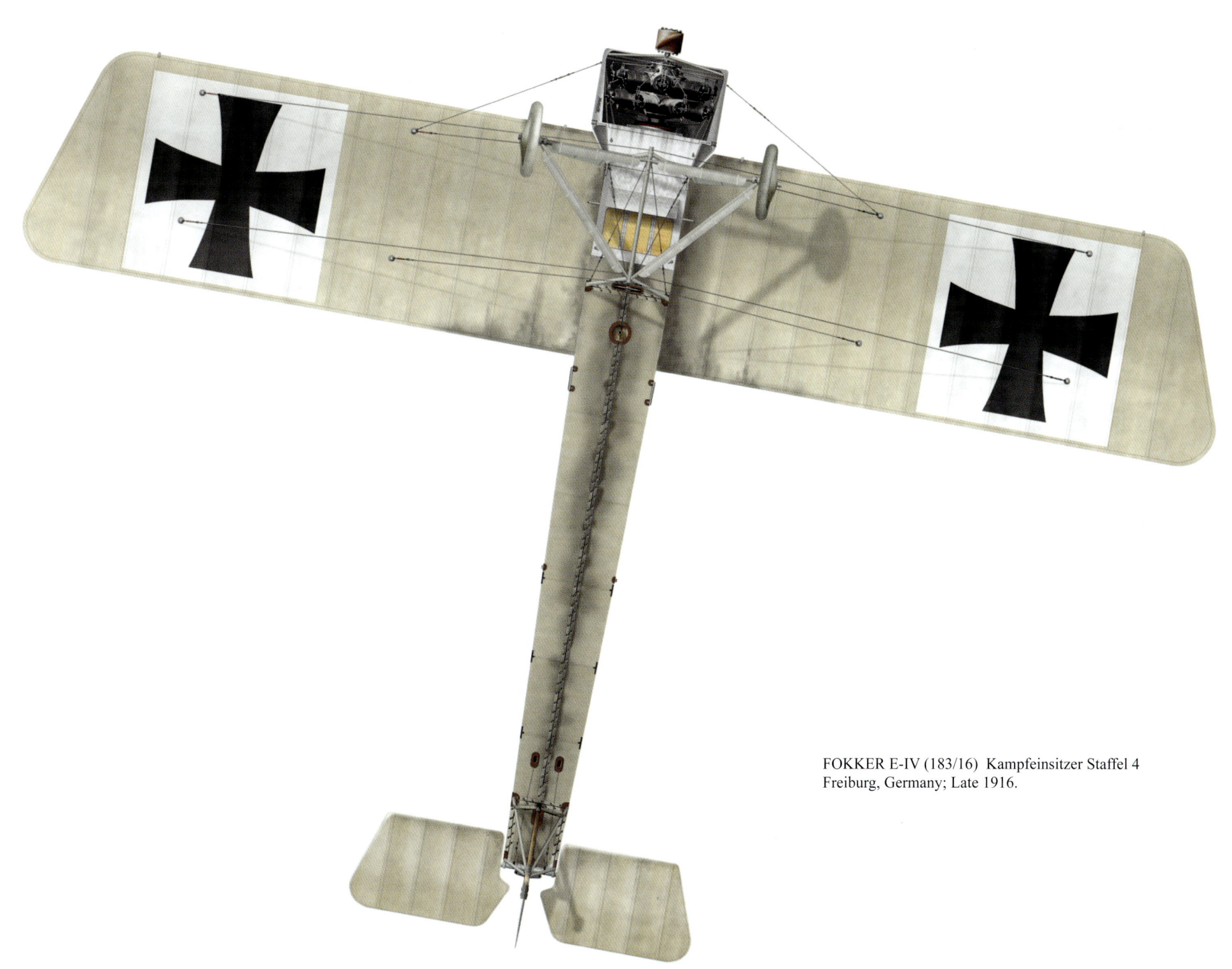

FOKKER E-IV (183/16) Kampfeinsitzer Staffel 4
Freiburg, Germany; Late 1916.

FOKKER E-IV (161/16) Ltn. Hans Müller, Kampfeinsitzer Staffel 6
Bonn-Hangelar, Germany; April 1917.

FOKKER E-IV (170/16) unit and location unknown.
Early 1917.
Unarmed trainer.

PFALZ E-I

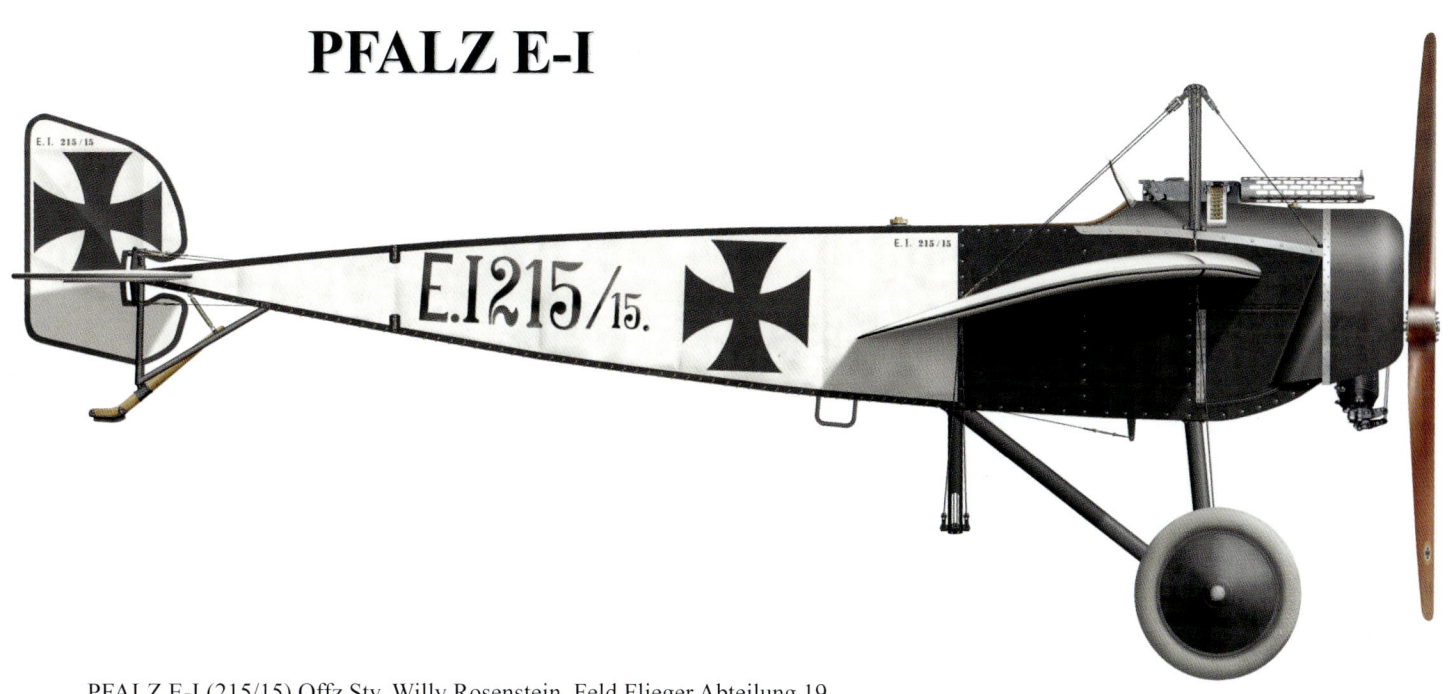

PFALZ E-I (215/15) Offz Stv. Willy Rosenstein, Feld Flieger Abteilung 19
Porcher, France; December 1915.

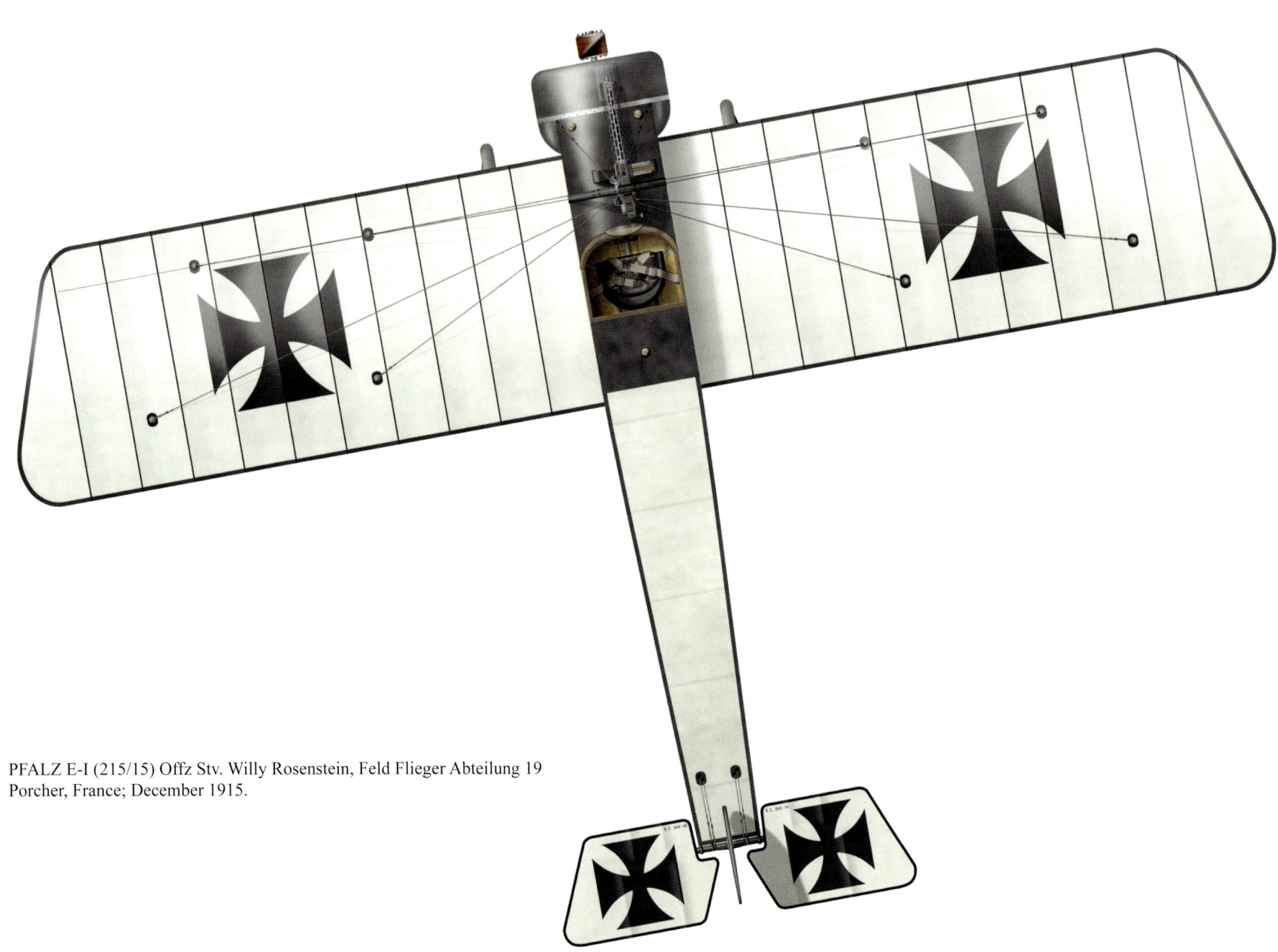

PFALZ E-I (215/15) Offz Stv. Willy Rosenstein, Feld Flieger Abteilung 19
Porcher, France; December 1915.

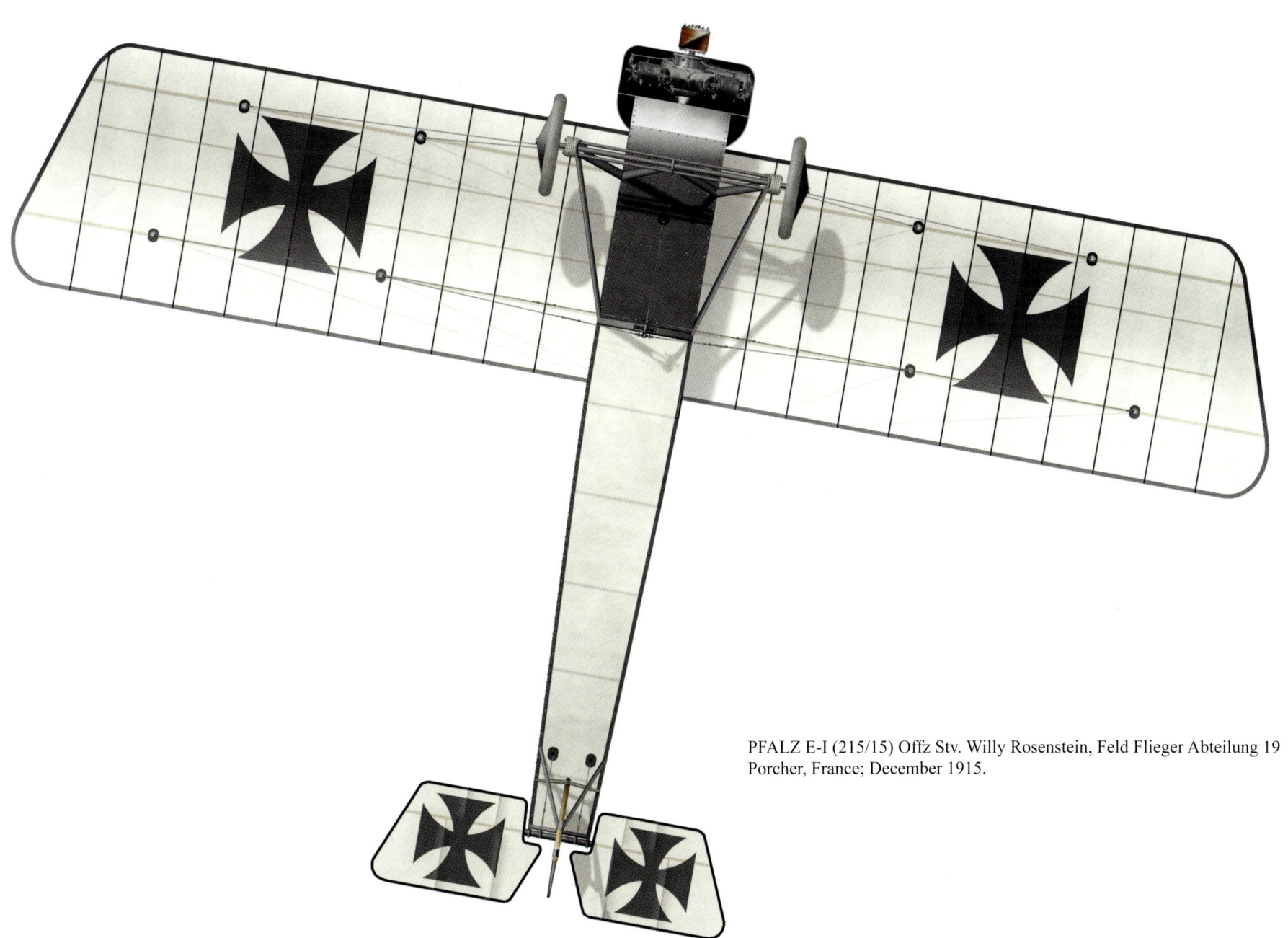

PFALZ E-I (215/15) Offz Stv. Willy Rosenstein, Feld Flieger Abteilung 19
Porcher, France; December 1915.

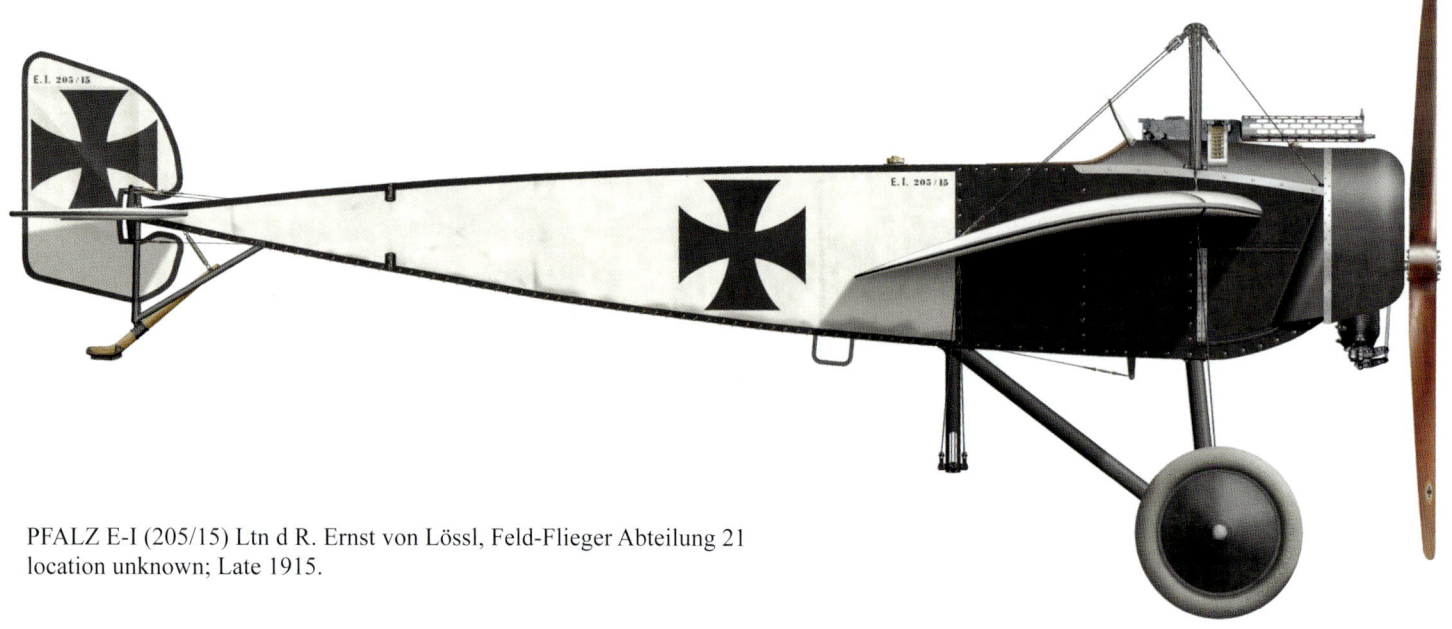

PFALZ E-I (205/15) Ltn d R. Ernst von Lössl, Feld-Flieger Abteilung 21
location unknown; Late 1915.

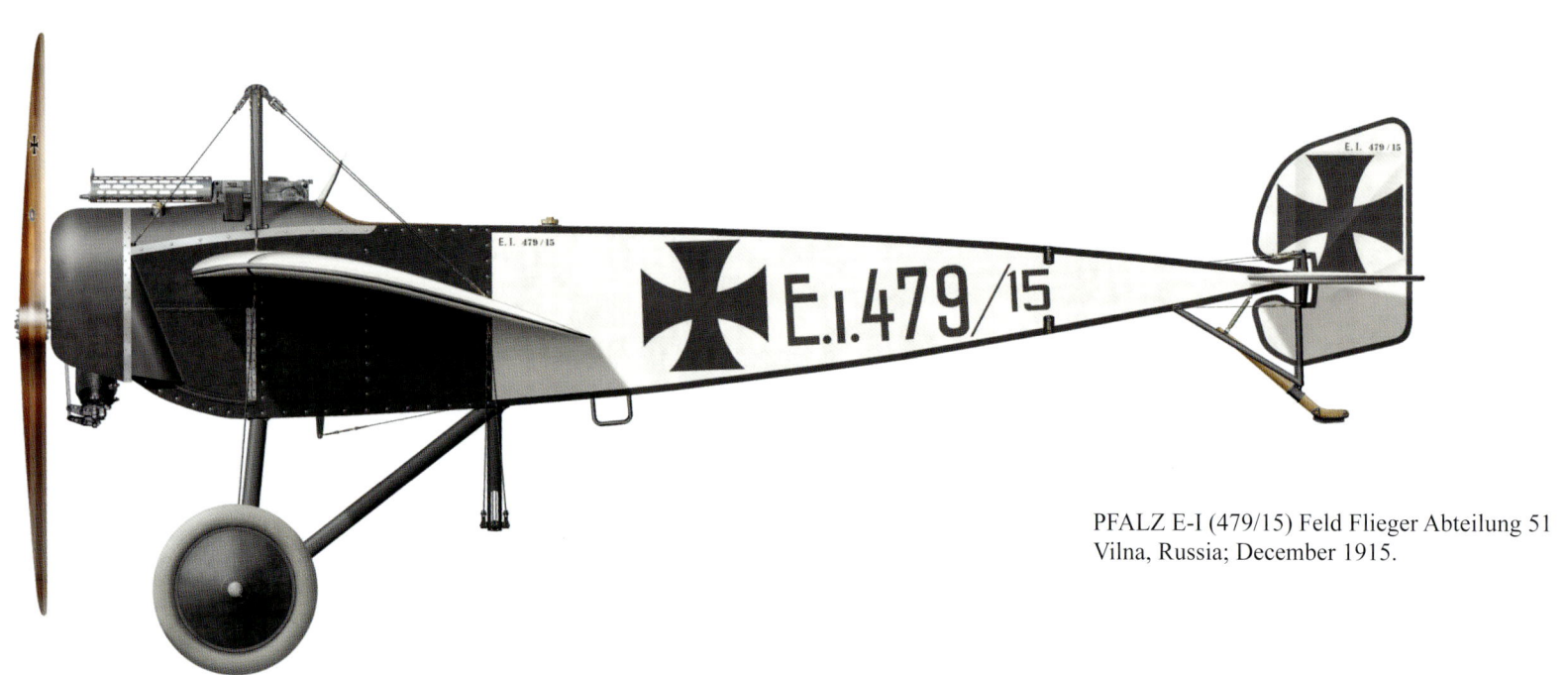

PFALZ E-I (479/15) Feld Flieger Abteilung 51
Vilna, Russia; December 1915.

PFALZ E-II

PFALZ E-II (278/15) Ltn. Walter von Bülow, Feld Flieger Abteilung 22
Metz, Germany; November 1915.

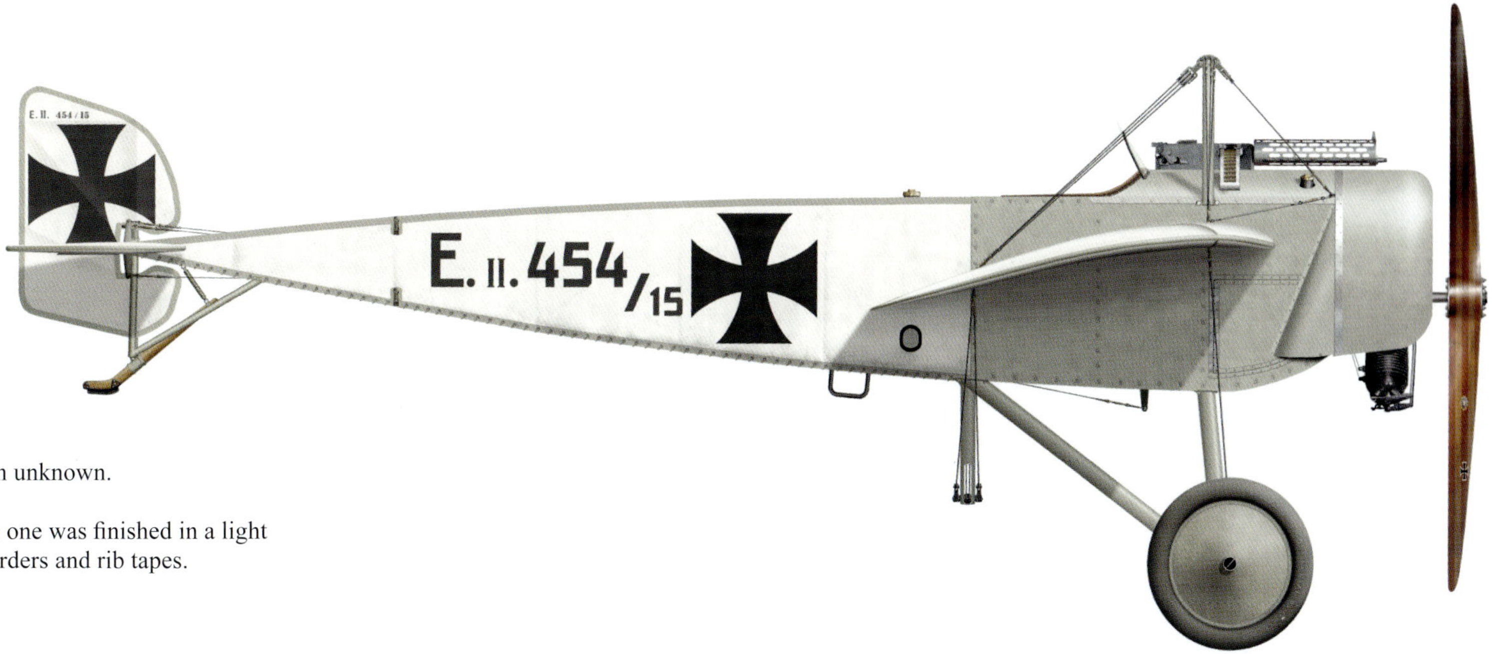

PFALZ E-II (454/15) unit and location unknown.
Early 1916.
Unlike most production machines this one was finished in a light colour on the cowling, metal parts, borders and rib tapes.

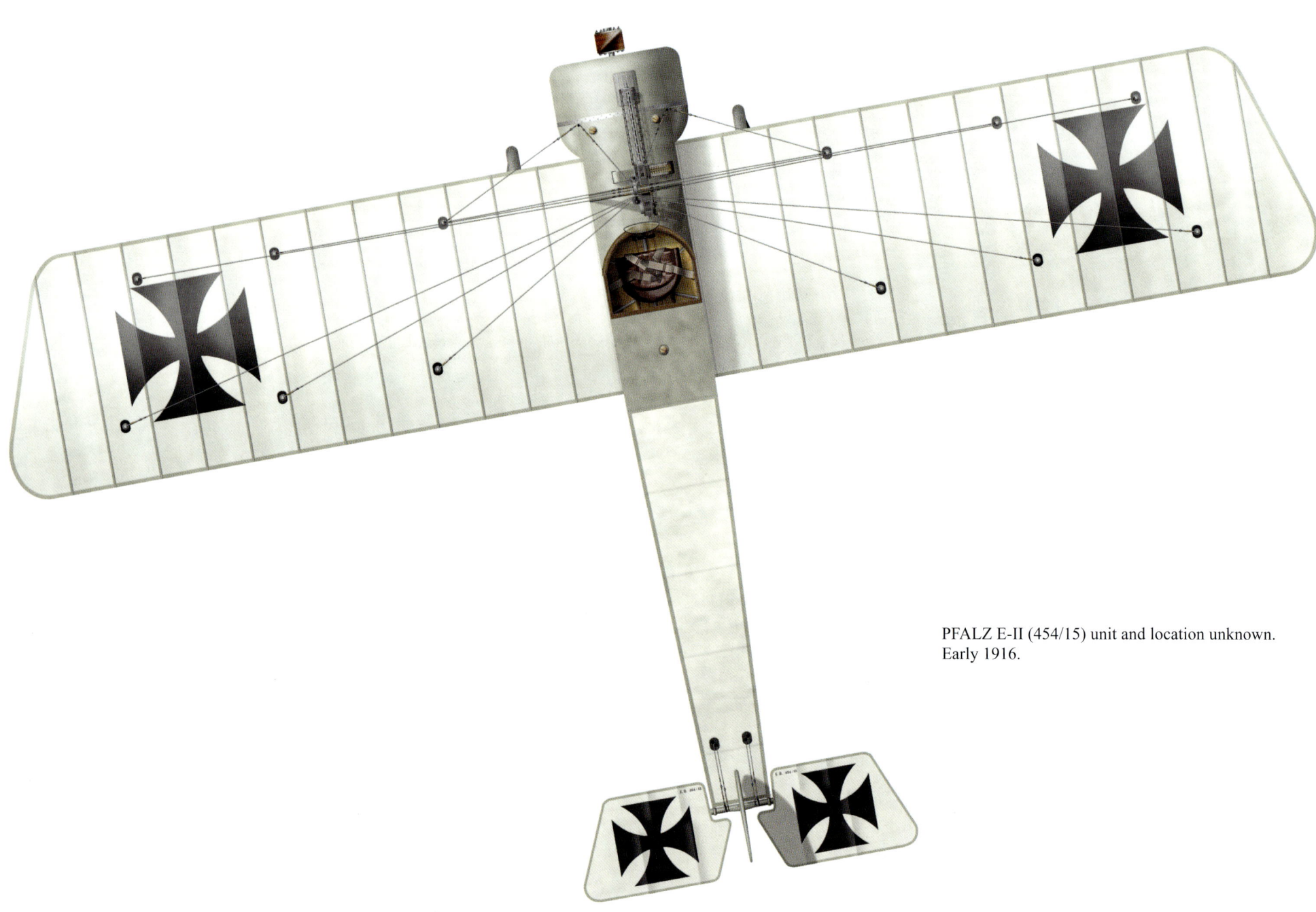

PFALZ E-II (454/15) unit and location unknown.
Early 1916.

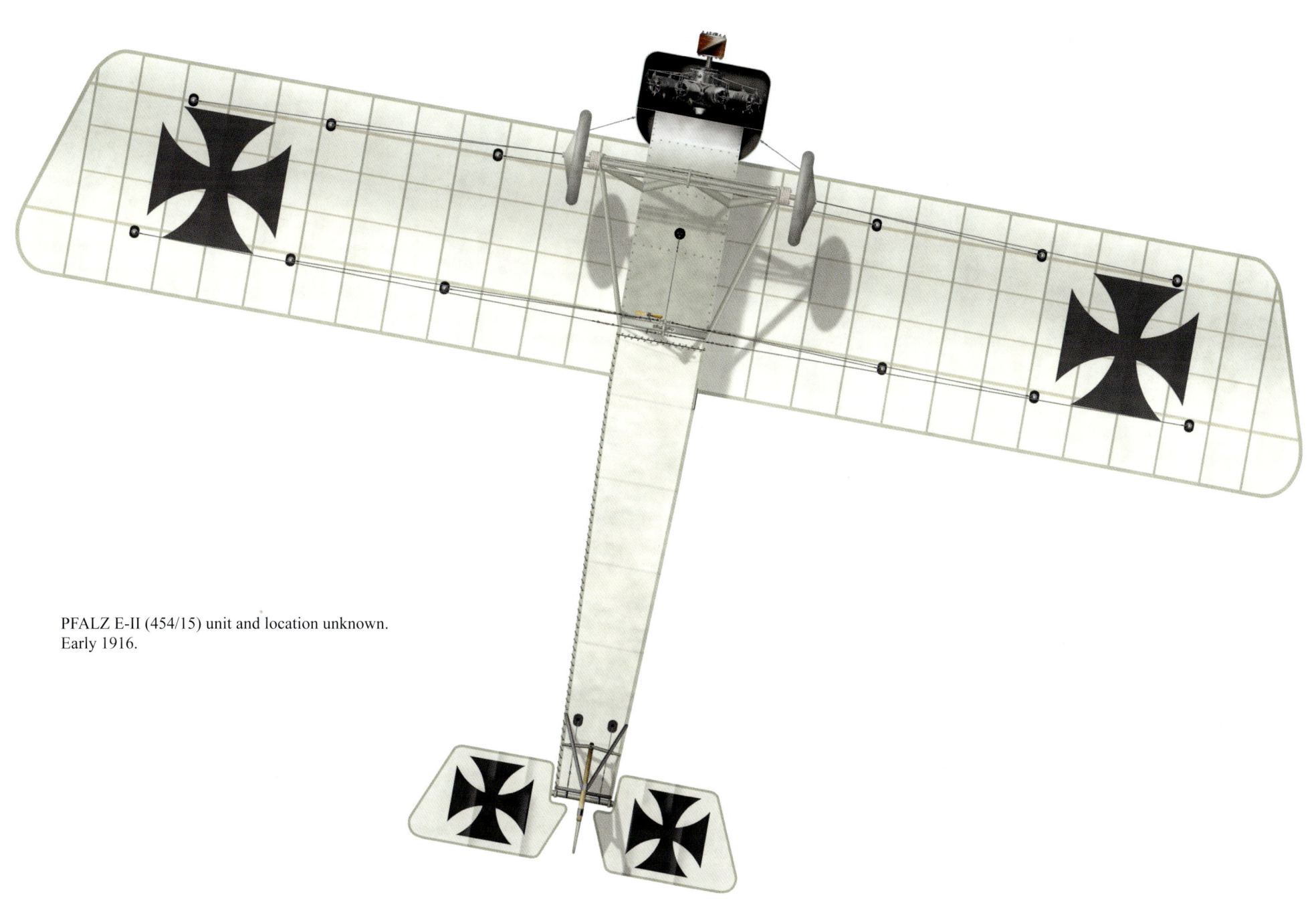

PFALZ E-II (454/15) unit and location unknown.
Early 1916.

PFALZ E-II (serial, unit and location unknown)
Early 1916.

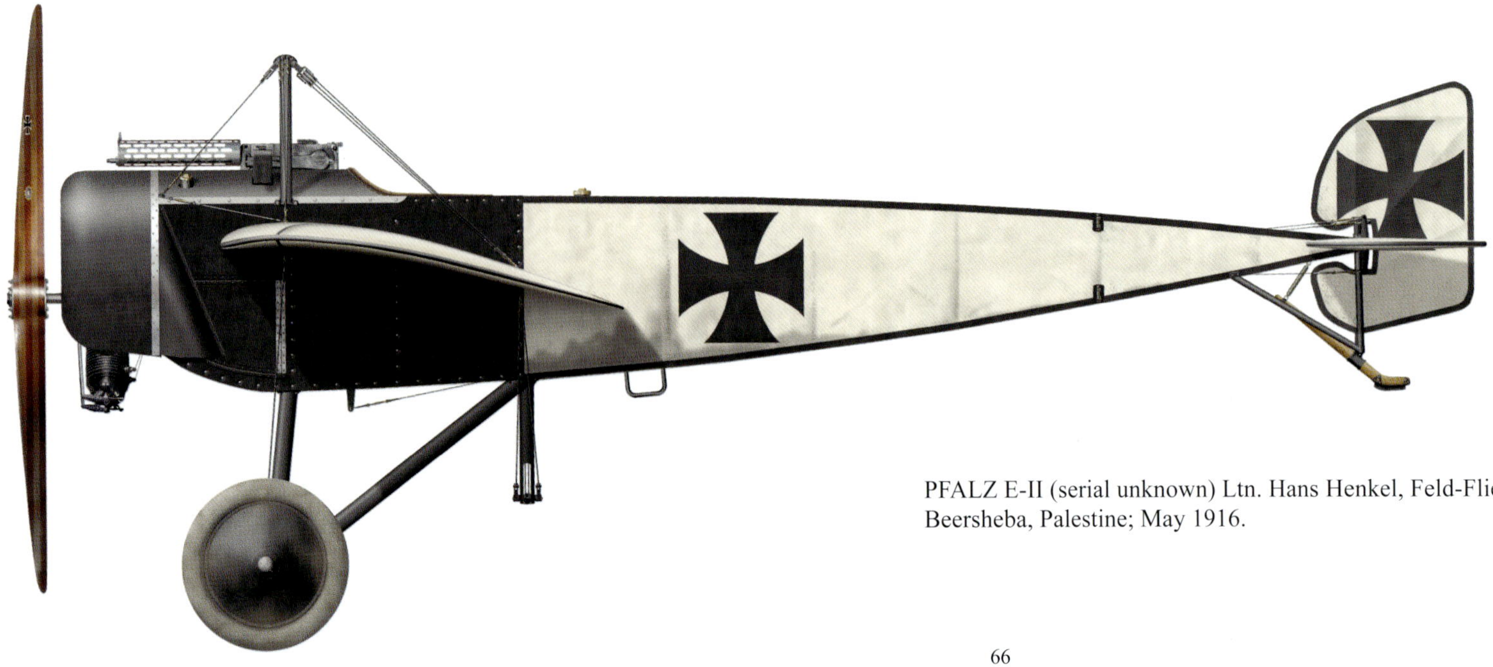

PFALZ E-II (serial unknown) Ltn. Hans Henkel, Feld-Flieger Abteilung 300 "Pascha"
Beersheba, Palestine; May 1916.

PFALZ E-IV

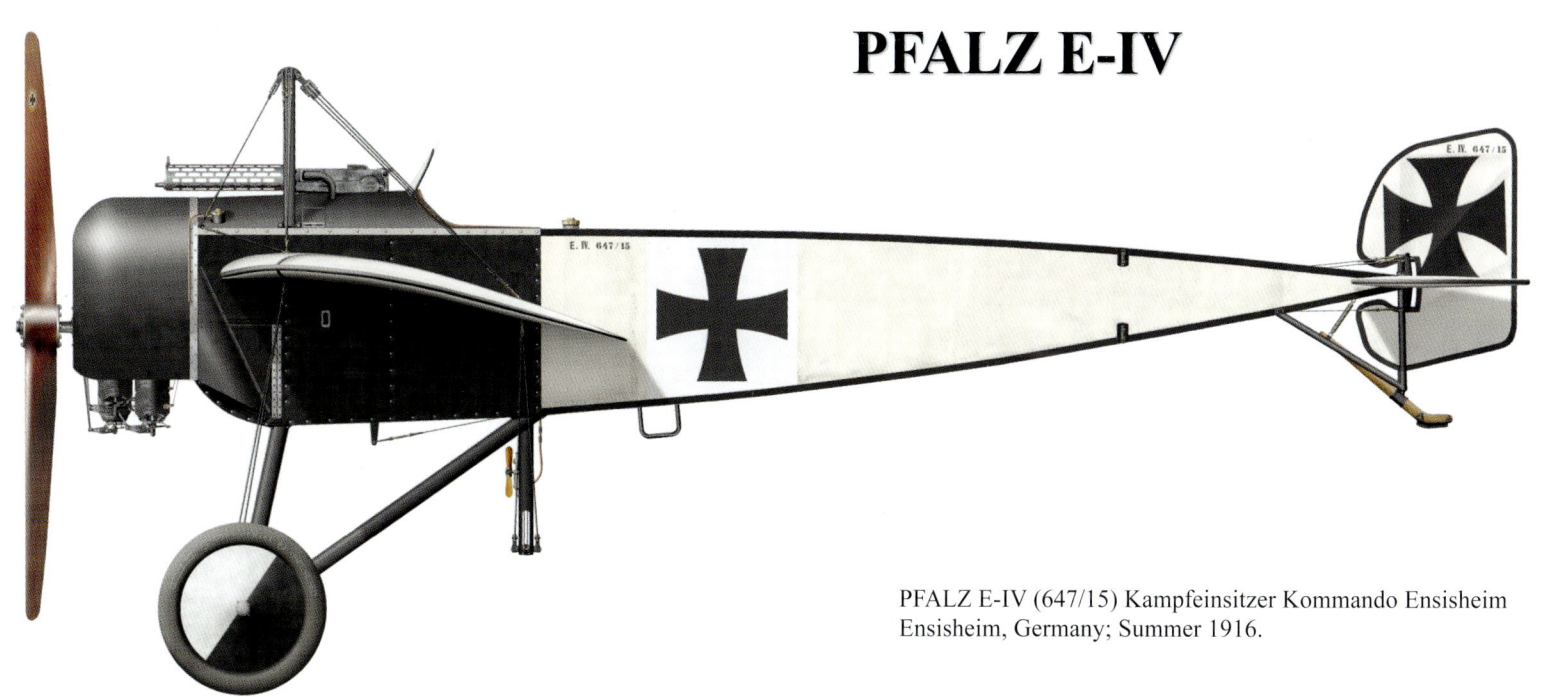

PFALZ E-IV (647/15) Kampfeinsitzer Kommando Ensisheim
Ensisheim, Germany; Summer 1916.

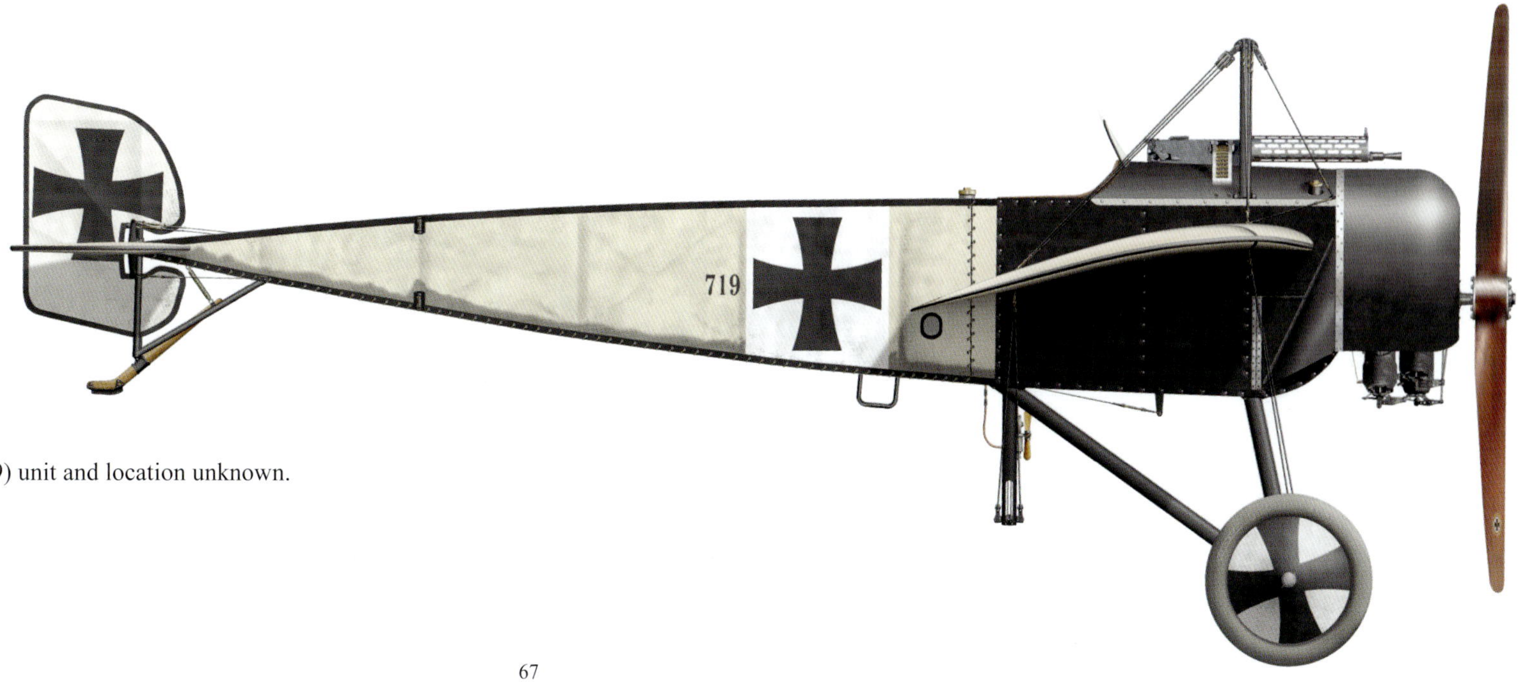

PFALZ E-IV (719) unit and location unknown.
Mid-1916.

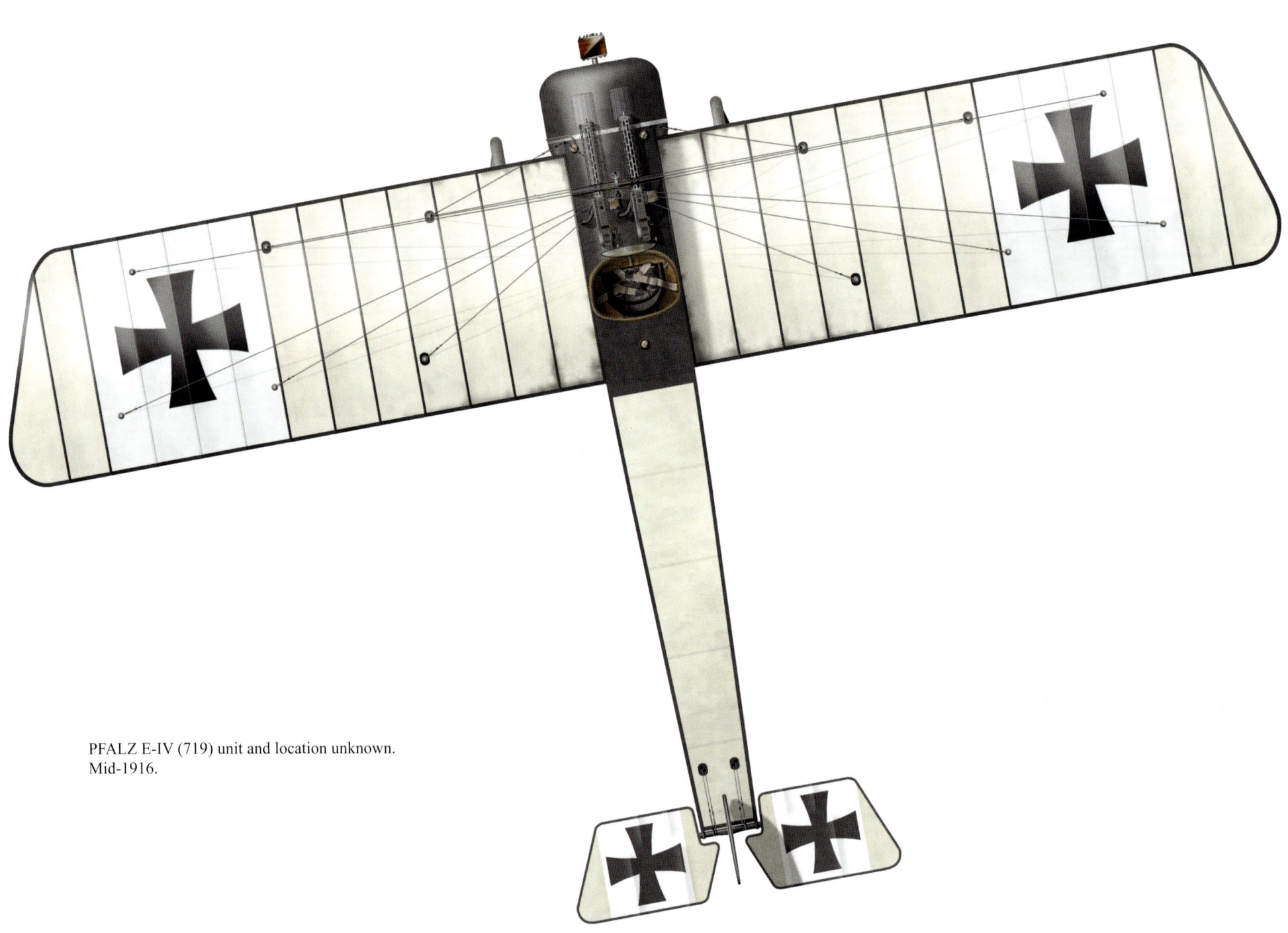

PFALZ E-IV (719) unit and location unknown.
Mid-1916.

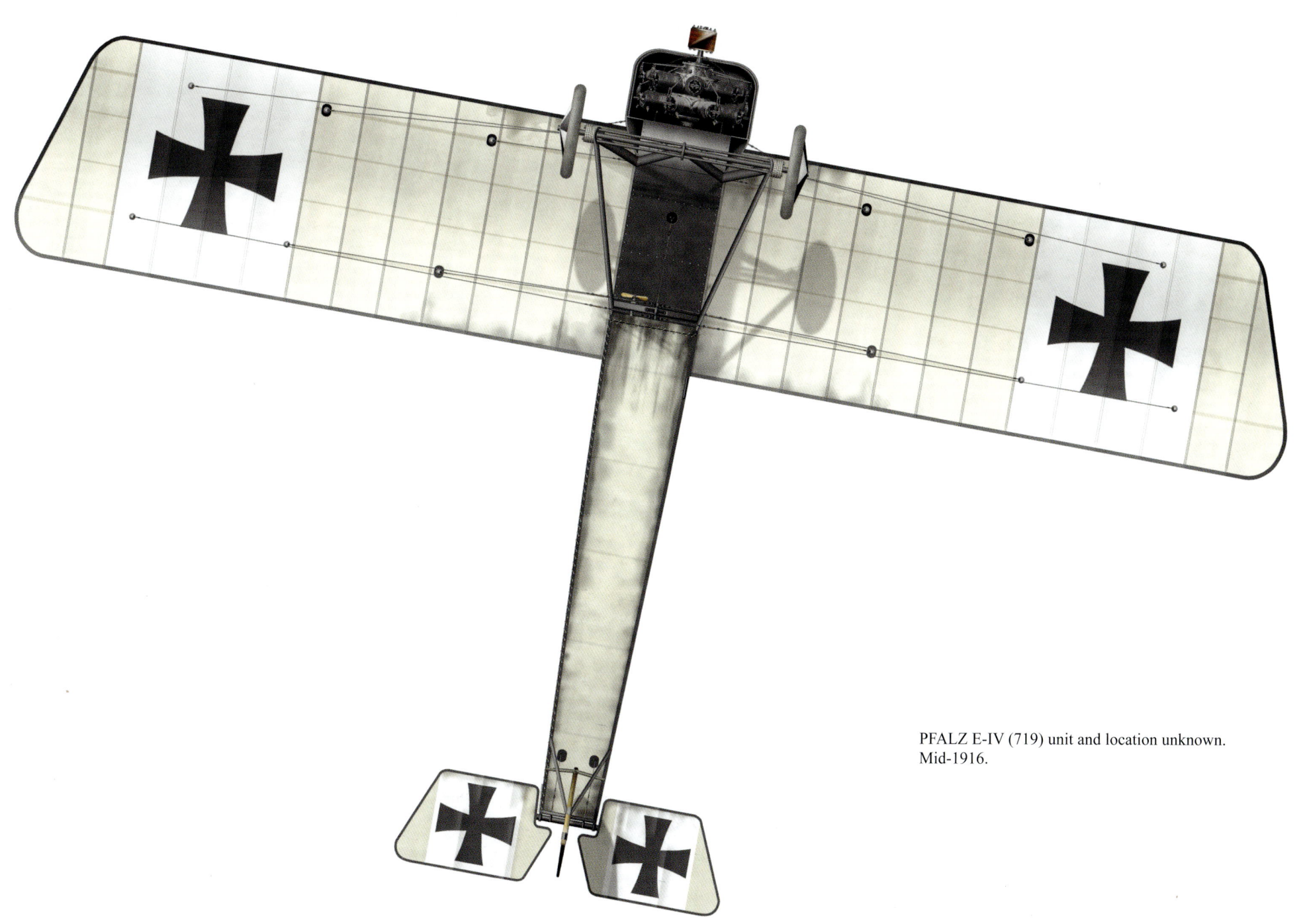

PFALZ E-IV (719) unit and location unknown.
Mid-1916.

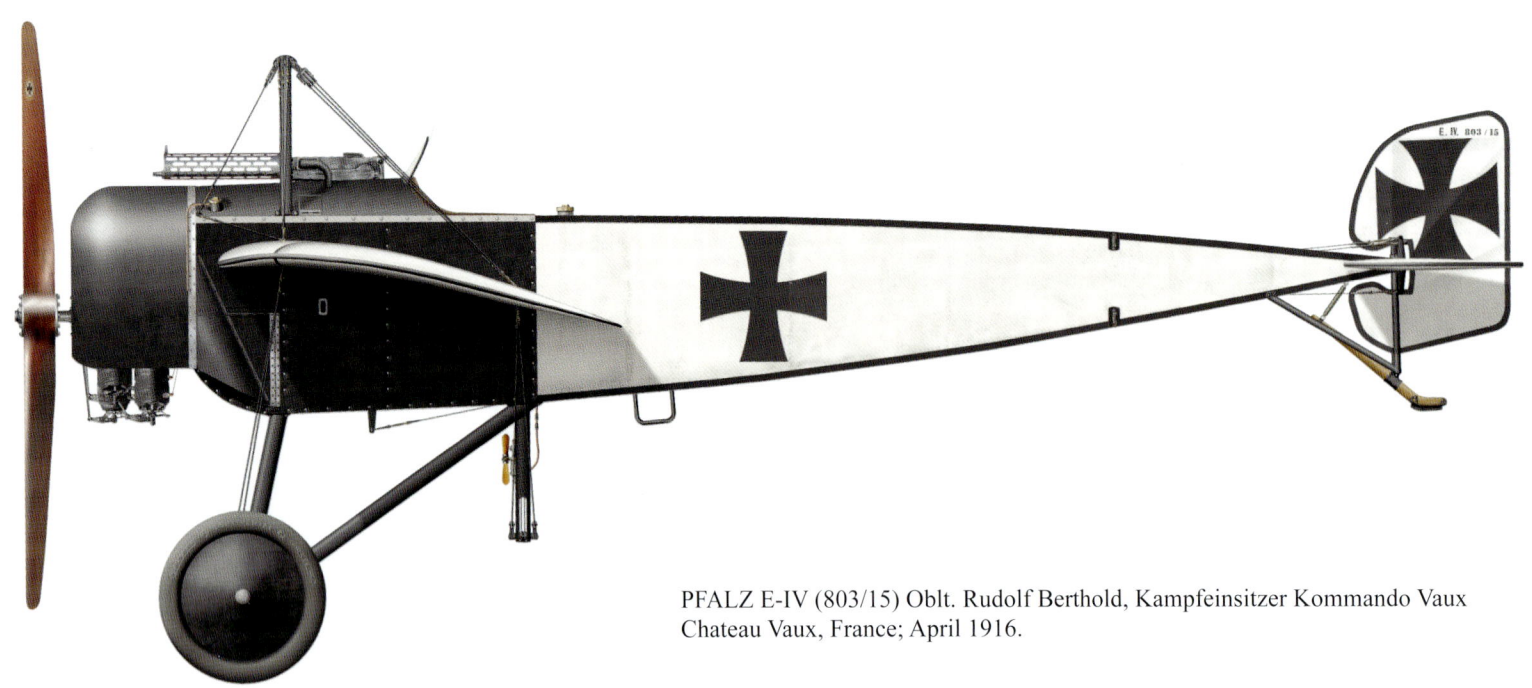

PFALZ E-IV (803/15) Oblt. Rudolf Berthold, Kampfeinsitzer Kommando Vaux
Chateau Vaux, France; April 1916.

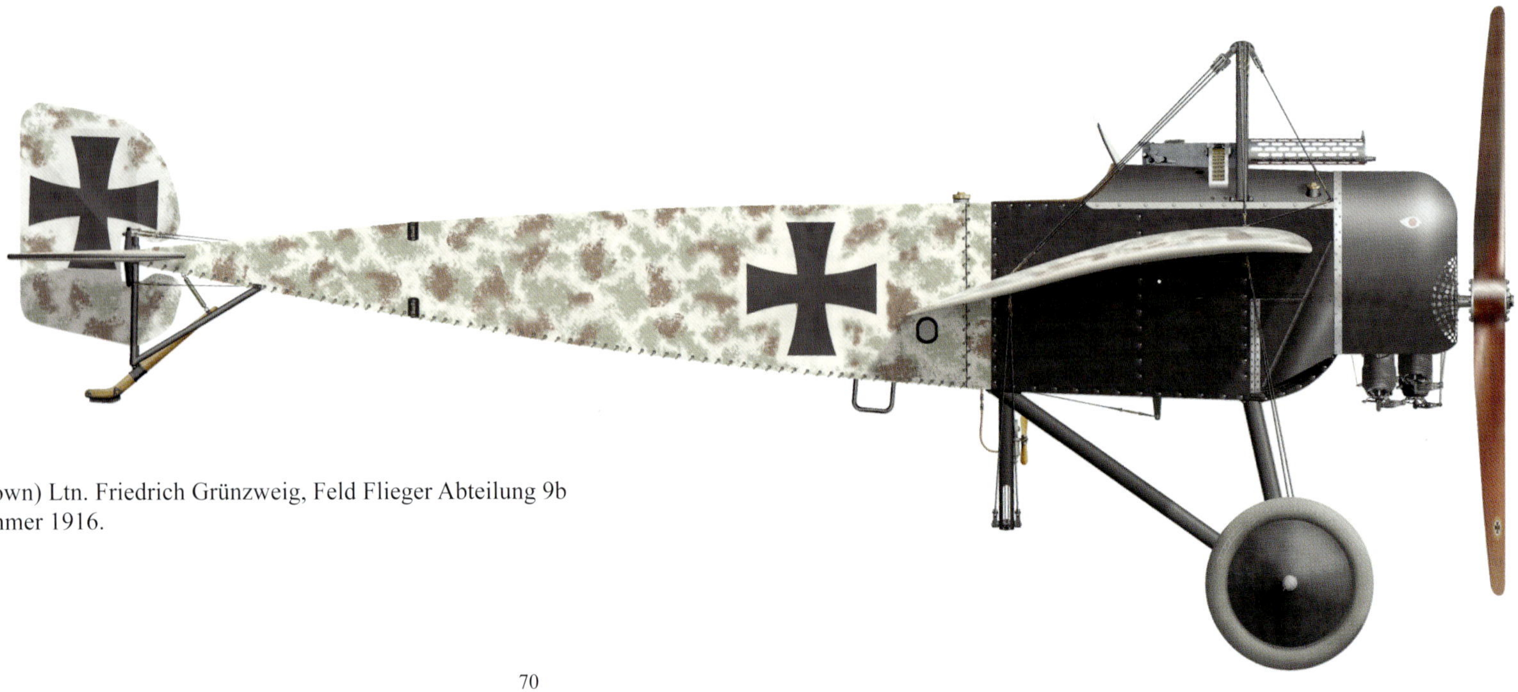

PFALZ E-IV (serial unknown) Ltn. Friedrich Grünzweig, Feld Flieger Abteilung 9b
Ensisheim, Germany; Summer 1916.

FOKKER D-I

FOKKER D-I (151/16) Jagdstaffel 1
Bertincourt, France; August 1916.

FOKKER D-I (159/16) unit and location unknown; Late 1916. Upper surfaces painted in camouflage colours most probably in the field. (See top view on page 72.)
Inter-plane struts were a replacement as the *Werk Nummern* does not match the number on the rudder.

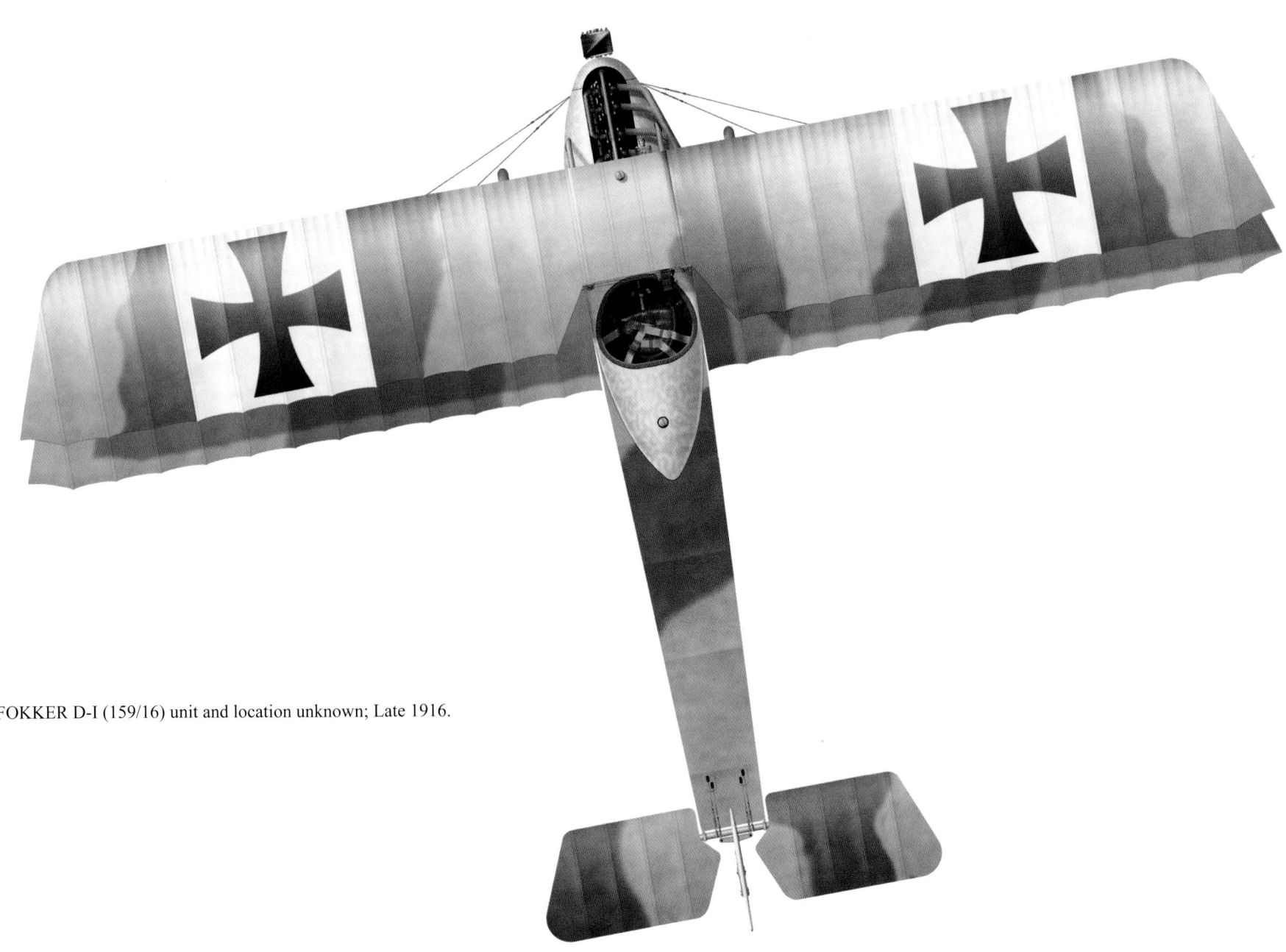

FOKKER D-I (159/16) unit and location unknown; Late 1916.

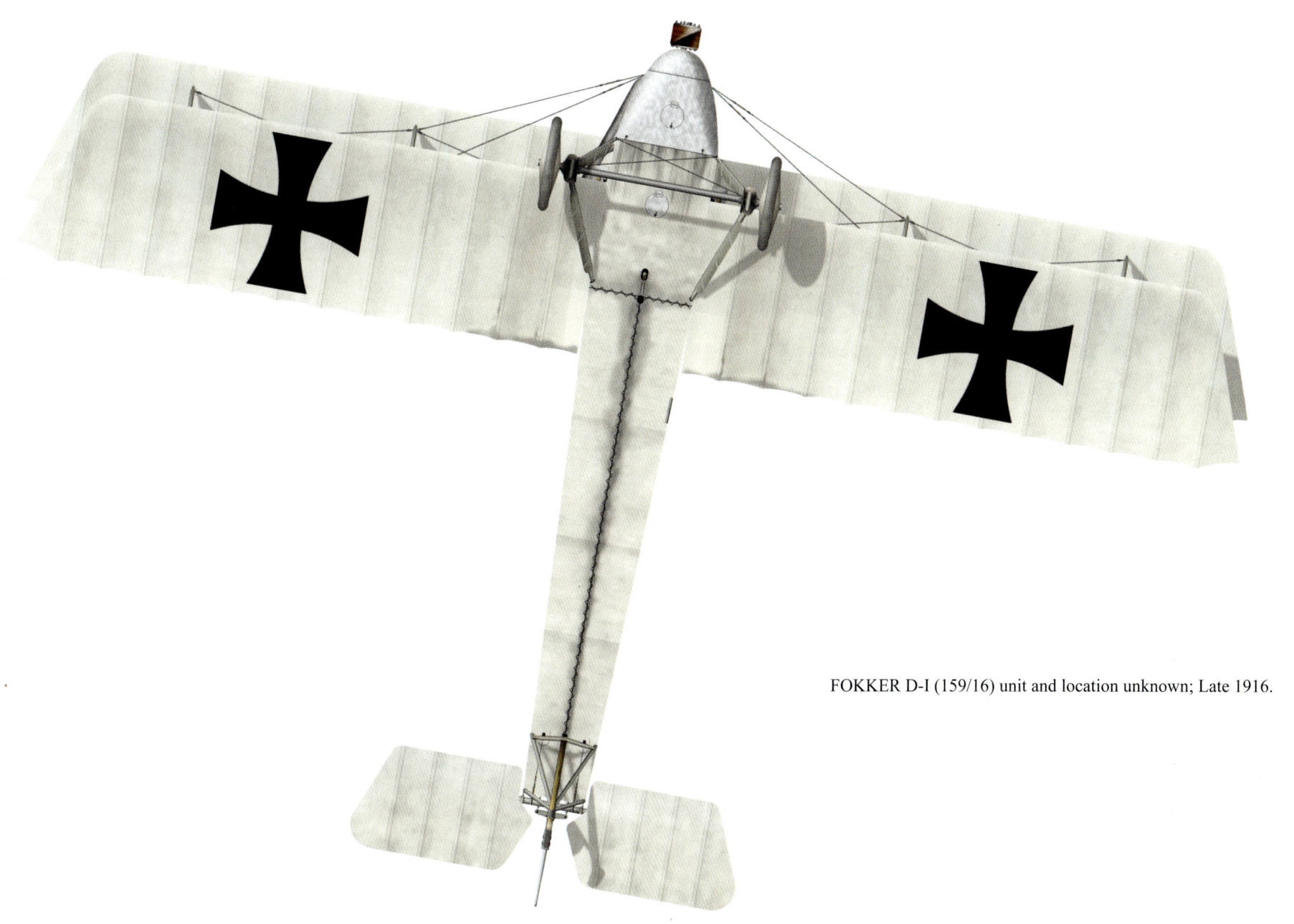
FOKKER D-I (159/16) unit and location unknown; Late 1916.

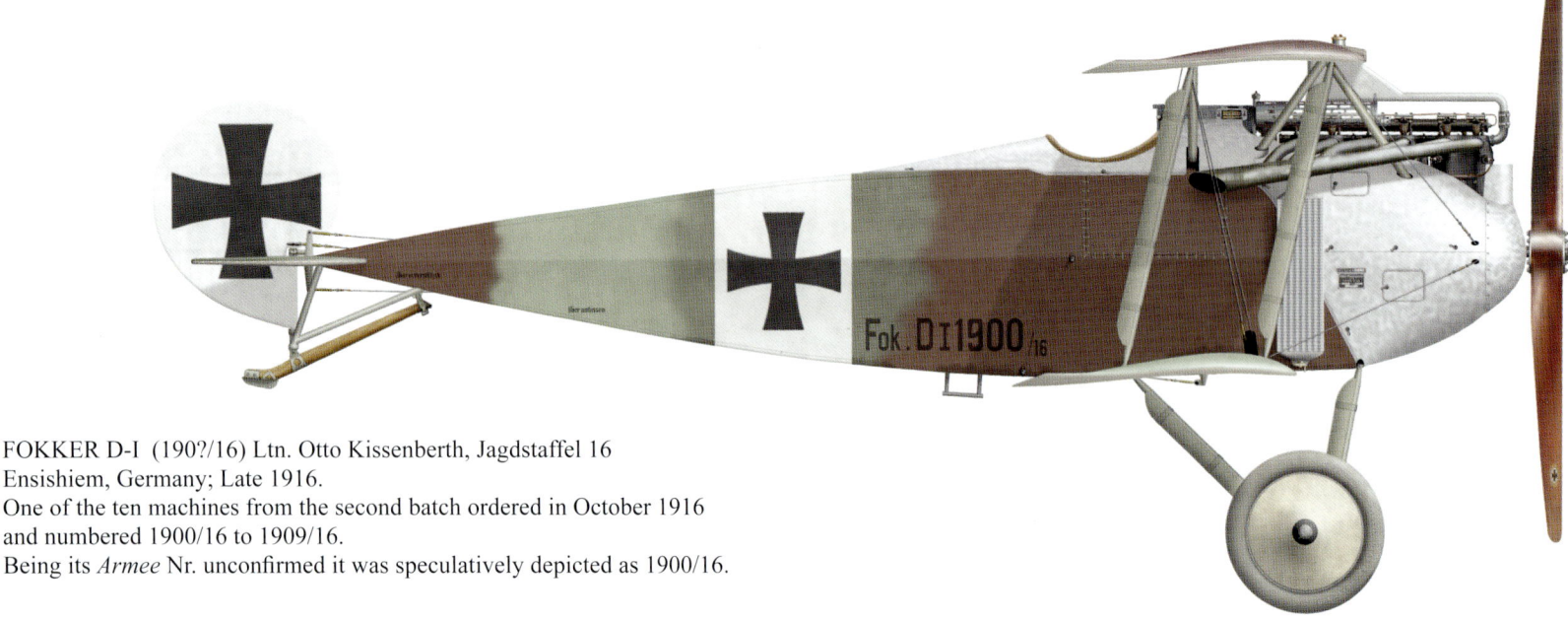

FOKKER D-I (190?/16) Ltn. Otto Kissenberth, Jagdstaffel 16
Ensisheim, Germany; Late 1916.
One of the ten machines from the second batch ordered in October 1916
and numbered 1900/16 to 1909/16.
Being its *Armee* Nr. unconfirmed it was speculatively depicted as 1900/16.

FOKKER D-I (208/16) Beobachter Schule Cöln
Cöln, Germany; September 1917.
Unarmed trainer.

FOKKER D-II

FOKKER D-II (541/16) Ltn d R. Josef Jacobs, Fokkerstaffel West
Chauny, France; September 1916.

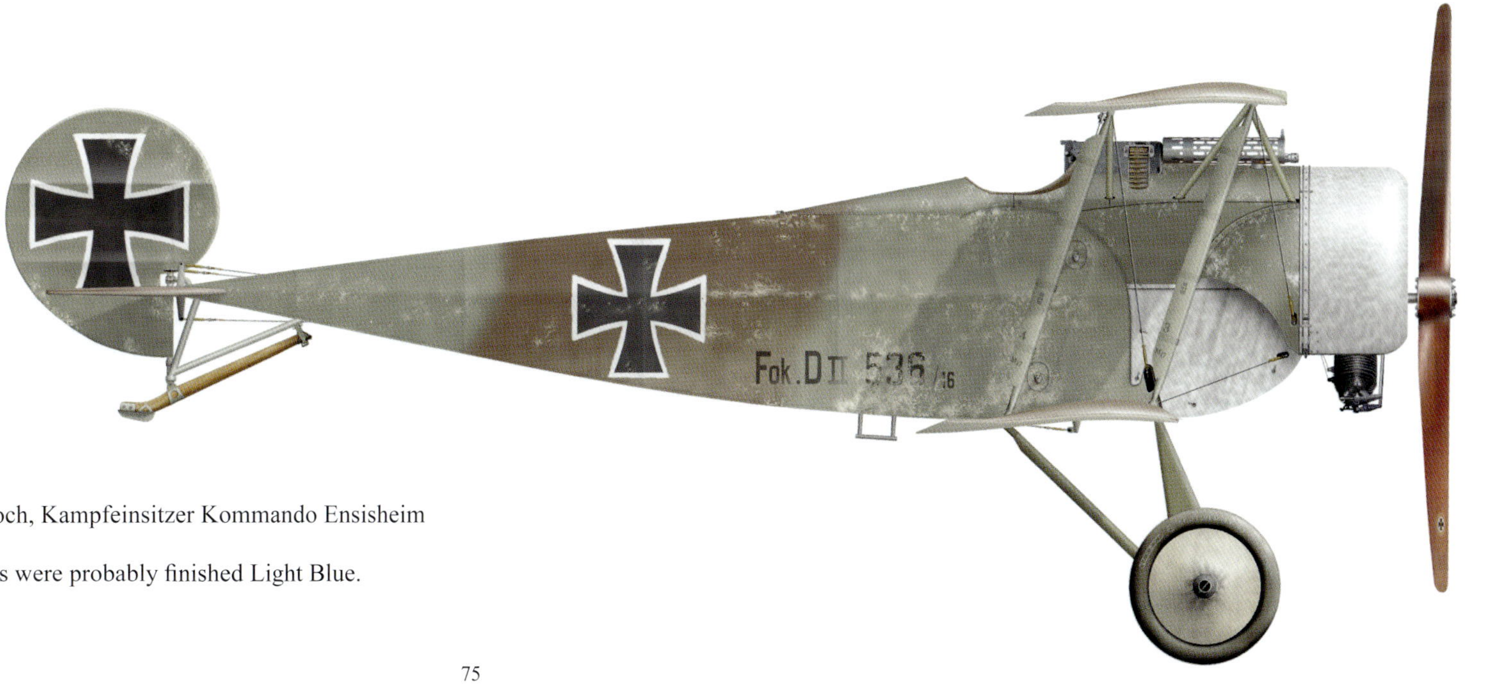

FOKKER D-II (536/16) Ltn. Otto Dessloch, Kampfeinsitzer Kommando Ensisheim
Ensisheim, Germany; October 1916.
According to some reports, undersurfaces were probably finished Light Blue.

FOKKER O-II (549/16) Ltn. Eberhard Baier, Jagdstaffel 16
Ensisheim, Germany; November 1916.

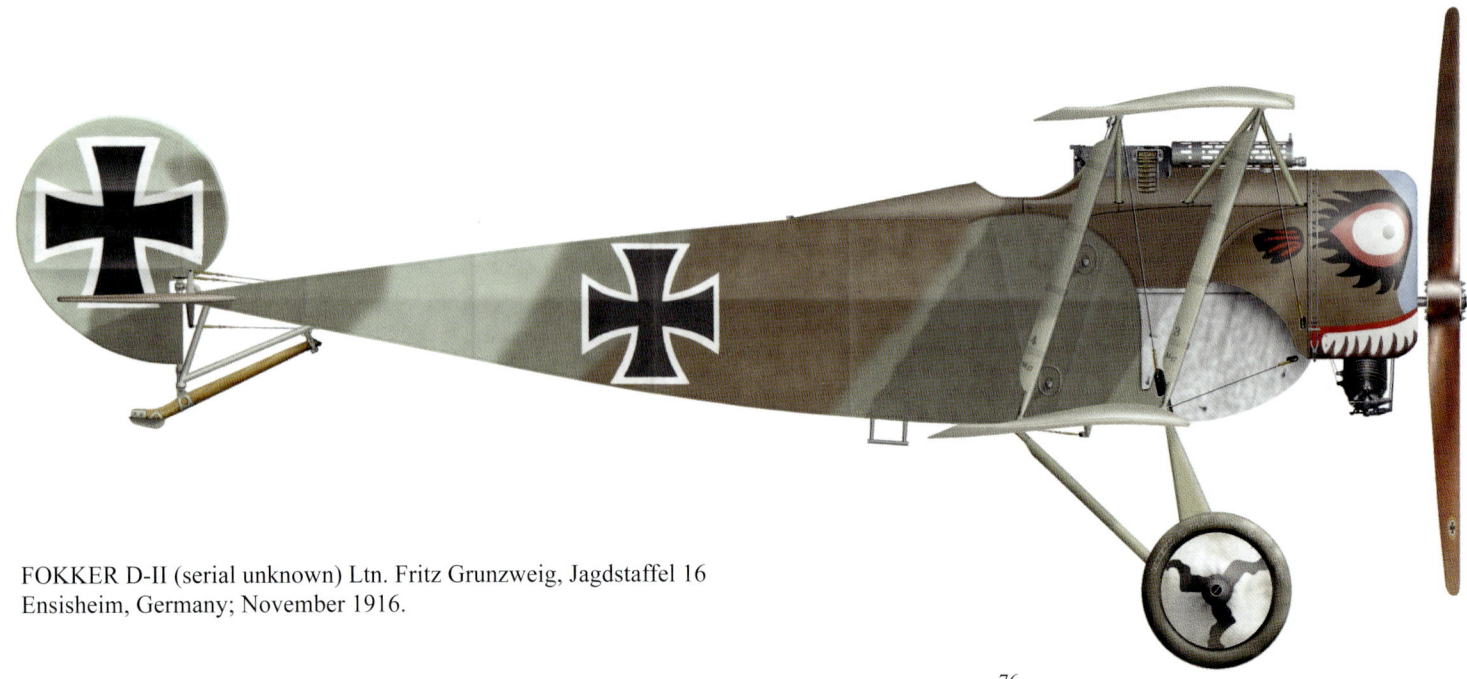

FOKKER D-II (serial unknown) Ltn. Fritz Grunzweig, Jagdstaffel 16
Ensisheim, Germany; November 1916.

FOKKER D-II (serial unknown) Ltn. Otto Kissenberth, Jagdstaffel 16
Ensisheim, Germany; November 1916.

FOKKER D-II (serial unknown) Ltn. Otto Kissenberth, Jagdstaffel 16
Ensisheim, Germany; November 1916.

FOKKER D-II (serial unknown) Ltn. Otto Kissenberth, Jagdstaffel 16
Ensisheim, Germany; November 1916.

FOKKER D-II (2393/16) Kampfeinsitzer Staffel 4
Freiburg, Germany; Late 1916.

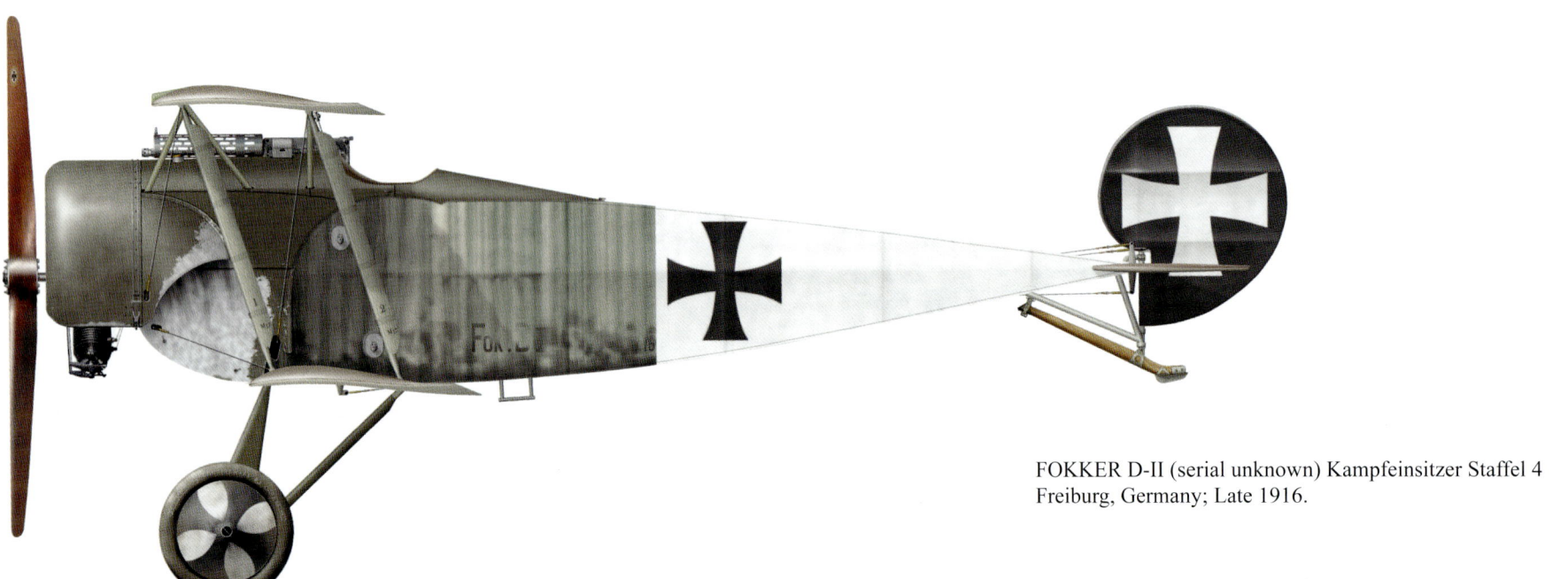

FOKKER D-II (serial unknown) Kampfeinsitzer Staffel 4
Freiburg, Germany; Late 1916.

FOKKER D-III

FOKKER D-III (352/16) Haptm. Oswald Boelcke, Jagdstaffel 2
Bertincourt, France; September 1916.

FOKKER D-III (368/16) Vzfw. Ernst Udet, Jagdstaffel 15
Habsheim, Germany; October 1916.
Notice the silhouette of an observer made of tin plate to deceive
the enemy pilots into thinking they were attacking a two-seater.

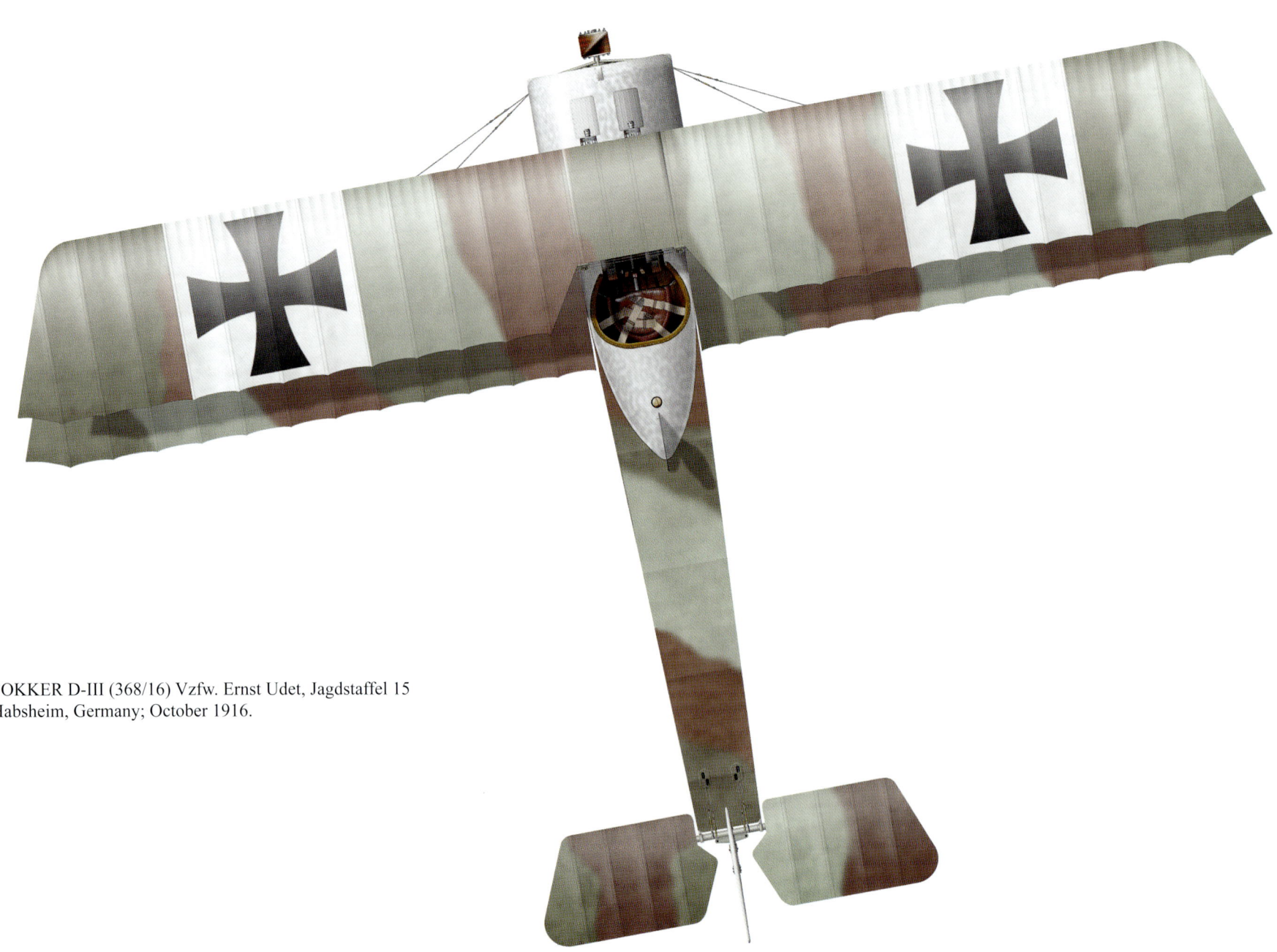

FOKKER D-III (368/16) Vzfw. Ernst Udet, Jagdstaffel 15
Habsheim, Germany; October 1916.

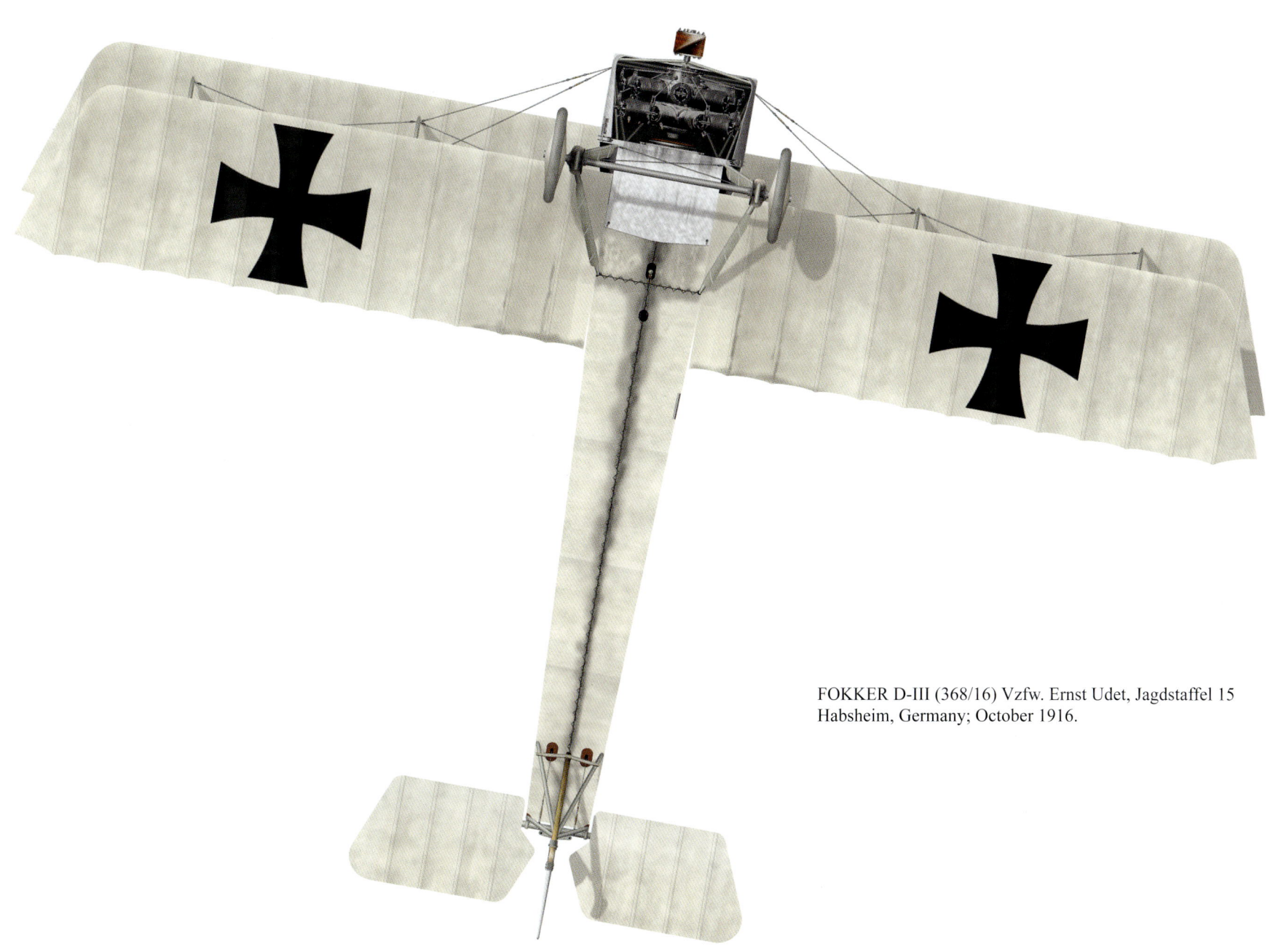

FOKKER D-III (368/16) Vzfw. Ernst Udet, Jagdstaffel 15
Habsheim, Germany; October 1916.

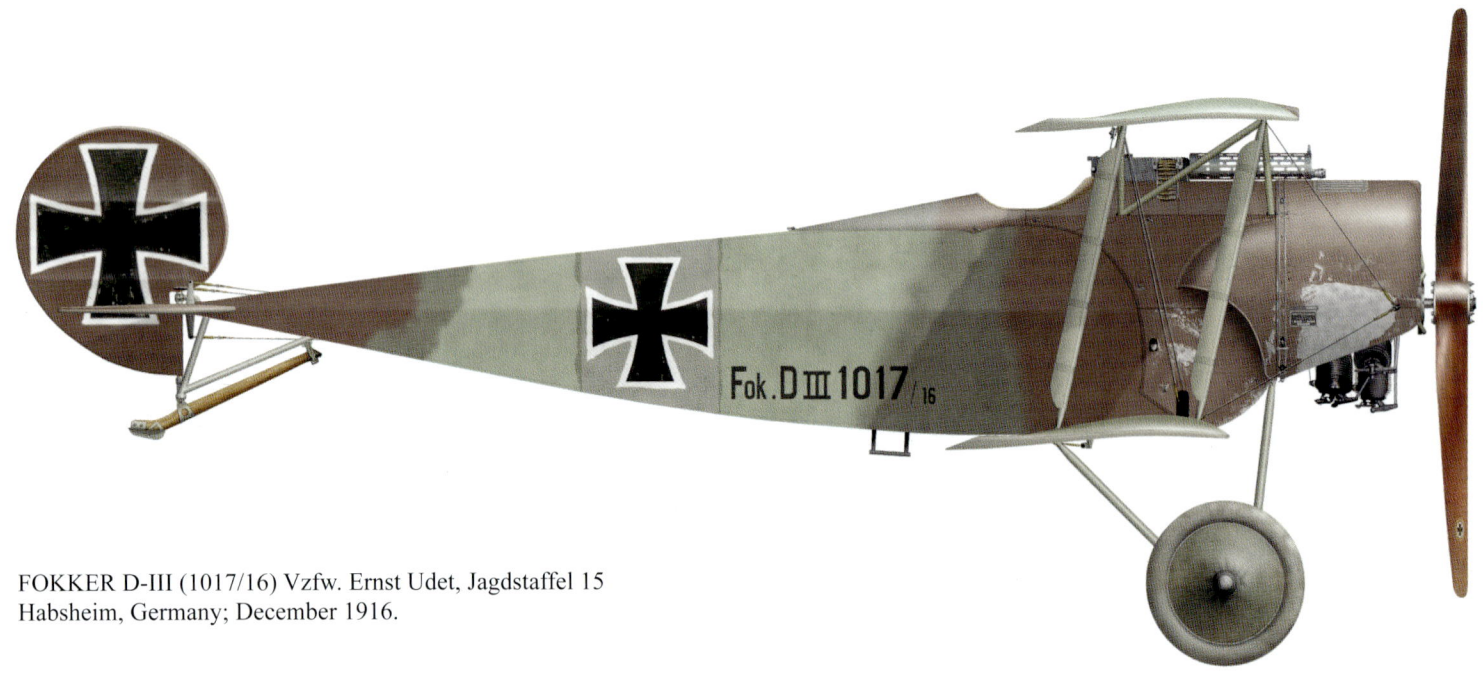

FOKKER D-III (1017/16) Vzfw. Ernst Udet, Jagdstaffel 15
Habsheim, Germany; December 1916.

FOKKER D-III (1612/16) Kampfeinsitzer Staffel 4
Freiburg, Germany; Late 1916.

HALBERSTADT D-II

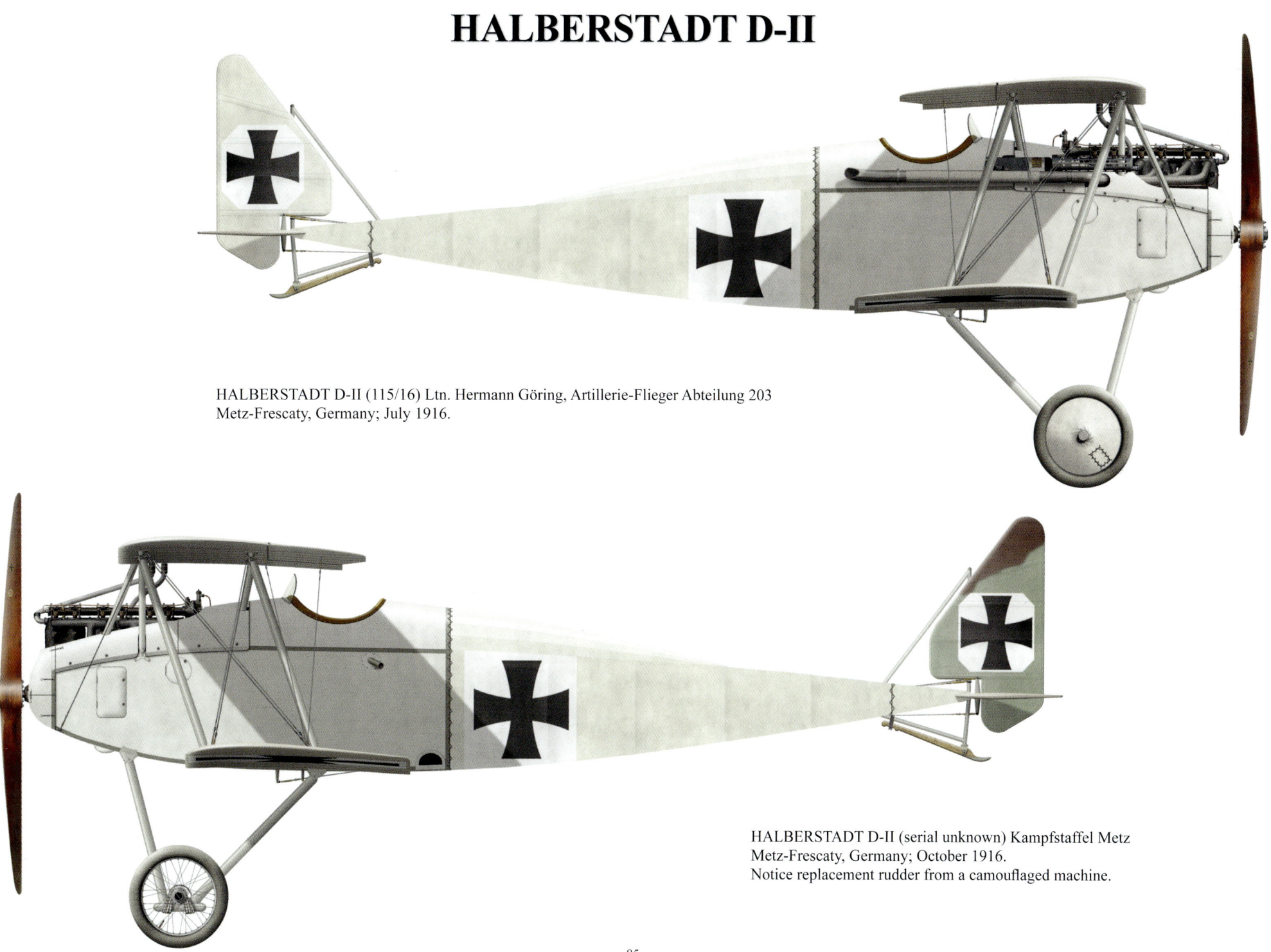

HALBERSTADT D-II (115/16) Ltn. Hermann Göring, Artillerie-Flieger Abteilung 203
Metz-Frescaty, Germany; July 1916.

HALBERSTADT D-II (serial unknown) Kampfstaffel Metz
Metz-Frescaty, Germany; October 1916.
Notice replacement rudder from a camouflaged machine.

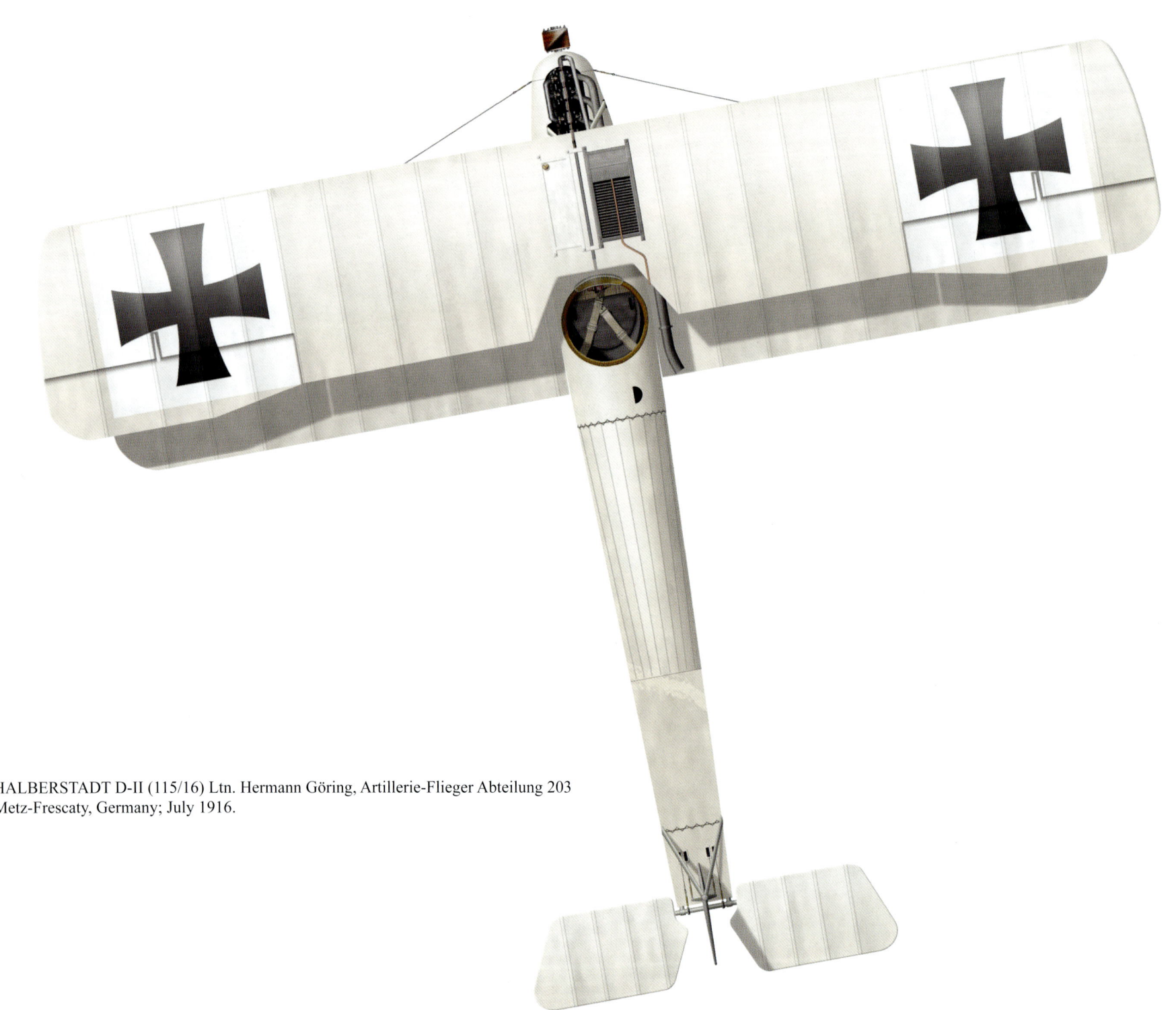

HALBERSTADT D-II (115/16) Ltn. Hermann Göring, Artillerie-Flieger Abteilung 203
Metz-Frescaty, Germany; July 1916.

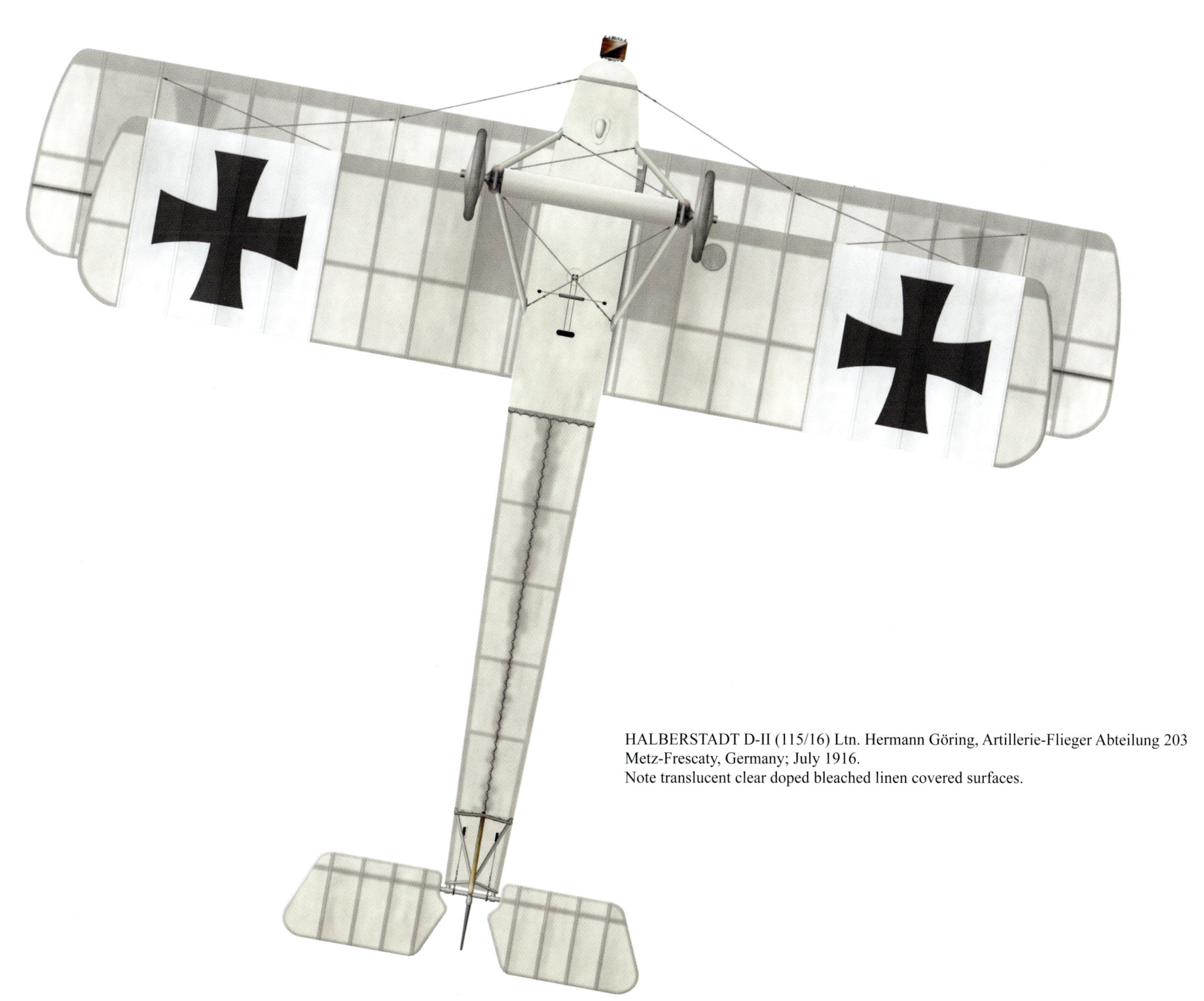

HALBERSTADT D-II (115/16) Ltn. Hermann Göring, Artillerie-Flieger Abteilung 203
Metz-Frescaty, Germany; July 1916.
Note translucent clear doped bleached linen covered surfaces.

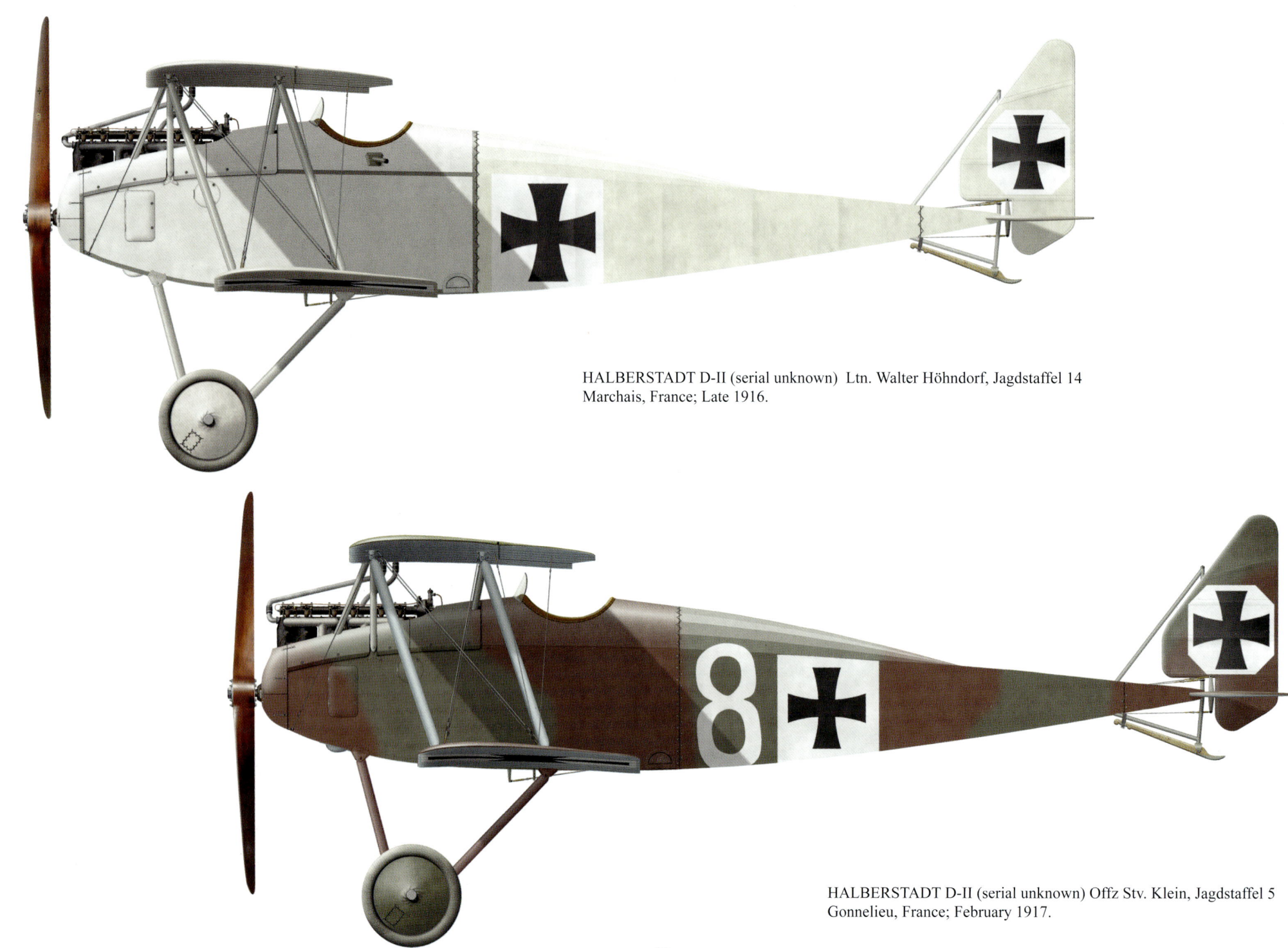

HALBERSTADT D-II (serial unknown) Ltn. Walter Höhndorf, Jagdstaffel 14
Marchais, France; Late 1916.

HALBERSTADT D-II (serial unknown) Offz Stv. Klein, Jagdstaffel 5
Gonnelieu, France; February 1917.

HALBERSTADT D-II (Han) & (Av)

HALBERSTADT D-II (Han) (818/16) Halbgeschwader I
Hudova, Macedonia; Late 1916.

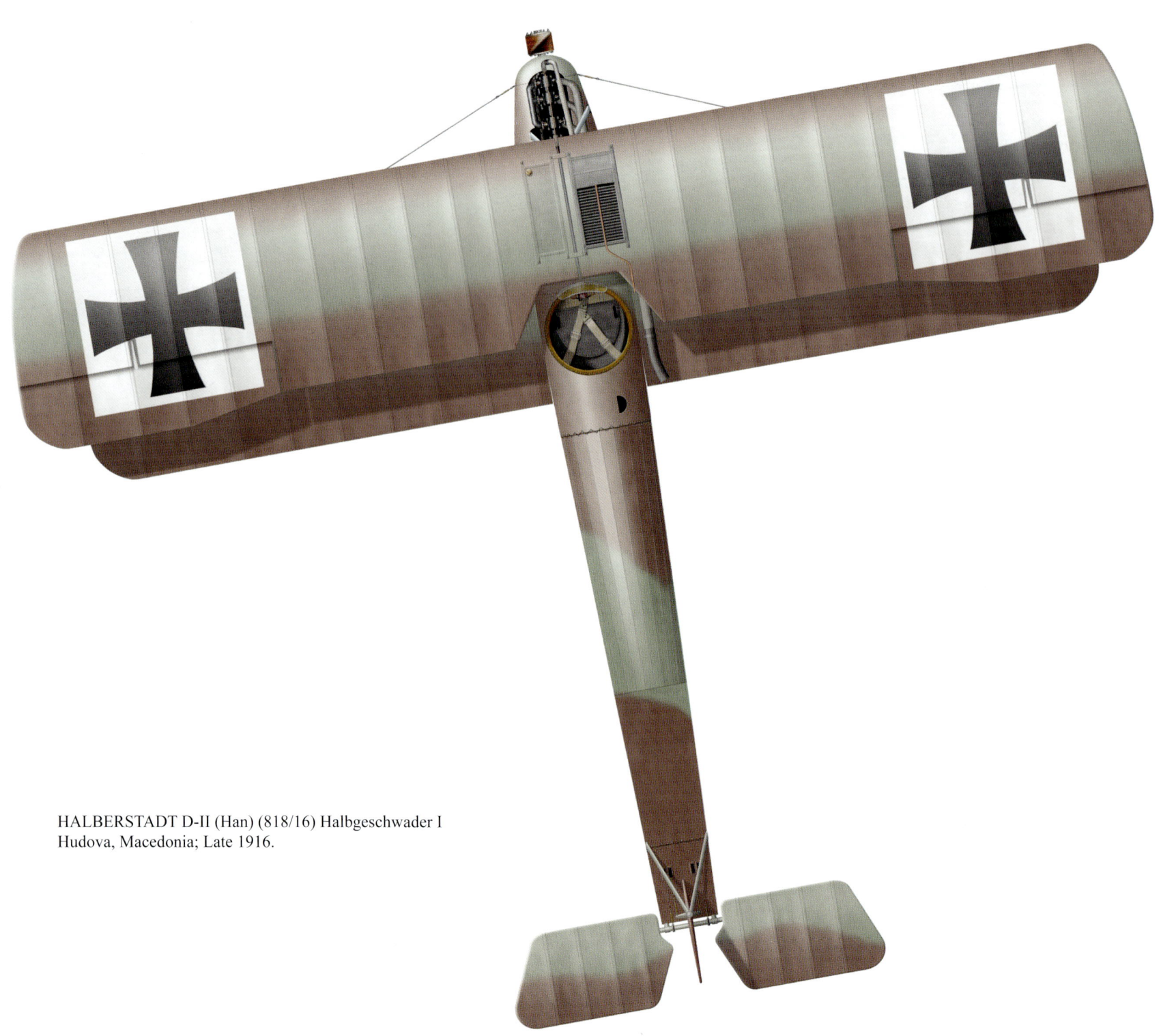

HALBERSTADT D-II (Han) (818/16) Halbgeschwader I
Hudova, Macedonia; Late 1916.

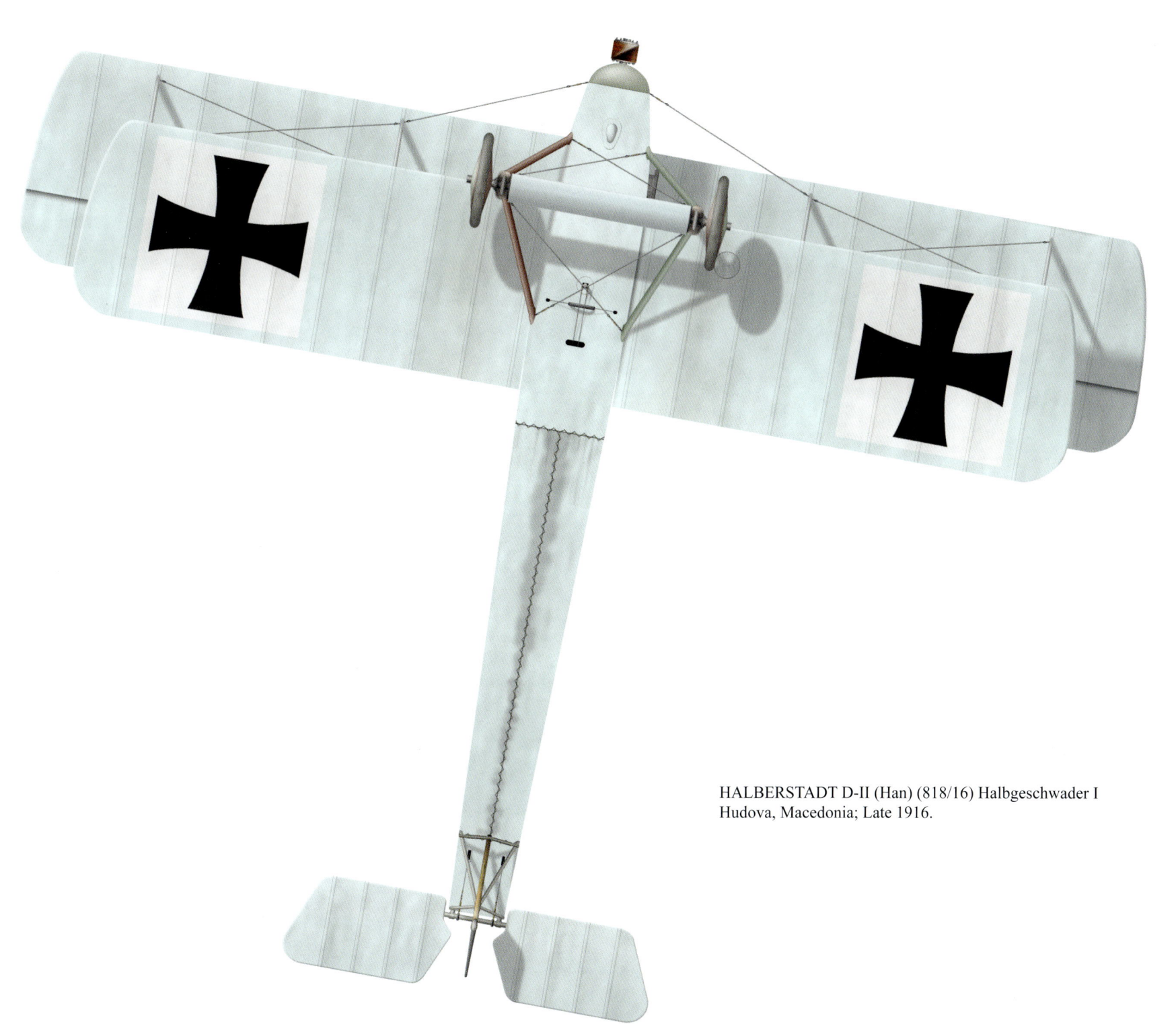

HALBERSTADT D-II (Han) (818/16) Halbgeschwader I
Hudova, Macedonia; Late 1916.

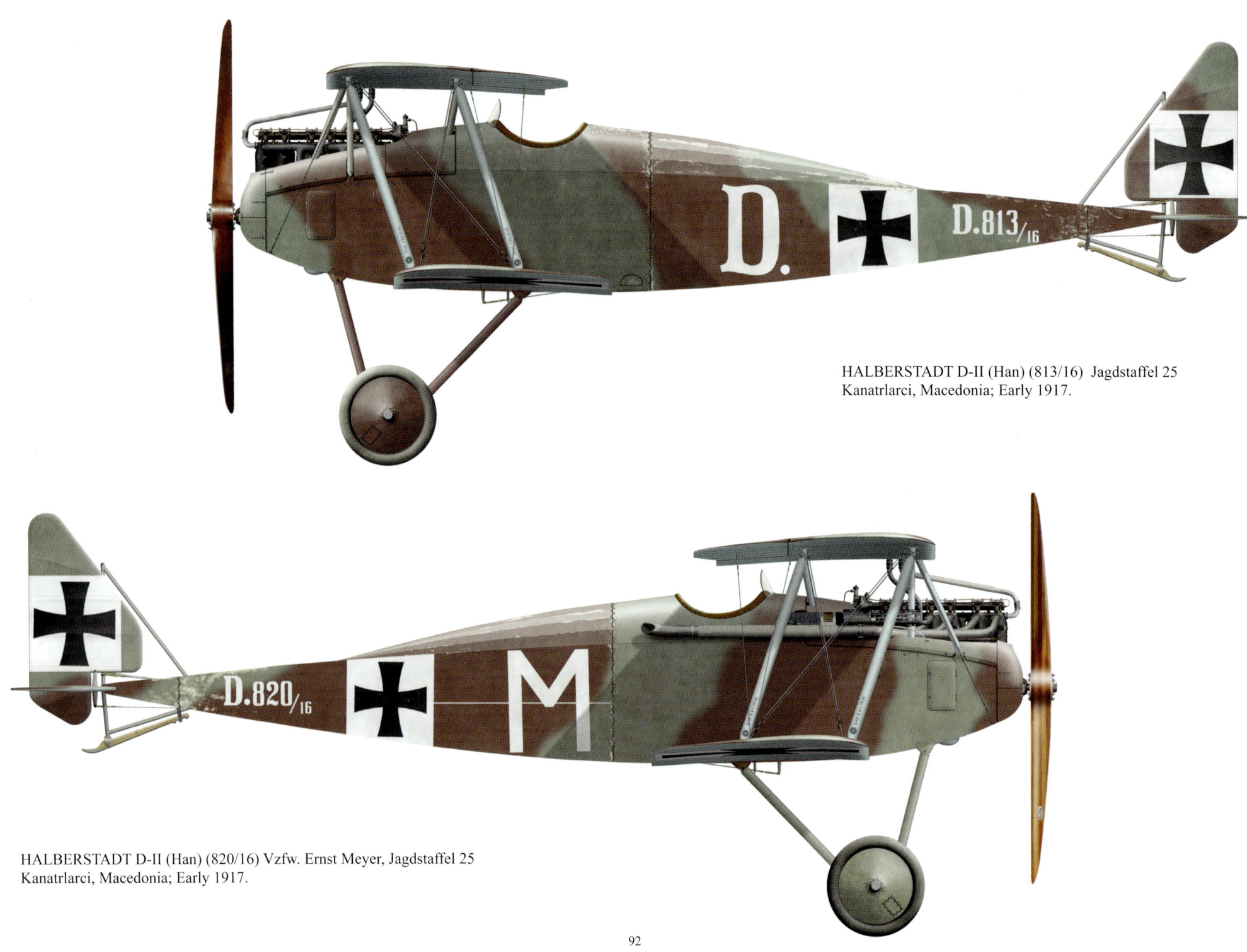

HALBERSTADT D-II (Han) (813/16) Jagdstaffel 25
Kanatrlarci, Macedonia; Early 1917.

HALBERSTADT D-II (Han) (820/16) Vzfw. Ernst Meyer, Jagdstaffel 25
Kanatrlarci, Macedonia; Early 1917.

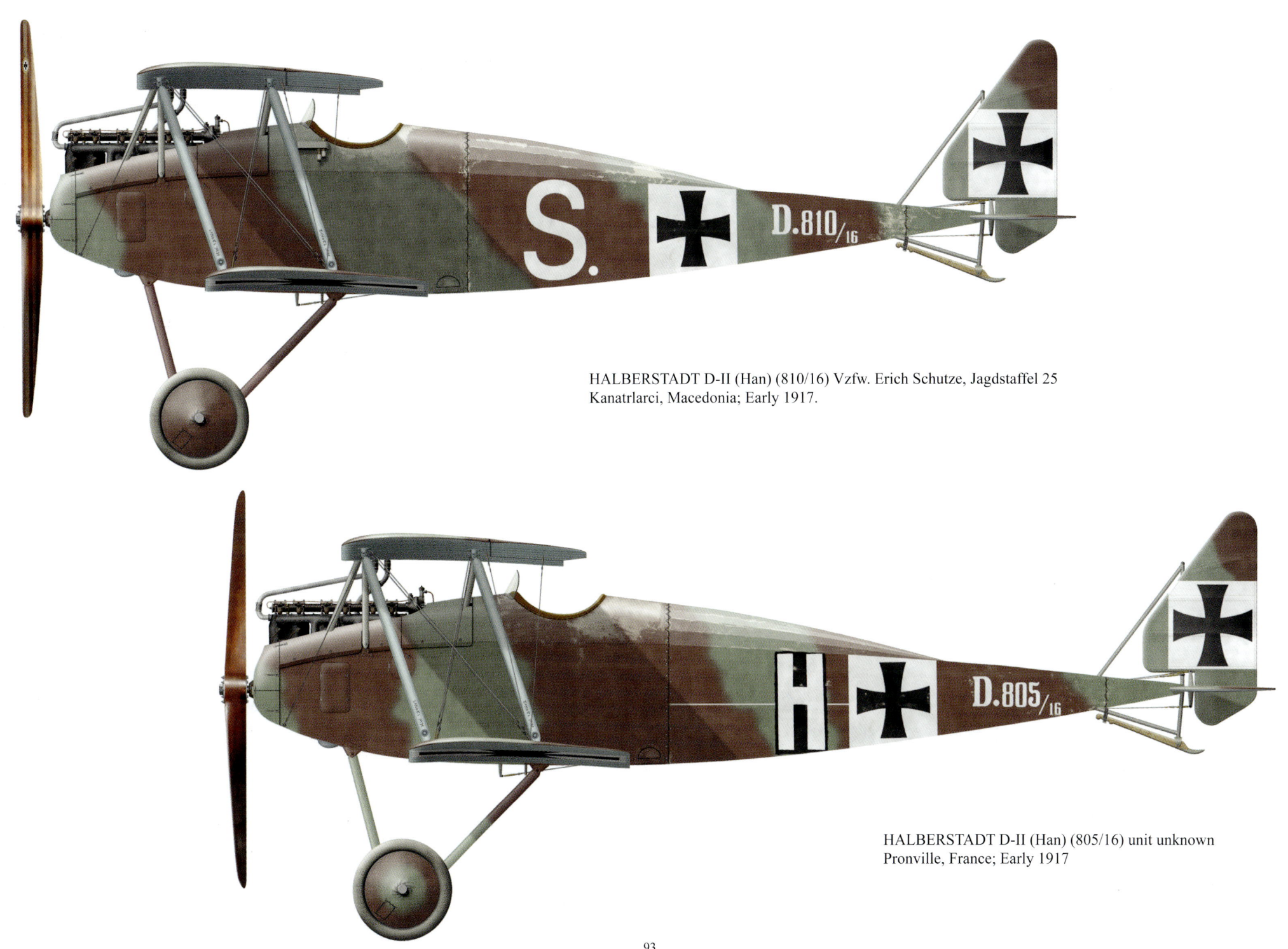

HALBERSTADT D-II (Han) (810/16) Vzfw. Erich Schutze, Jagdstaffel 25
Kanatrlarci, Macedonia; Early 1917.

HALBERSTADT D-II (Han) (805/16) unit unknown
Pronville, France; Early 1917

HALBERSTADT D-II (Av) (588/16) Ltn. Rolf von Lersner, Jastaschule 1
Valenciennes, France; Early 1917.

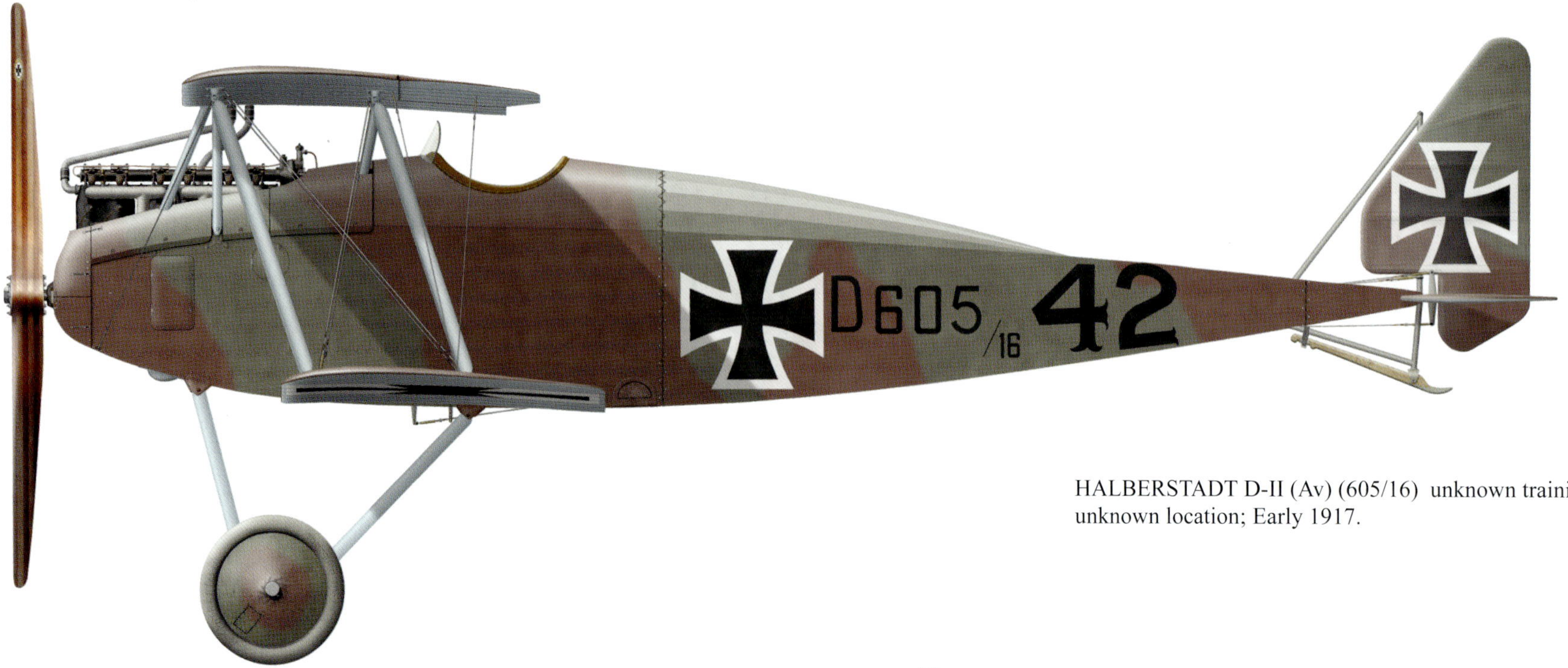

HALBERSTADT D-II (Av) (605/16) unknown training unit
unknown location; Early 1917.

HALBERSTADT D-III

HALBERSTADT D-III (serial unknown) Ltn. Ernst von Althaus, Jagdstaffel 4
Vaux, France; August 1916.

HALBERSTADT D-III (serial unknown) Ltn. Hans von Keudell, Jagdstaffel 1
Bertincourt, France; September 1916.

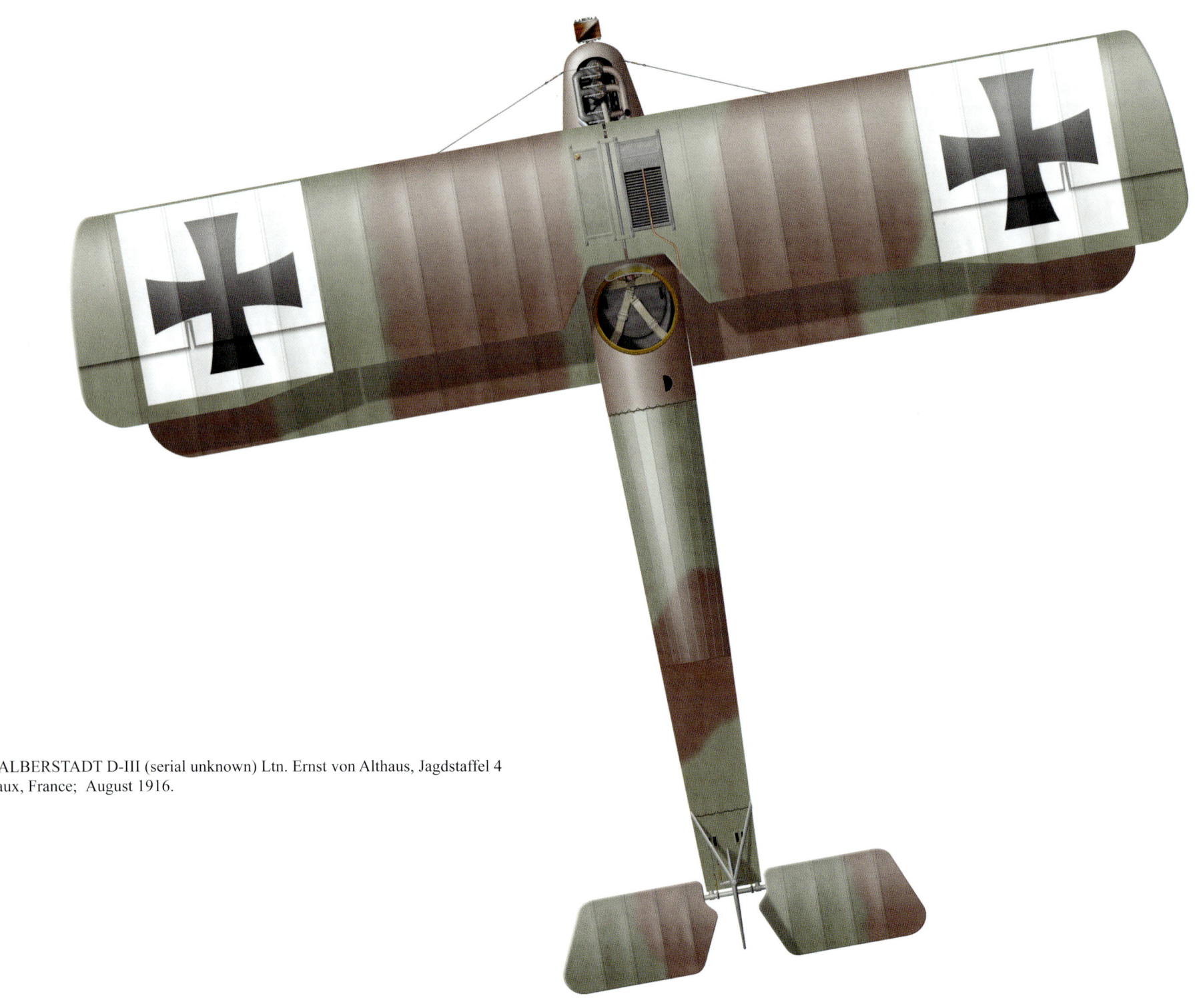

HALBERSTADT D-III (serial unknown) Ltn. Ernst von Althaus, Jagdstaffel 4
Vaux, France; August 1916.

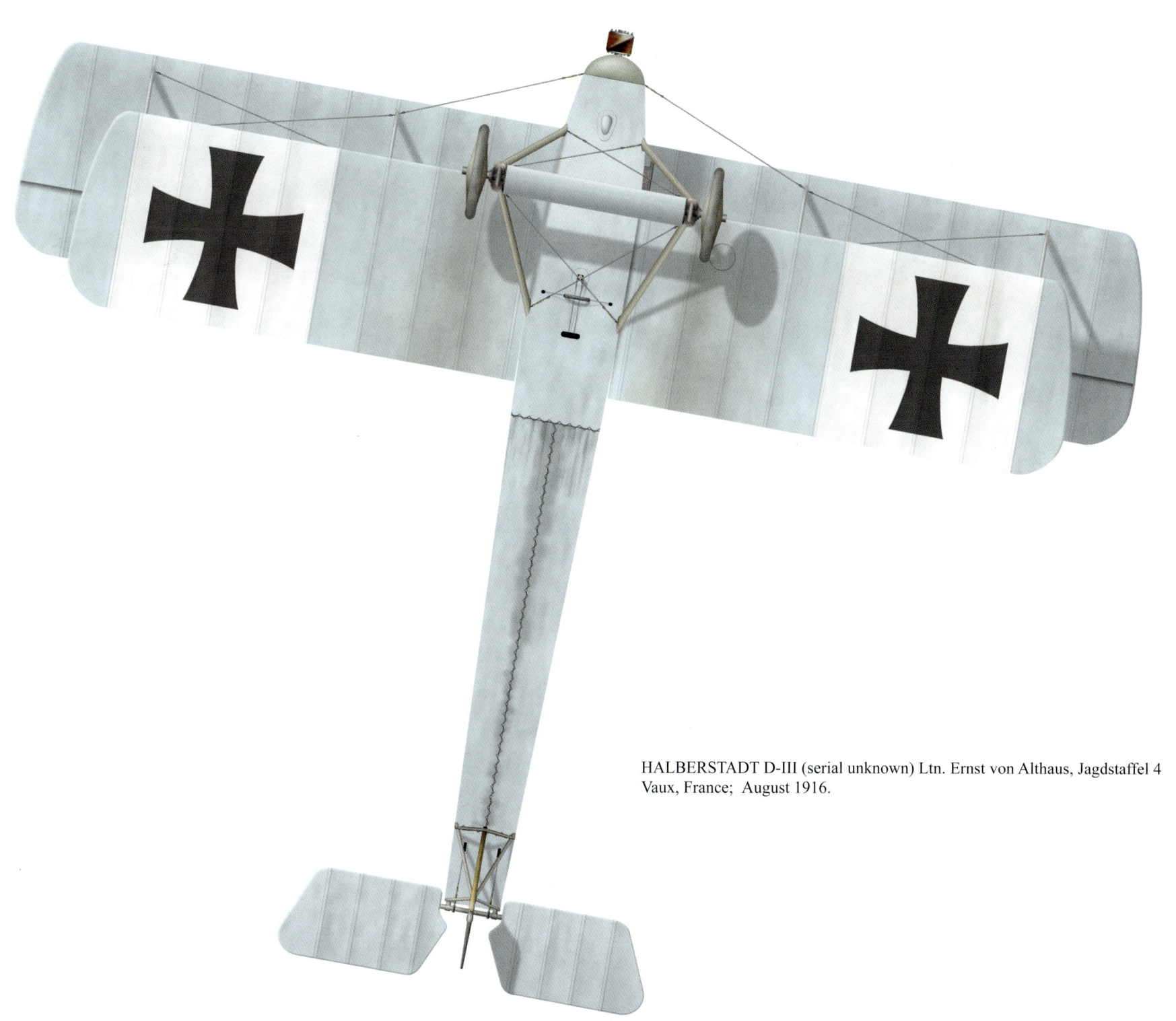

HALBERSTADT D-III (serial unknown) Ltn. Ernst von Althaus, Jagdstaffel 4
Vaux, France; August 1916.

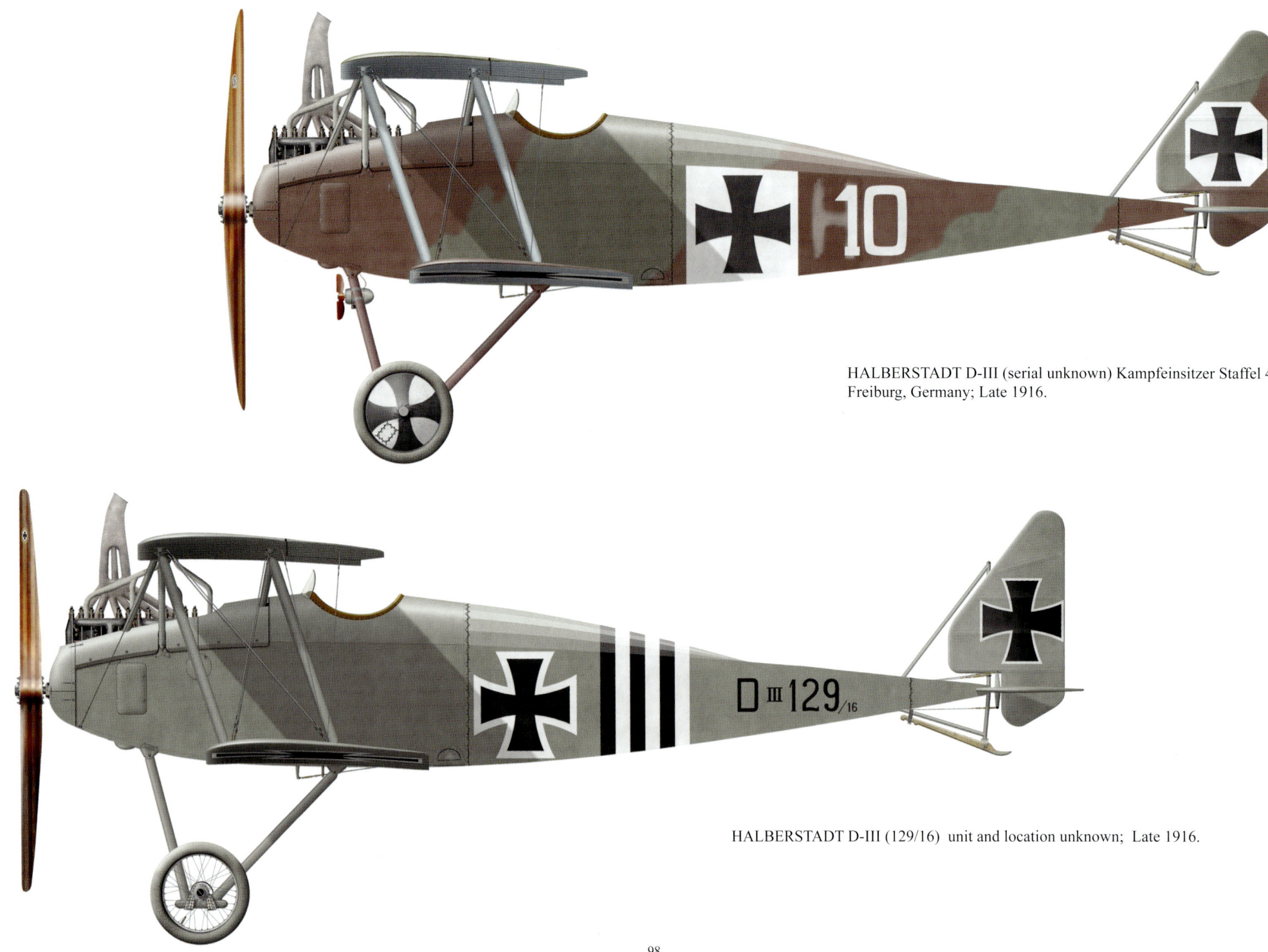

HALBERSTADT D-III (serial unknown) Kampfeinsitzer Staffel 4 Freiburg, Germany; Late 1916.

HALBERSTADT D-III (129/16) unit and location unknown; Late 1916.

HALBERSTADT D-V

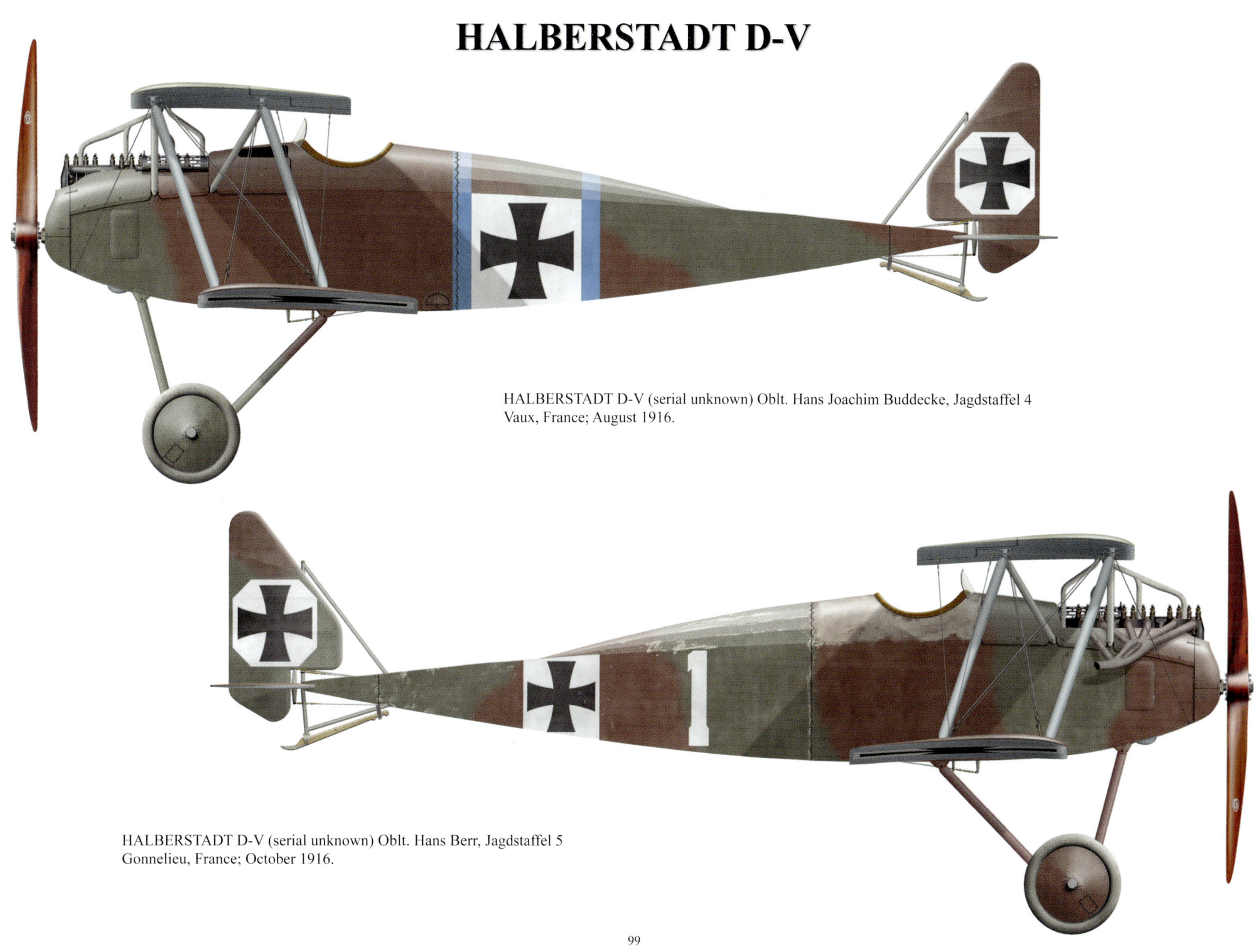

HALBERSTADT D-V (serial unknown) Oblt. Hans Joachim Buddecke, Jagdstaffel 4
Vaux, France; August 1916.

HALBERSTADT D-V (serial unknown) Oblt. Hans Berr, Jagdstaffel 5
Gonnelieu, France; October 1916.

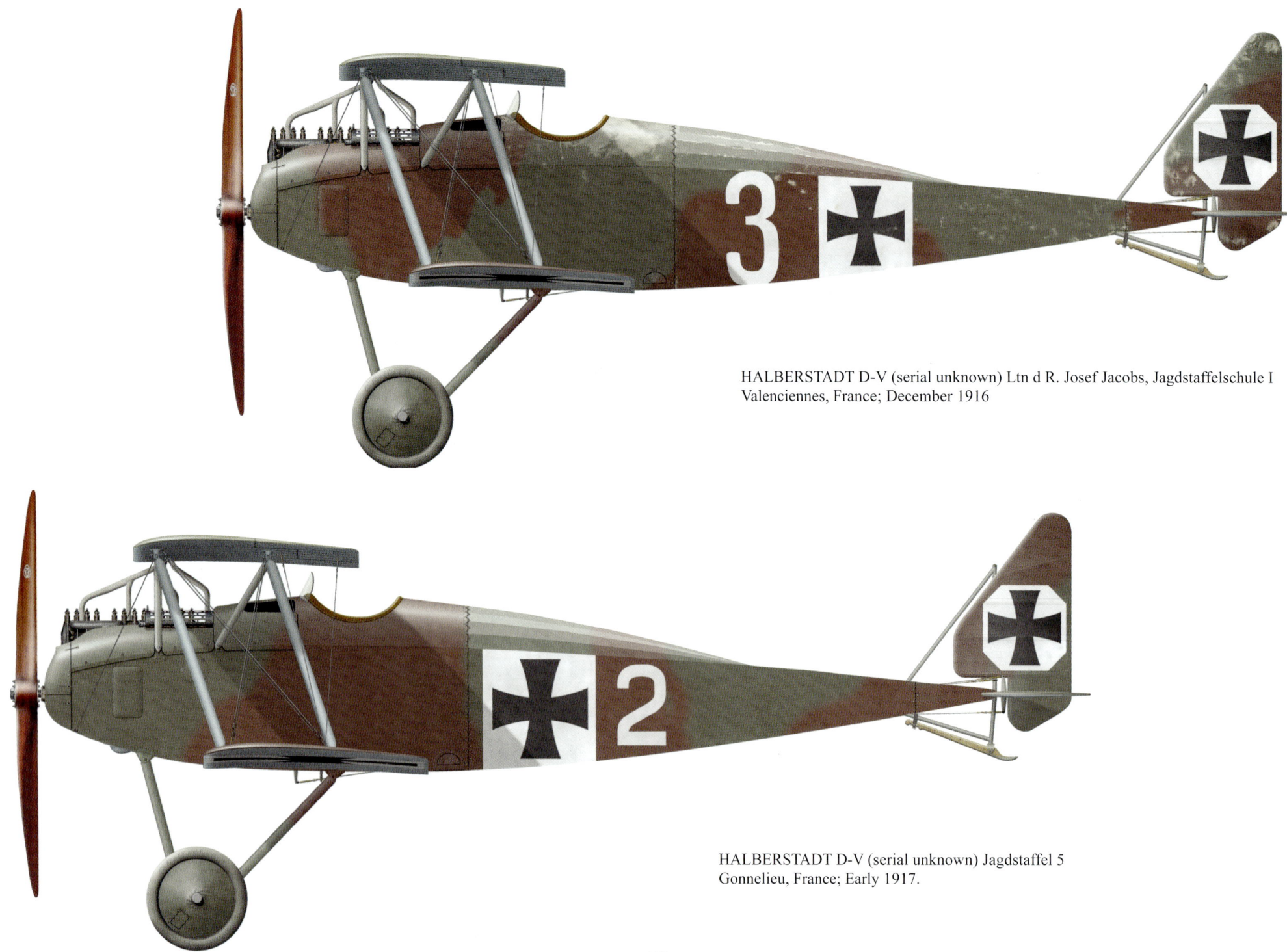

HALBERSTADT D-V (serial unknown) Ltn d R. Josef Jacobs, Jagdstaffelschule I Valenciennes, France; December 1916

HALBERSTADT D-V (serial unknown) Jagdstaffel 5 Gonnelieu, France; Early 1917.

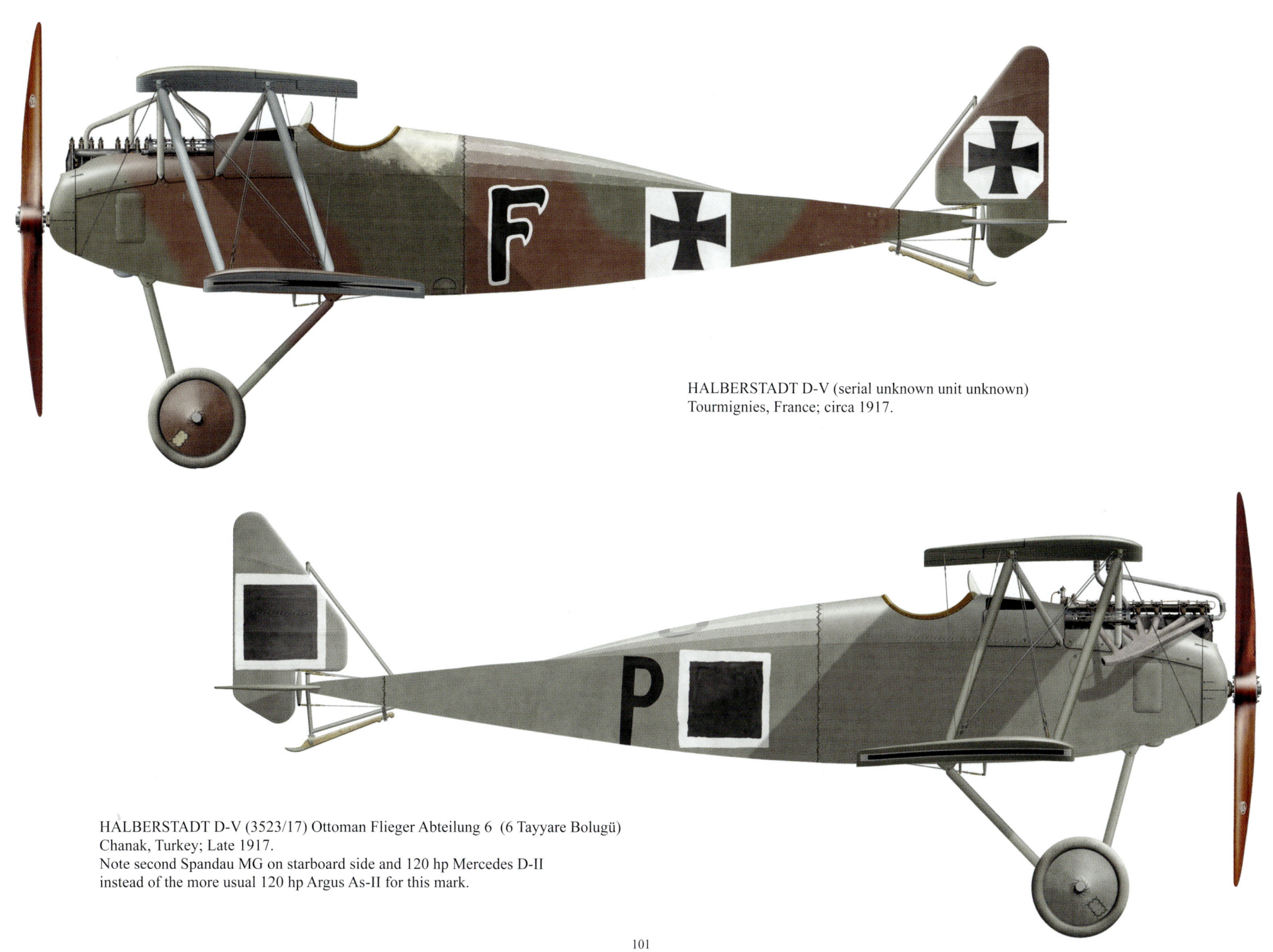

HALBERSTADT D-V (serial unknown unit unknown)
Tourmignies, France; circa 1917.

HALBERSTADT D-V (3523/17) Ottoman Flieger Abteilung 6 (6 Tayyare Bolugü)
Chanak, Turkey; Late 1917.
Note second Spandau MG on starboard side and 120 hp Mercedes D-II
instead of the more usual 120 hp Argus As-II for this mark.

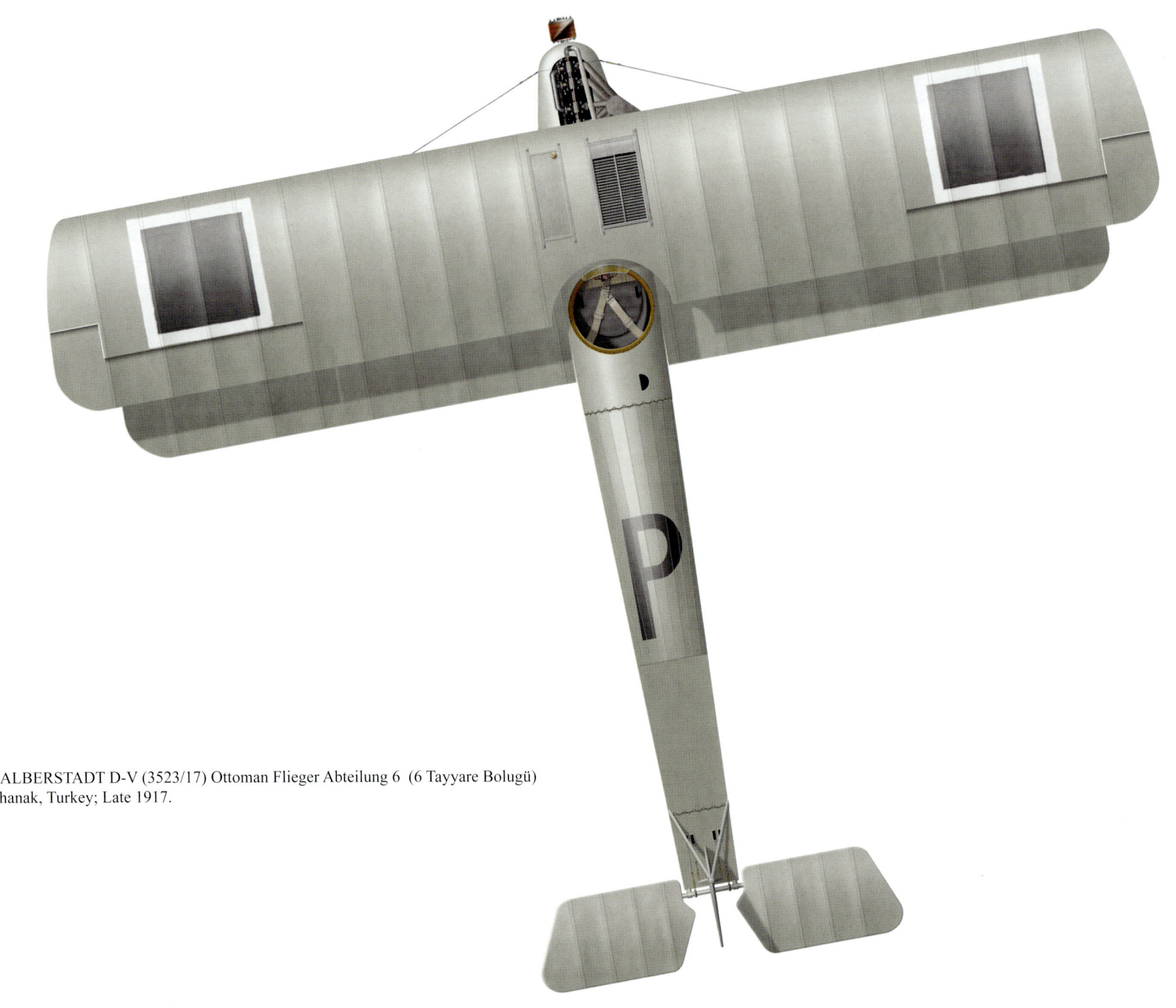

HALBERSTADT D-V (3523/17) Ottoman Flieger Abteilung 6 (6 Tayyare Bolugü)
Chanak, Turkey; Late 1917.

ALBATROS D-I

ALBATROS D-I (390/16) Ltn d R. Otto Höhne, Jagdstaffel 2
Lagnicourt, France; September 1916.

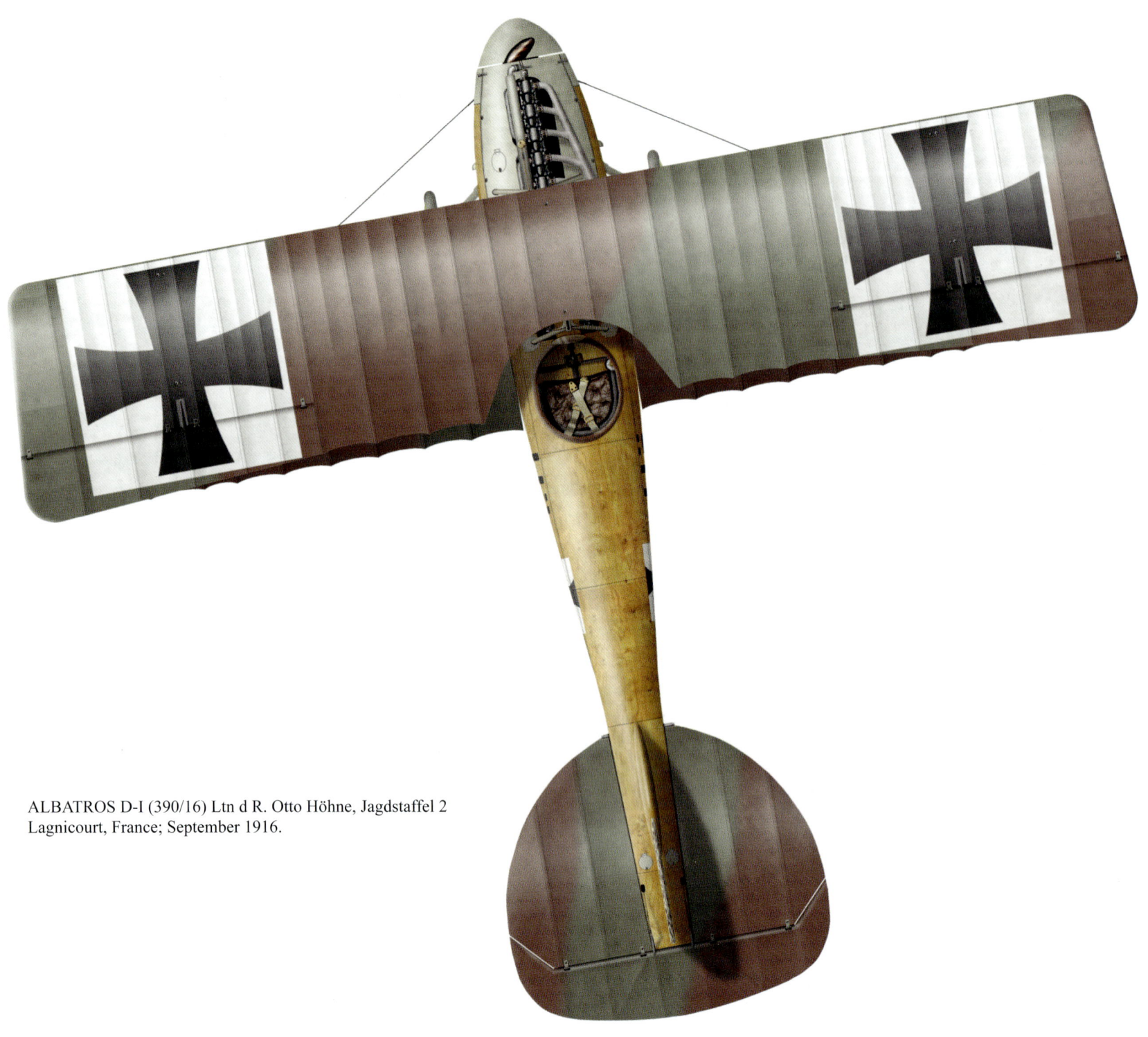

ALBATROS D-I (390/16) Ltn d R. Otto Höhne, Jagdstaffel 2
Lagnicourt, France; September 1916.

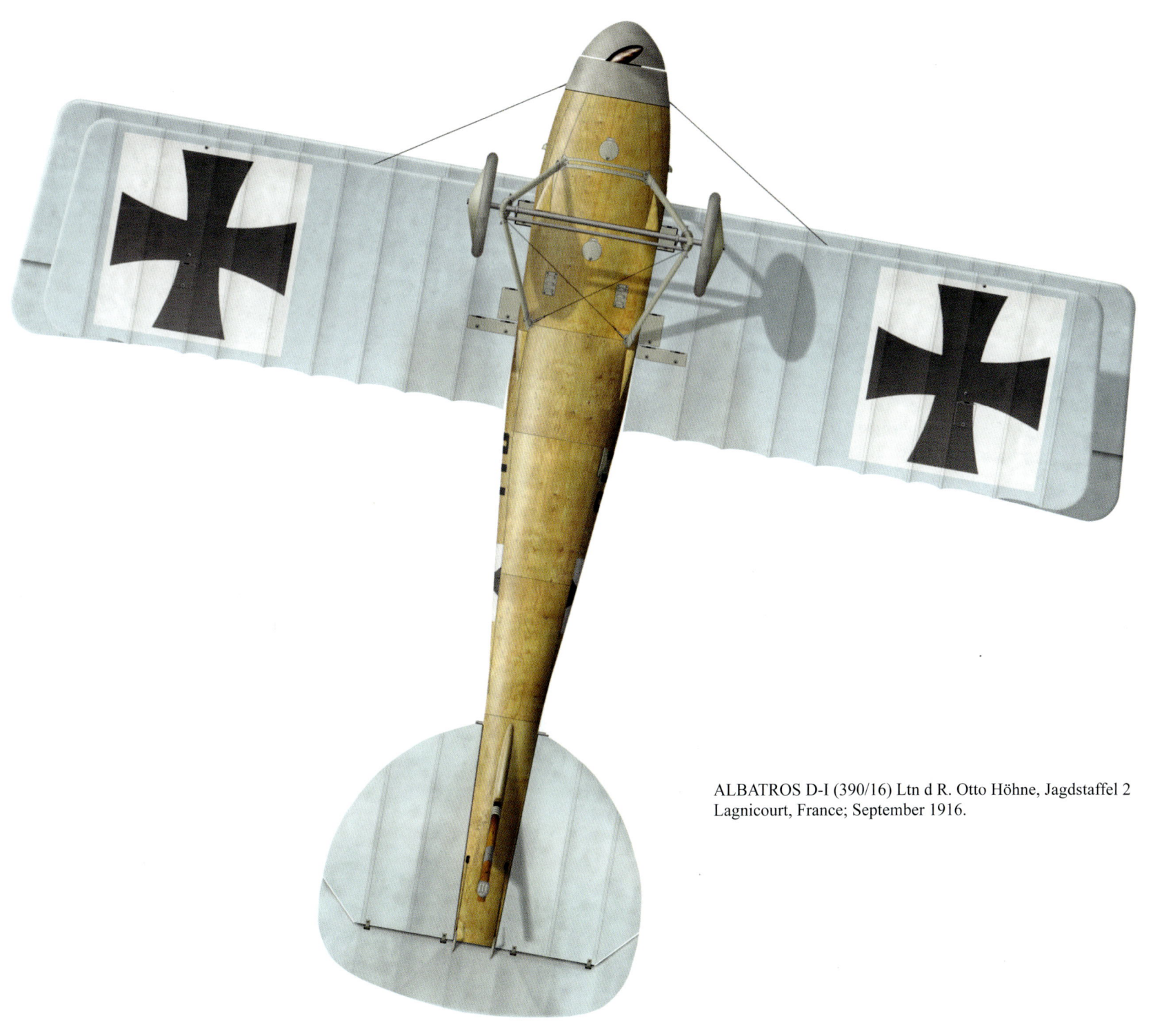

ALBATROS D-I (390/16) Ltn d R. Otto Höhne, Jagdstaffel 2
Lagnicourt, France; September 1916.

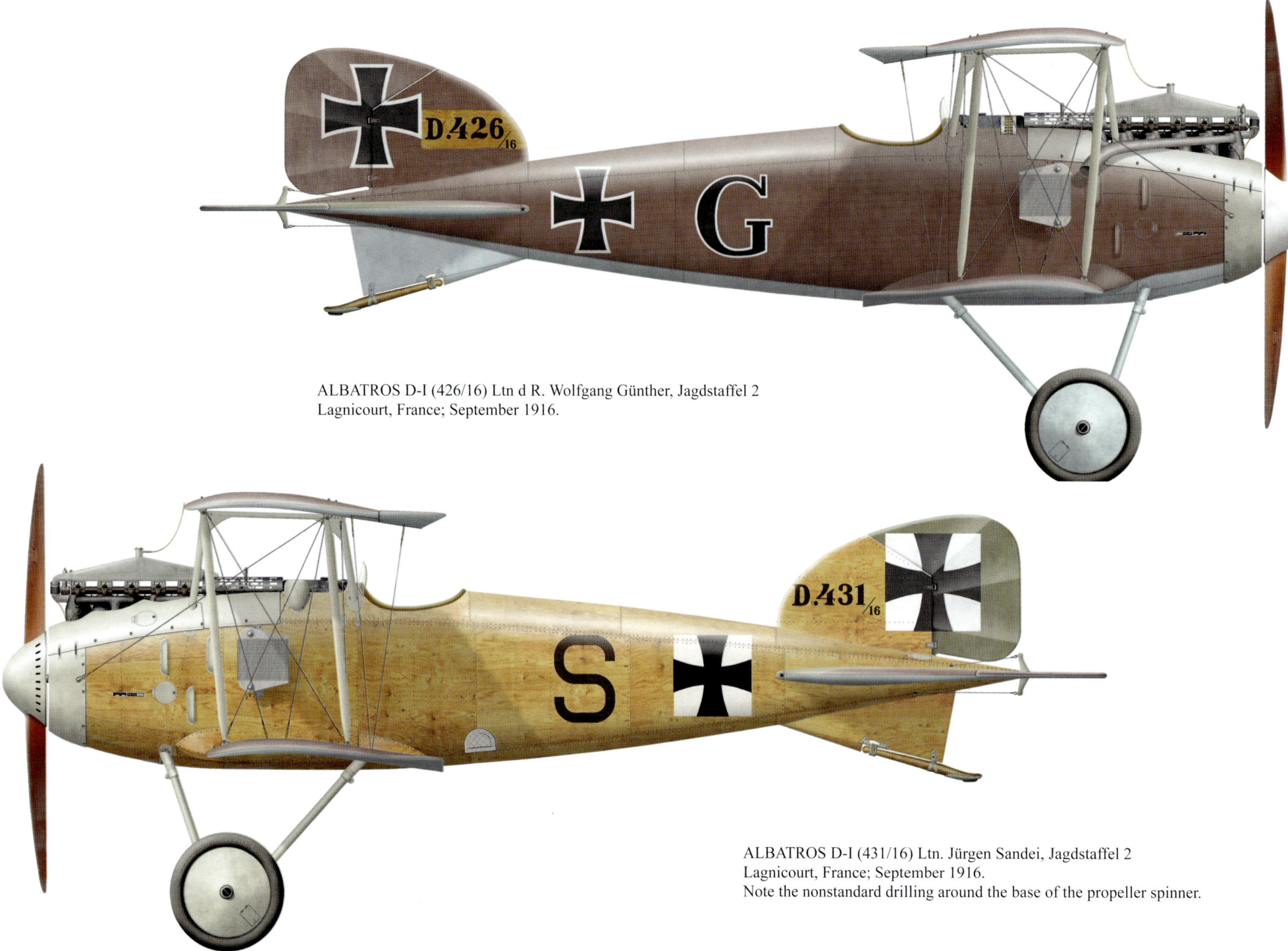

ALBATROS D-I (426/16) Ltn d R. Wolfgang Günther, Jagdstaffel 2
Lagnicourt, France; September 1916.

ALBATROS D-I (431/16) Ltn. Jürgen Sandei, Jagdstaffel 2
Lagnicourt, France; September 1916.
Note the nonstandard drilling around the base of the propeller spinner.

ALBATROS D-I (431/16) Ltn. Jürgen Sandei, Jagdstaffel 2
Lagnicourt, France; September 1916.

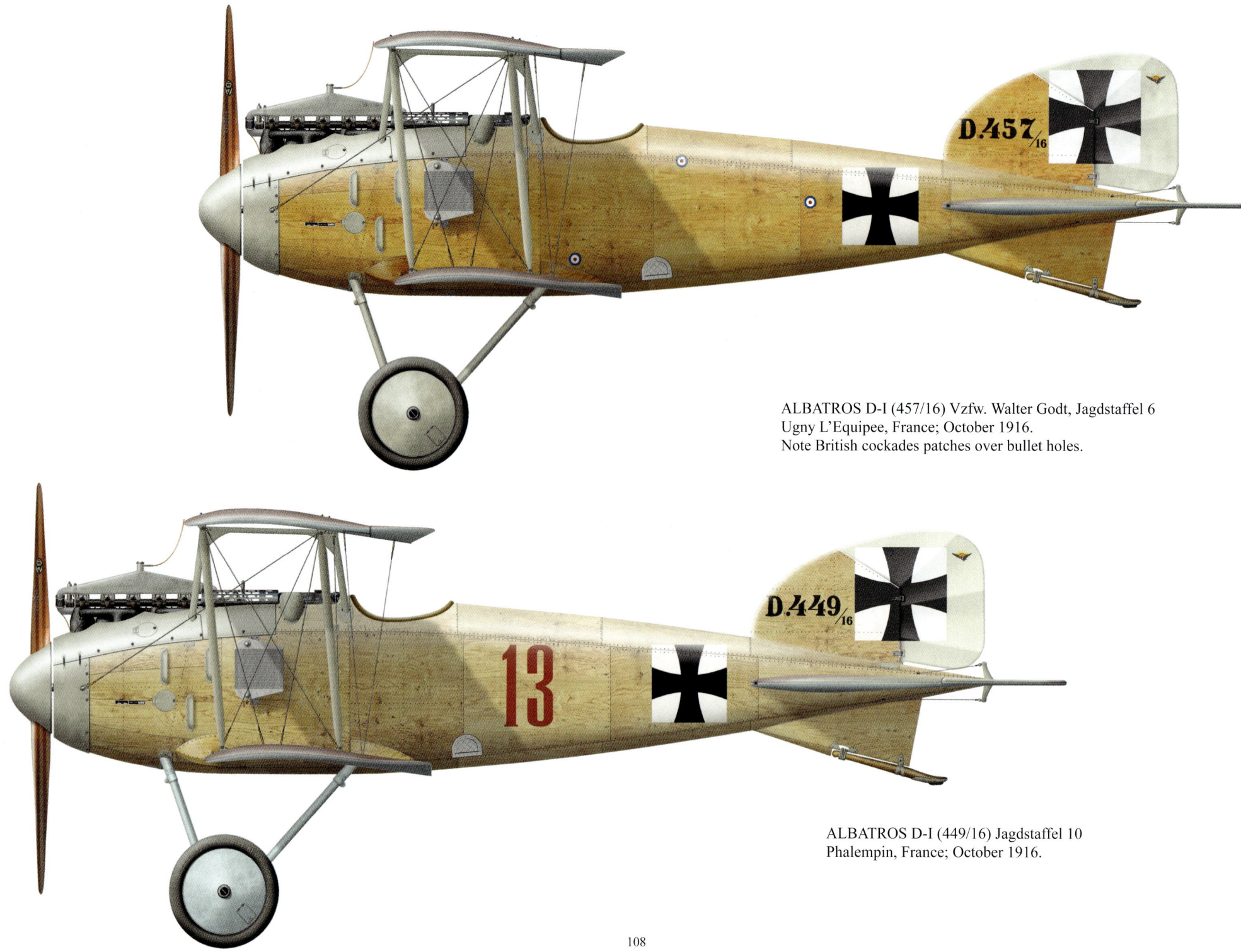

ALBATROS D-I (457/16) Vzfw. Walter Godt, Jagdstaffel 6
Ugny L'Equipee, France; October 1916.
Note British cockades patches over bullet holes.

ALBATROS D-I (449/16) Jagdstaffel 10
Phalempin, France; October 1916.

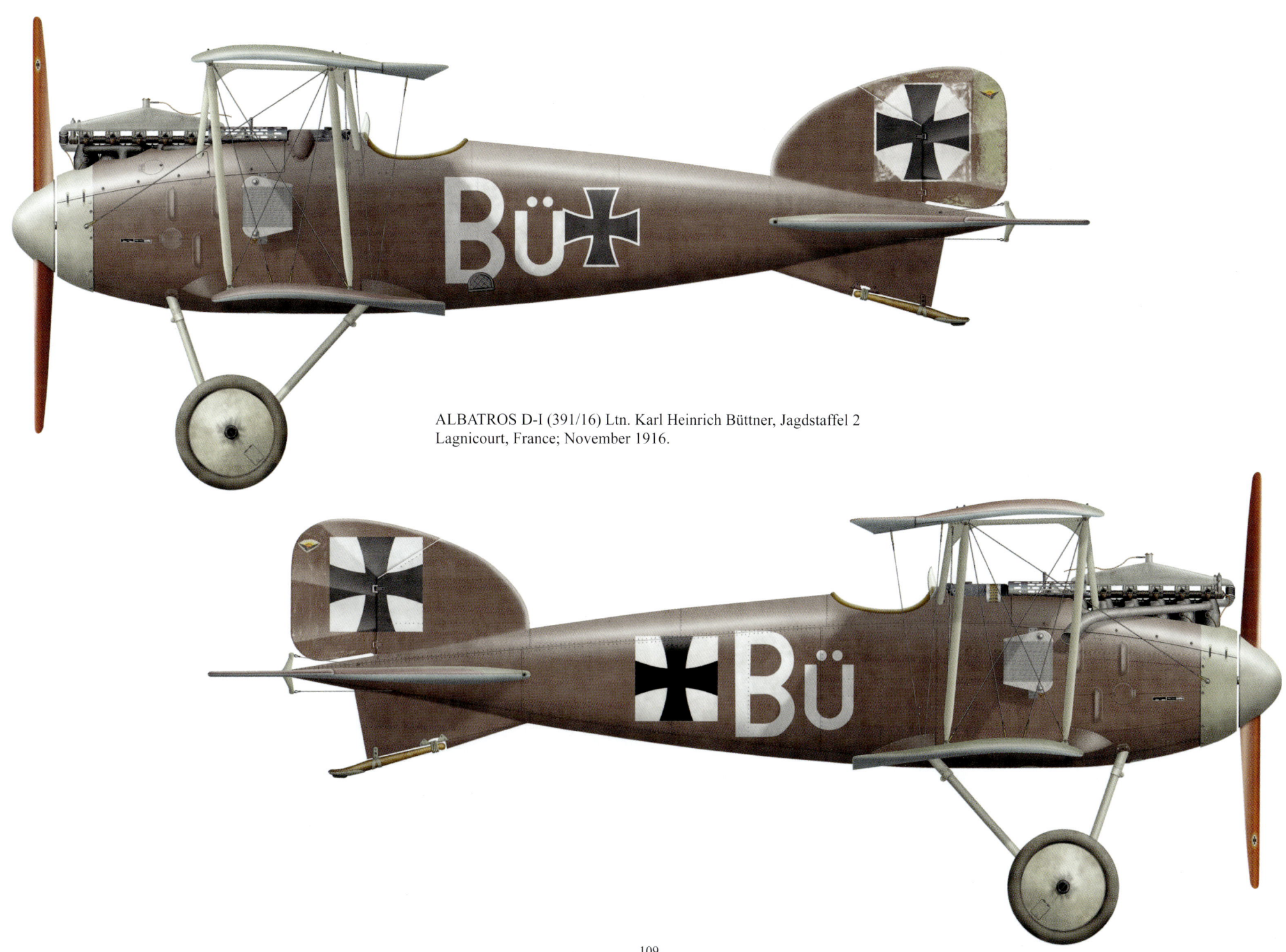

ALBATROS D-I (391/16) Ltn. Karl Heinrich Büttner, Jagdstaffel 2
Lagnicourt, France; November 1916.

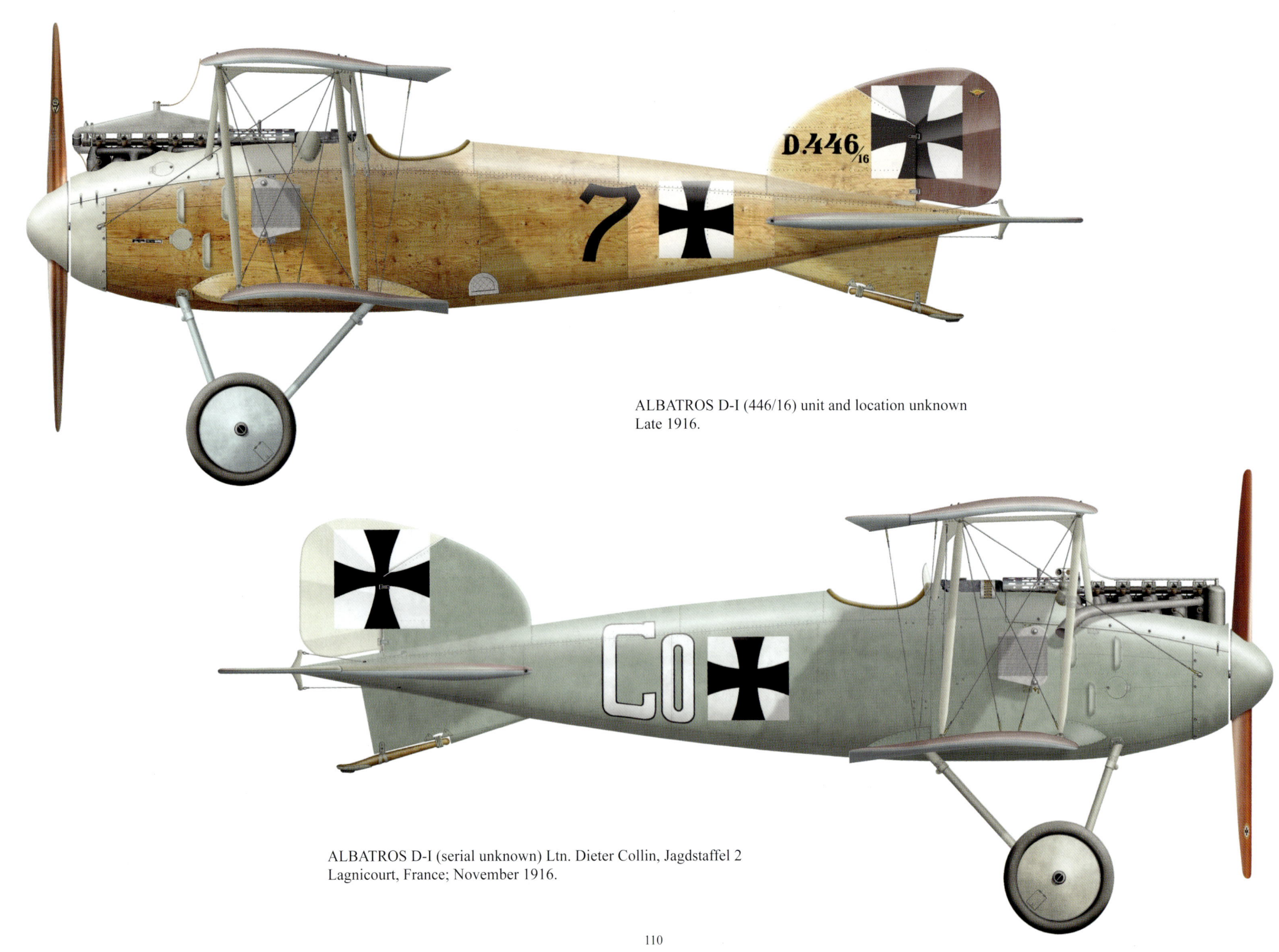

ALBATROS D-I (446/16) unit and location unknown
Late 1916.

ALBATROS D-I (serial unknown) Ltn. Dieter Collin, Jagdstaffel 2
Lagnicourt, France; November 1916.

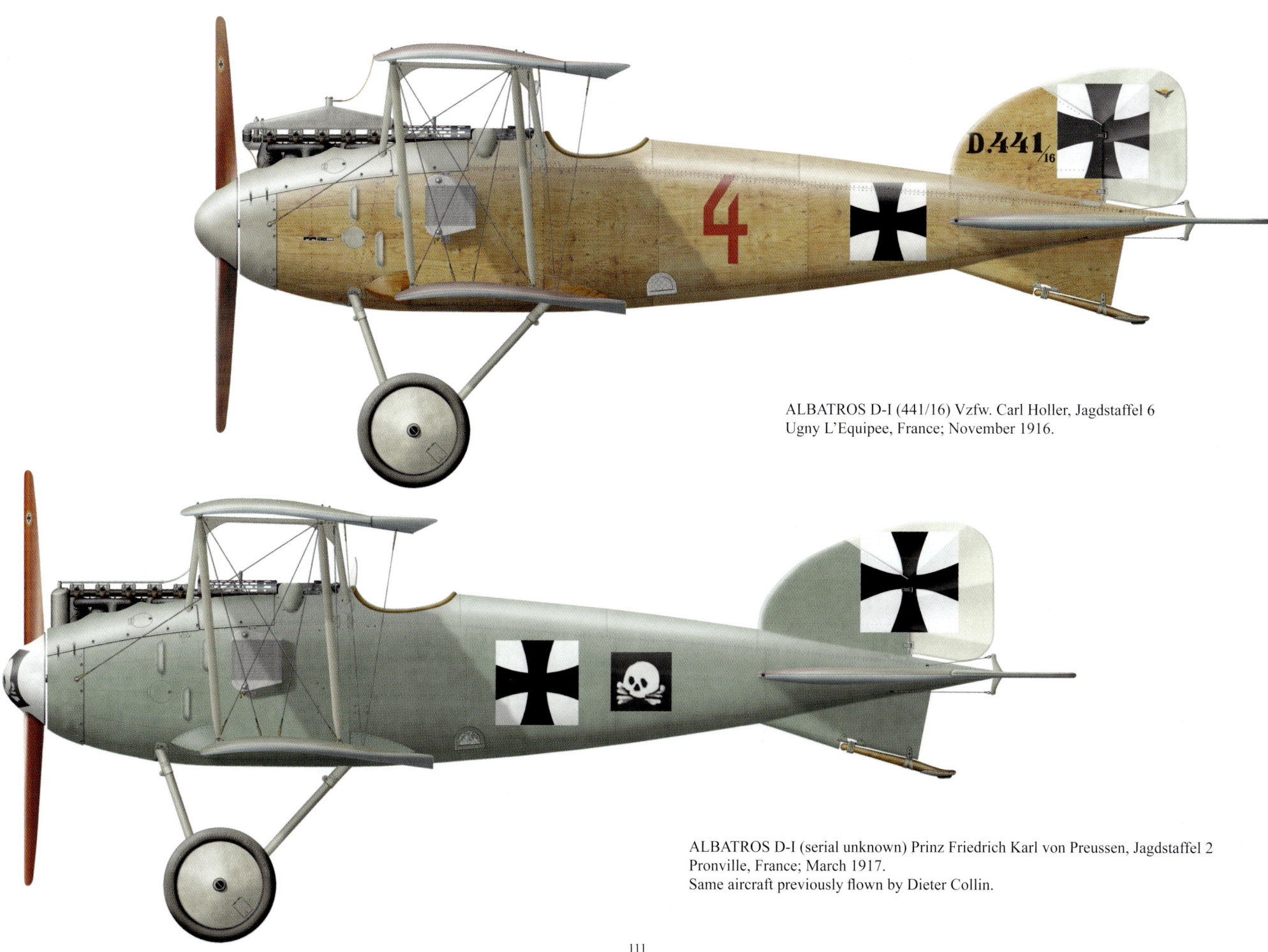

ALBATROS D-I (441/16) Vzfw. Carl Holler, Jagdstaffel 6
Ugny L'Equipee, France; November 1916.

ALBATROS D-I (serial unknown) Prinz Friedrich Karl von Preussen, Jagdstaffel 2
Pronville, France; March 1917.
Same aircraft previously flown by Dieter Collin.

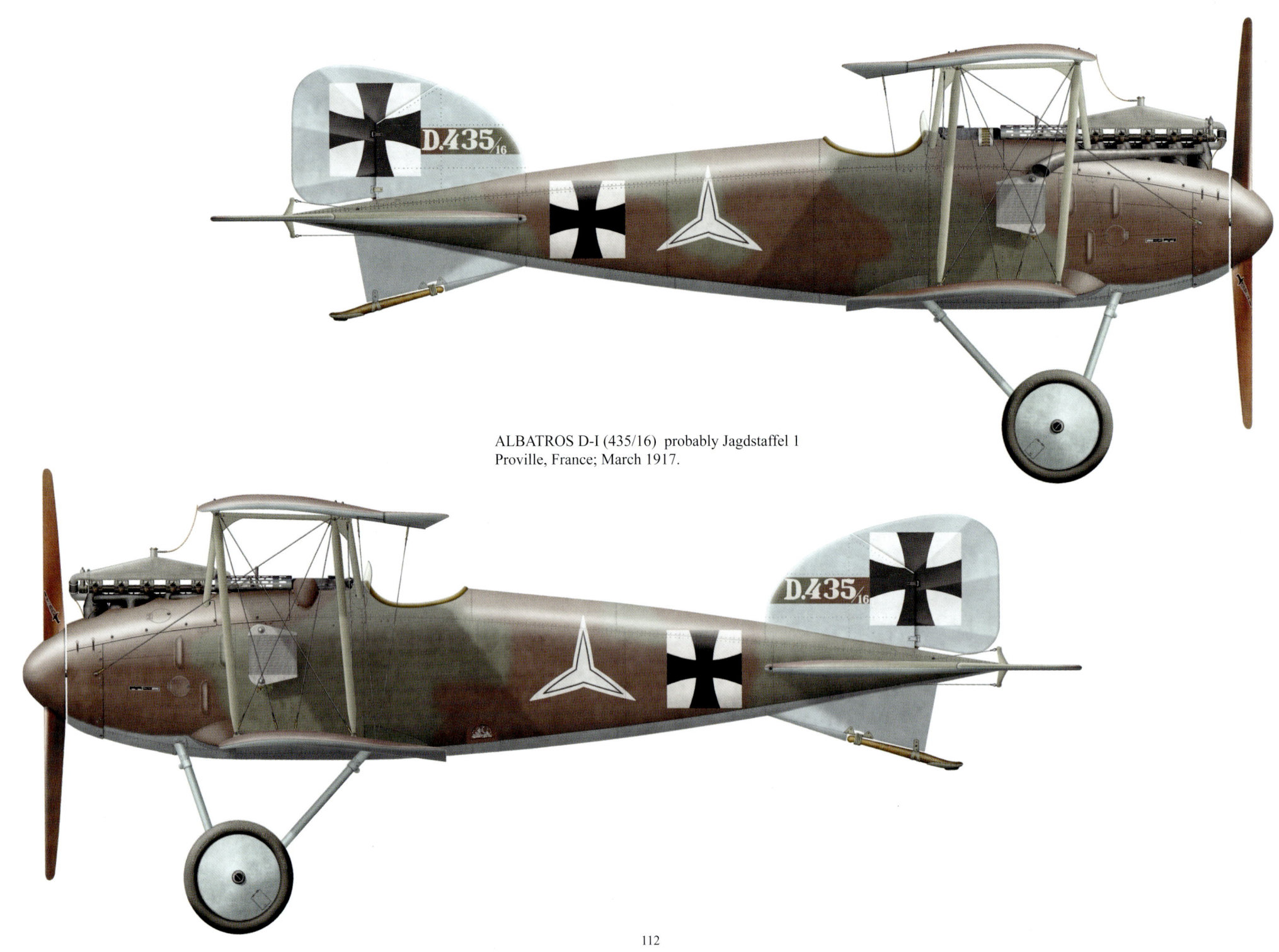

ALBATROS D-I (435/16) probably Jagdstaffel 1
Proville, France; March 1917.

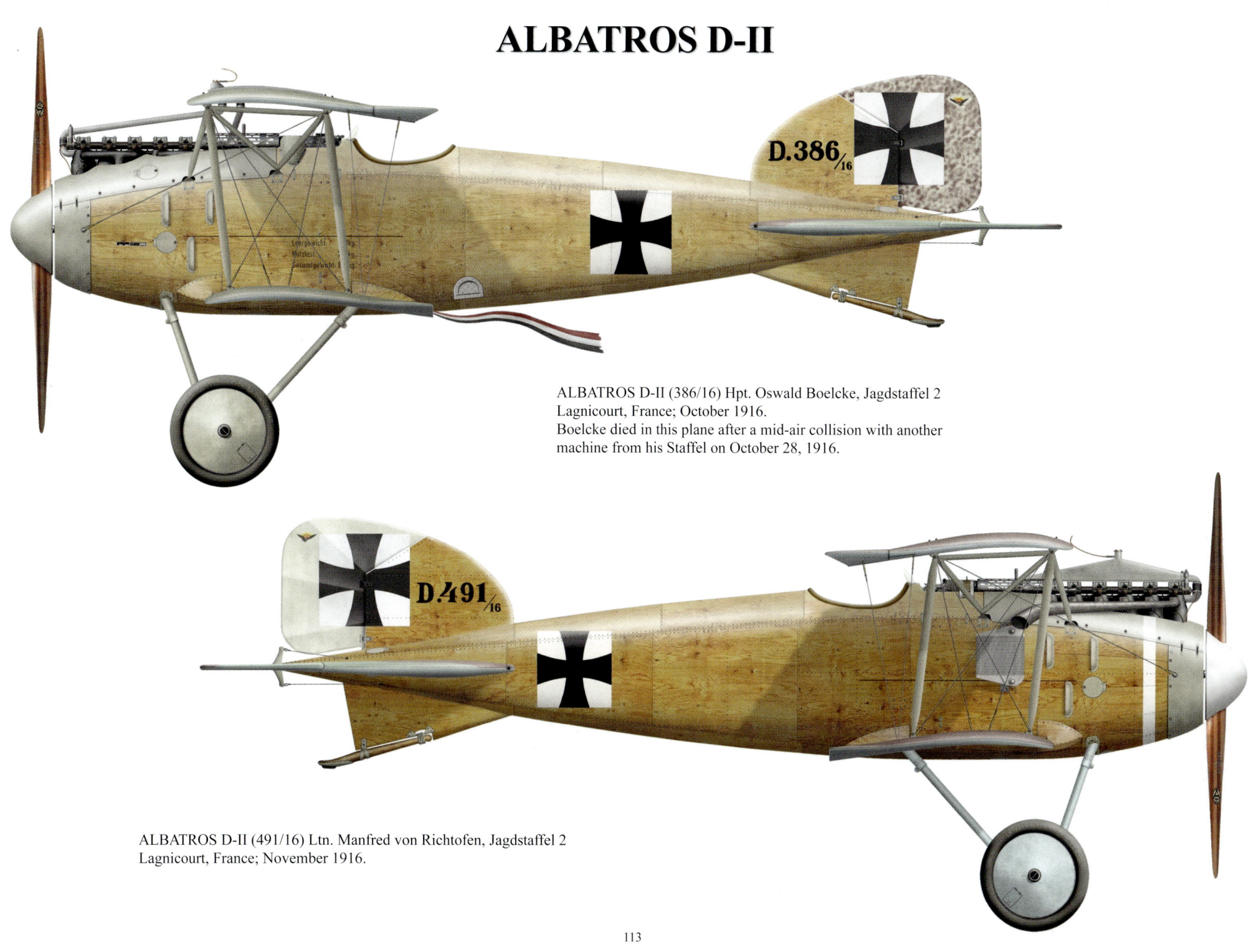

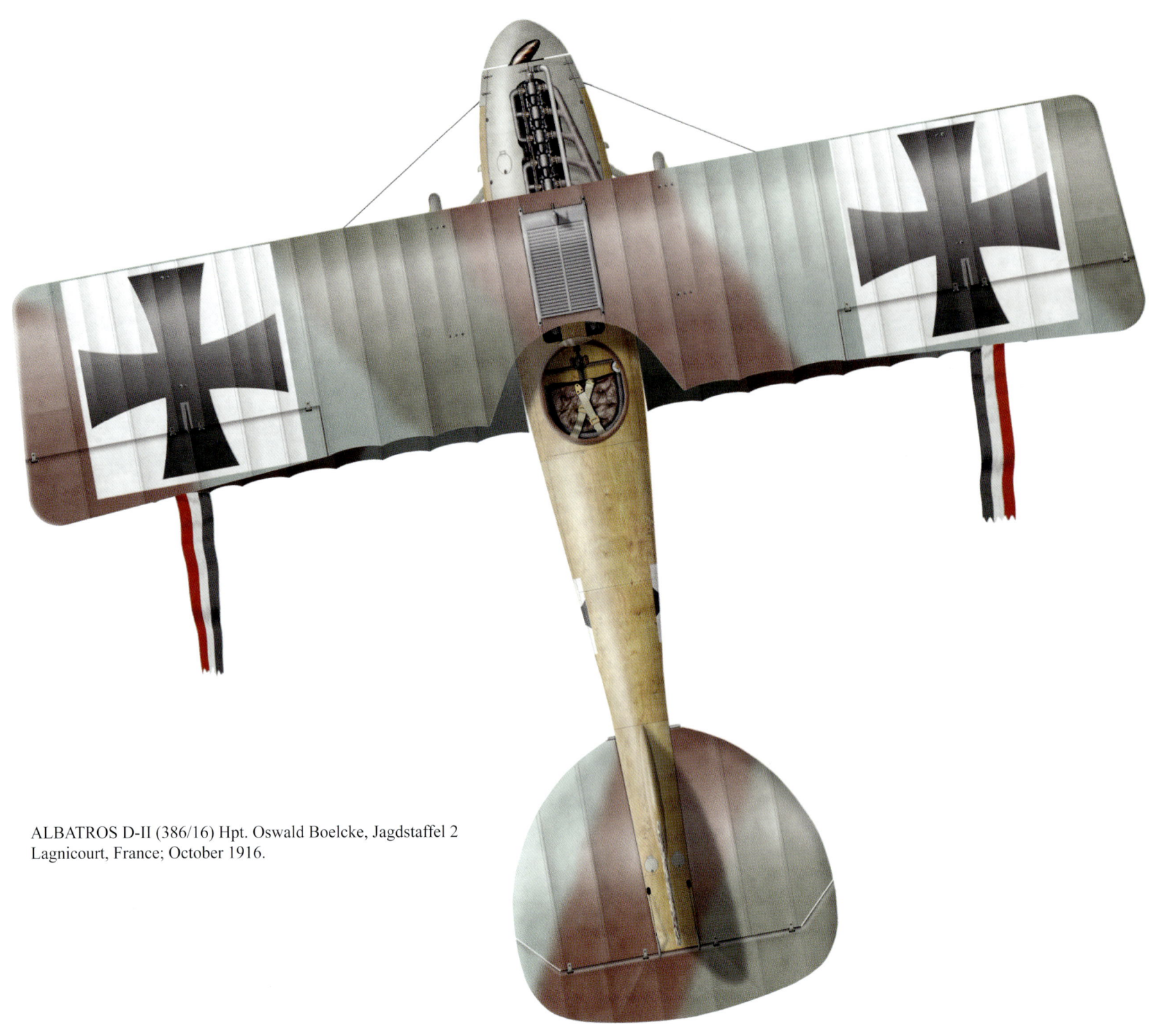

ALBATROS D-II (386/16) Hpt. Oswald Boelcke, Jagdstaffel 2
Lagnicourt, France; October 1916.

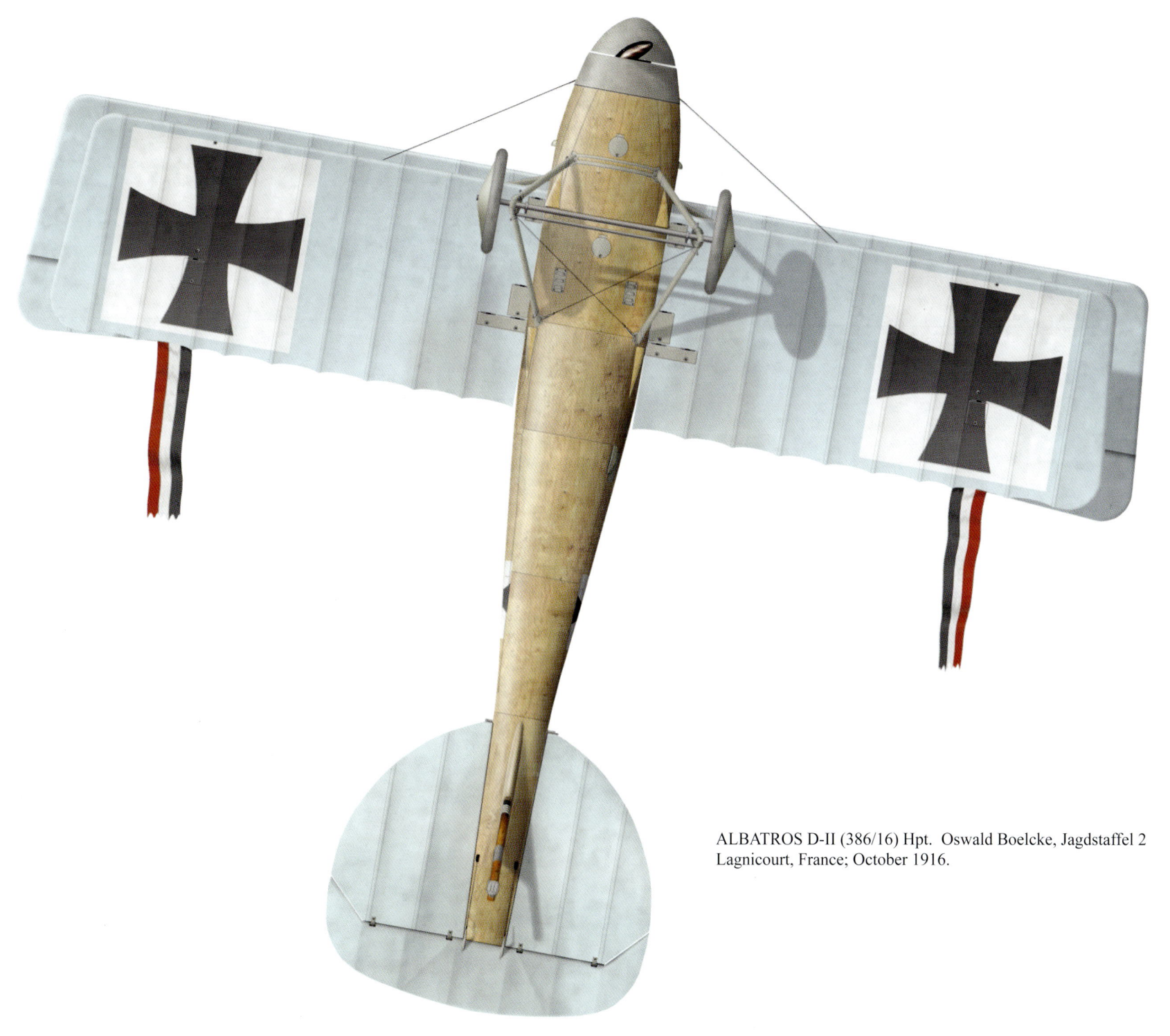

ALBATROS D-II (386/16) Hpt. Oswald Boelcke, Jagdstaffel 2
Lagnicourt, France; October 1916.

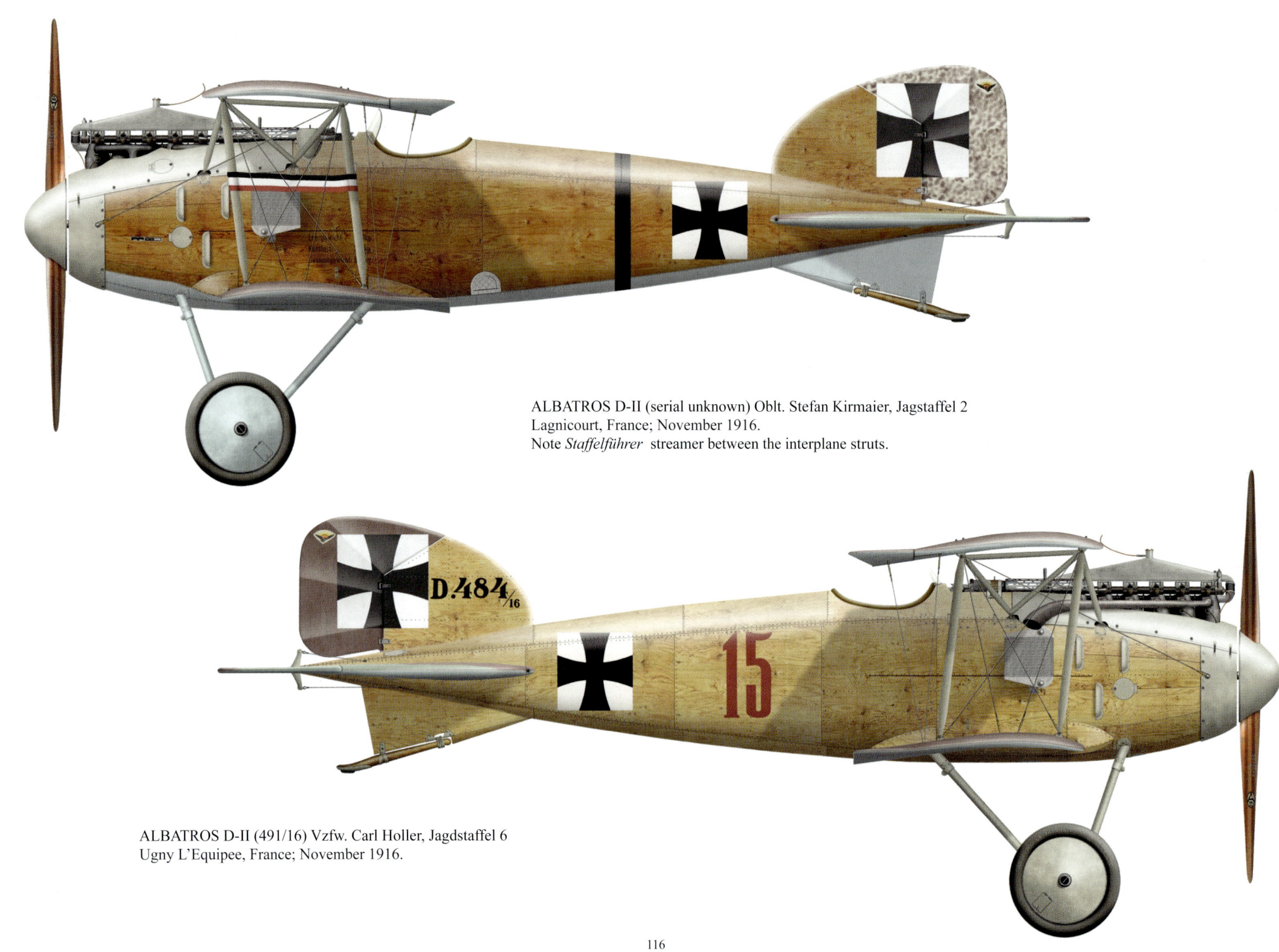

ALBATROS D-II (serial unknown) Oblt. Stefan Kirmaier, Jagstaffel 2
Lagnicourt, France; November 1916.
Note *Staffelführer* streamer between the interplane struts.

ALBATROS D-II (491/16) Vzfw. Carl Holler, Jagdstaffel 6
Ugny L'Equipee, France; November 1916.

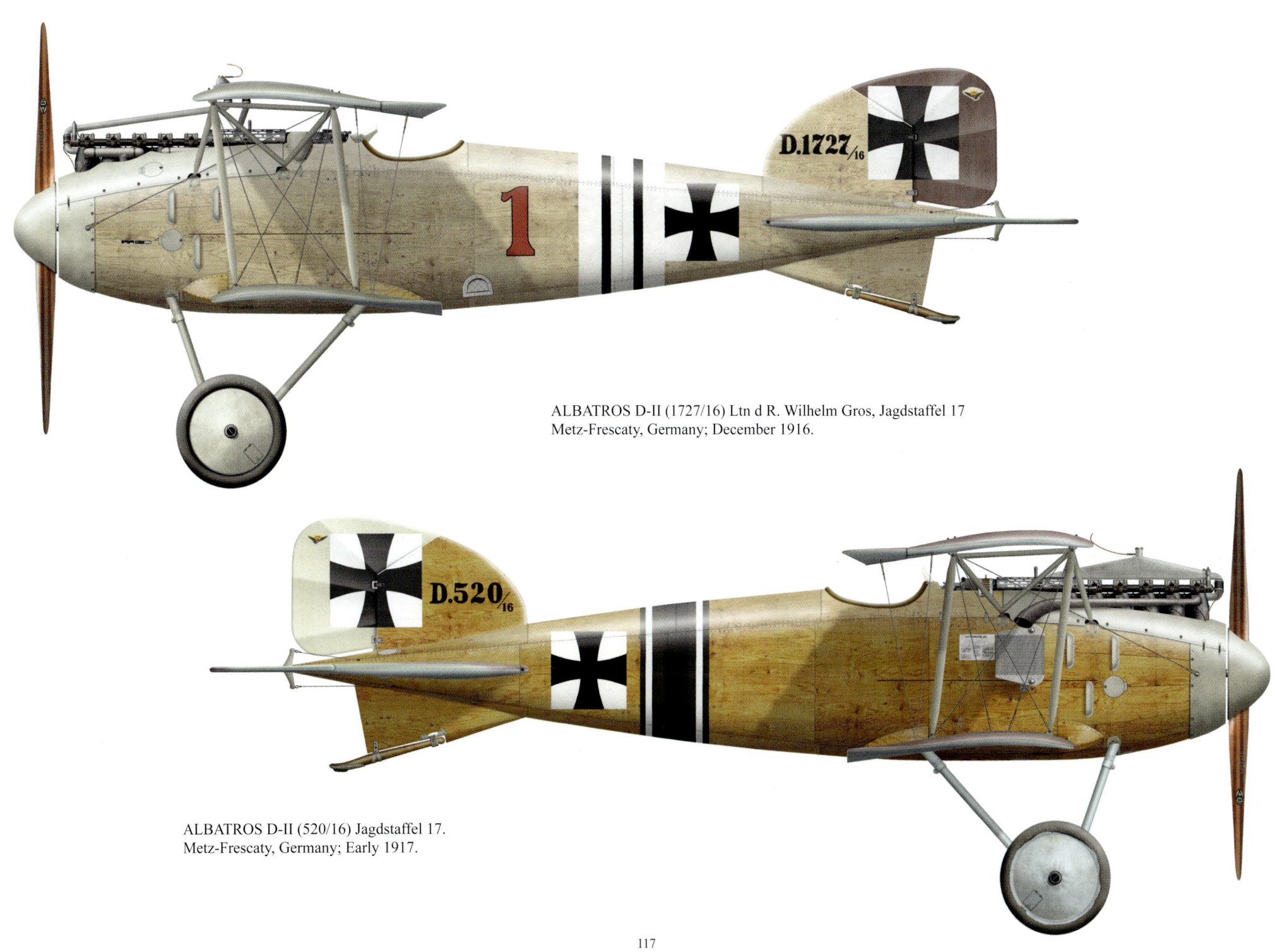

ALBATROS D-II (1727/16) Ltn d R. Wilhelm Gros, Jagdstaffel 17
Metz-Frescaty, Germany; December 1916.

ALBATROS D-II (520/16) Jagdstaffel 17.
Metz-Frescaty, Germany; Early 1917.

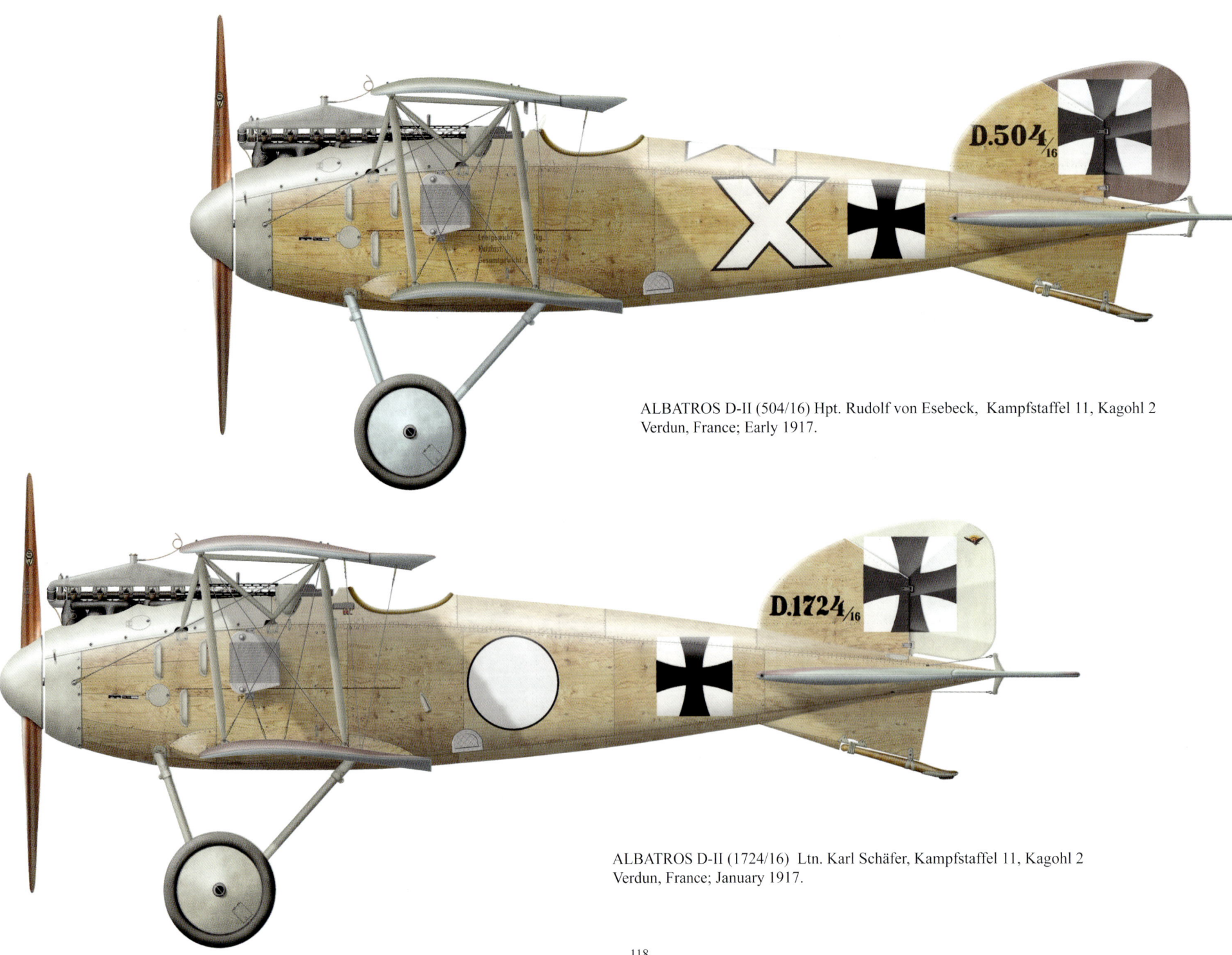

ALBATROS D-II (504/16) Hpt. Rudolf von Esebeck, Kampfstaffel 11, Kagohl 2
Verdun, France; Early 1917.

ALBATROS D-II (1724/16) Ltn. Karl Schäfer, Kampfstaffel 11, Kagohl 2
Verdun, France; January 1917.

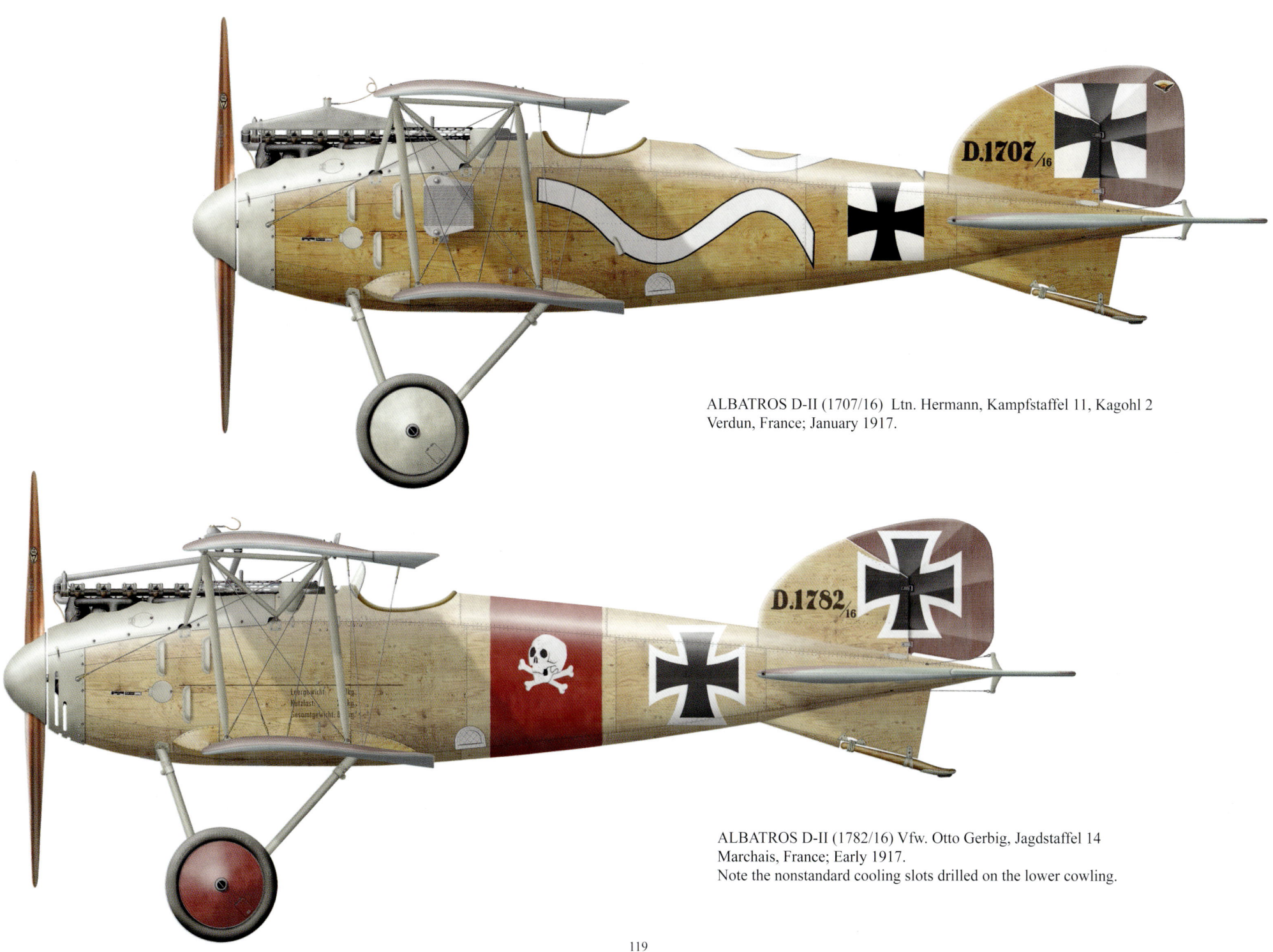

ALBATROS D-II (1707/16) Ltn. Hermann, Kampfstaffel 11, Kagohl 2
Verdun, France; January 1917.

ALBATROS D-II (1782/16) Vfw. Otto Gerbig, Jagdstaffel 14
Marchais, France; Early 1917.
Note the nonstandard cooling slots drilled on the lower cowling.

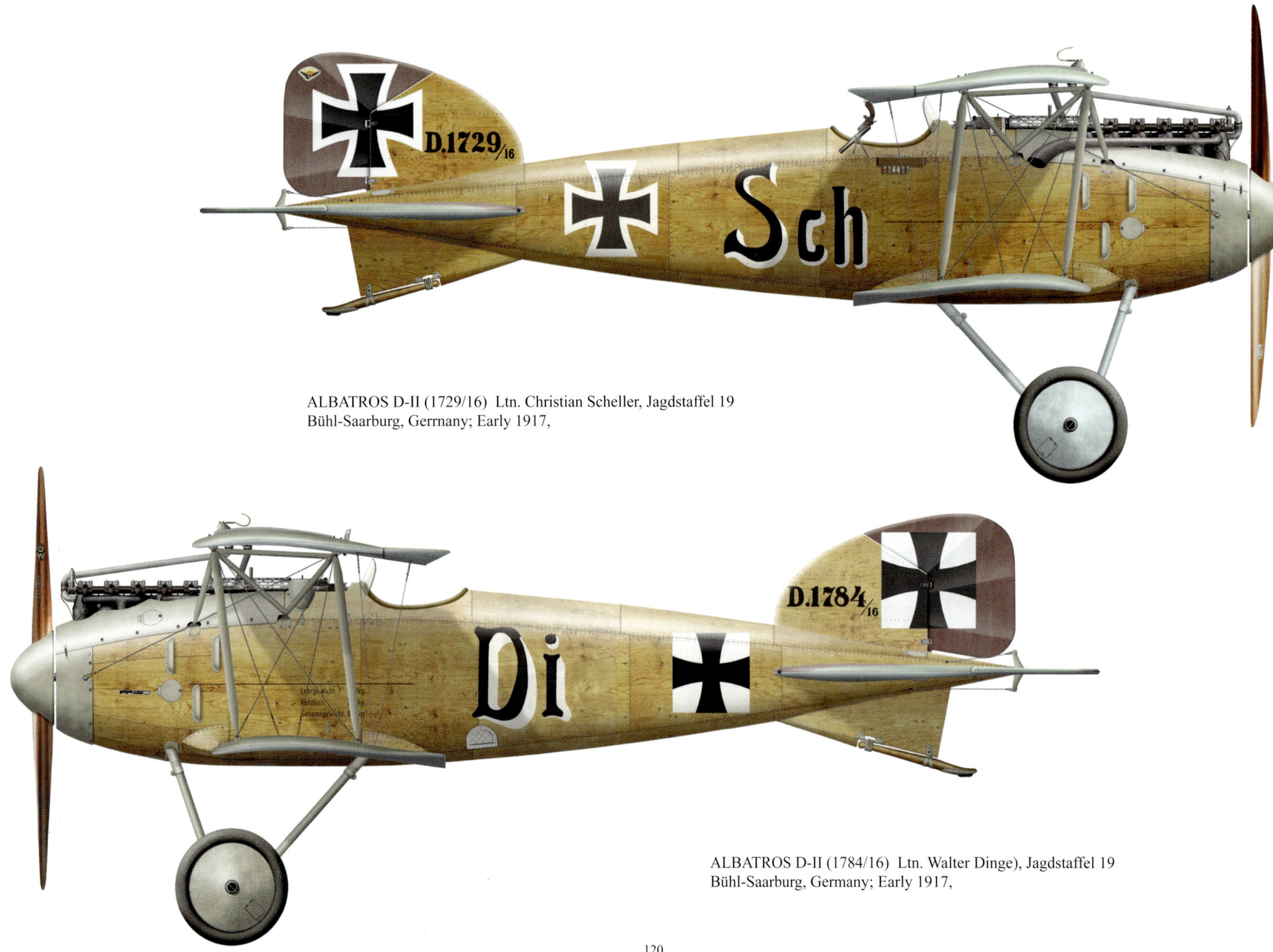

ALBATROS D-II (1729/16) Ltn. Christian Scheller, Jagdstaffel 19
Bühl-Saarburg, Gerrnany; Early 1917,

ALBATROS D-II (1784/16) Ltn. Walter Dinge), Jagdstaffel 19
Bühl-Saarburg, Germany; Early 1917,

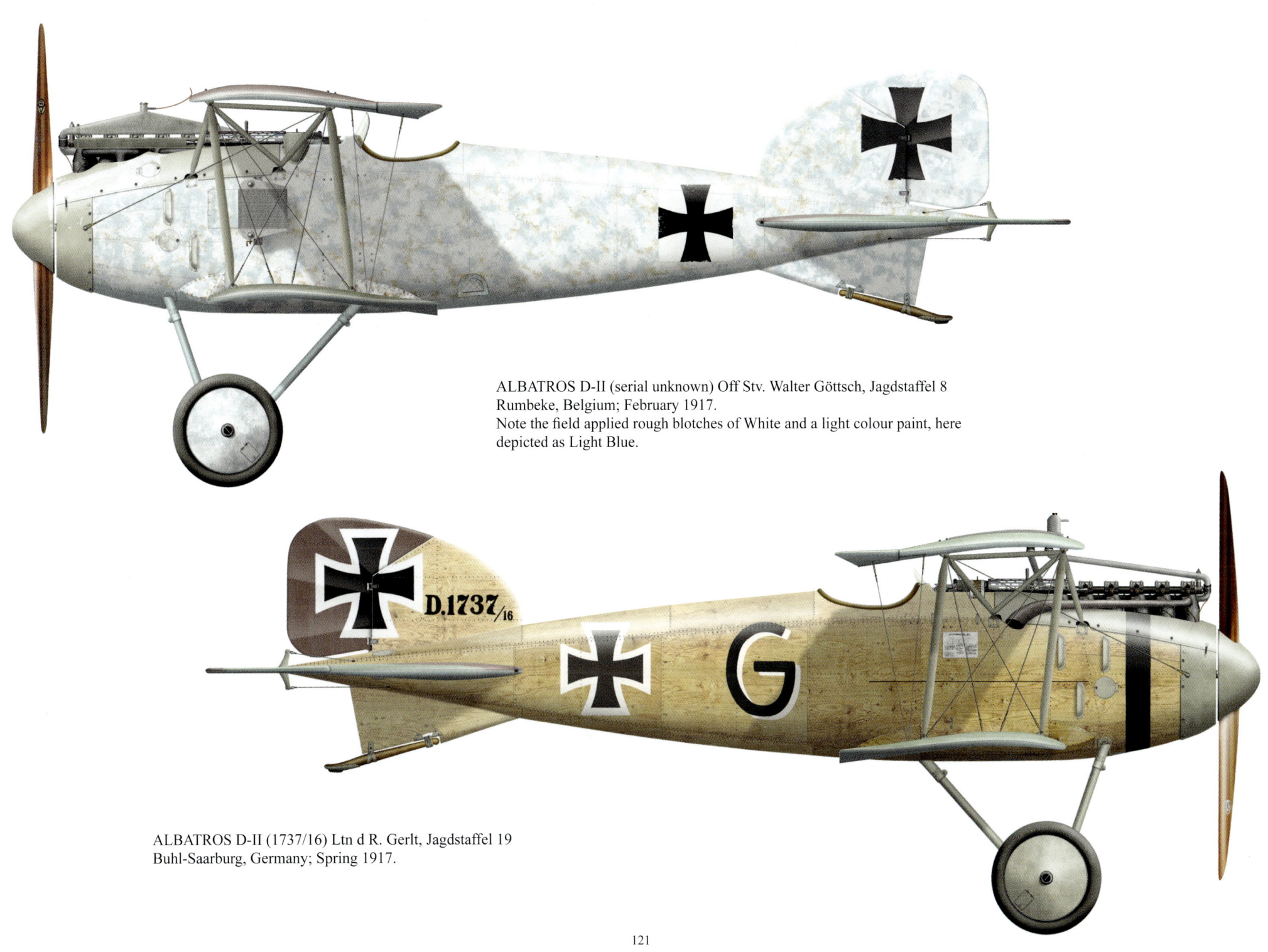

ALBATROS D-II (serial unknown) Off Stv. Walter Göttsch, Jagdstaffel 8
Rumbeke, Belgium; February 1917.
Note the field applied rough blotches of White and a light colour paint, here depicted as Light Blue.

ALBATROS D-II (1737/16) Ltn d R. Gerlt, Jagdstaffel 19
Buhl-Saarburg, Germany; Spring 1917.

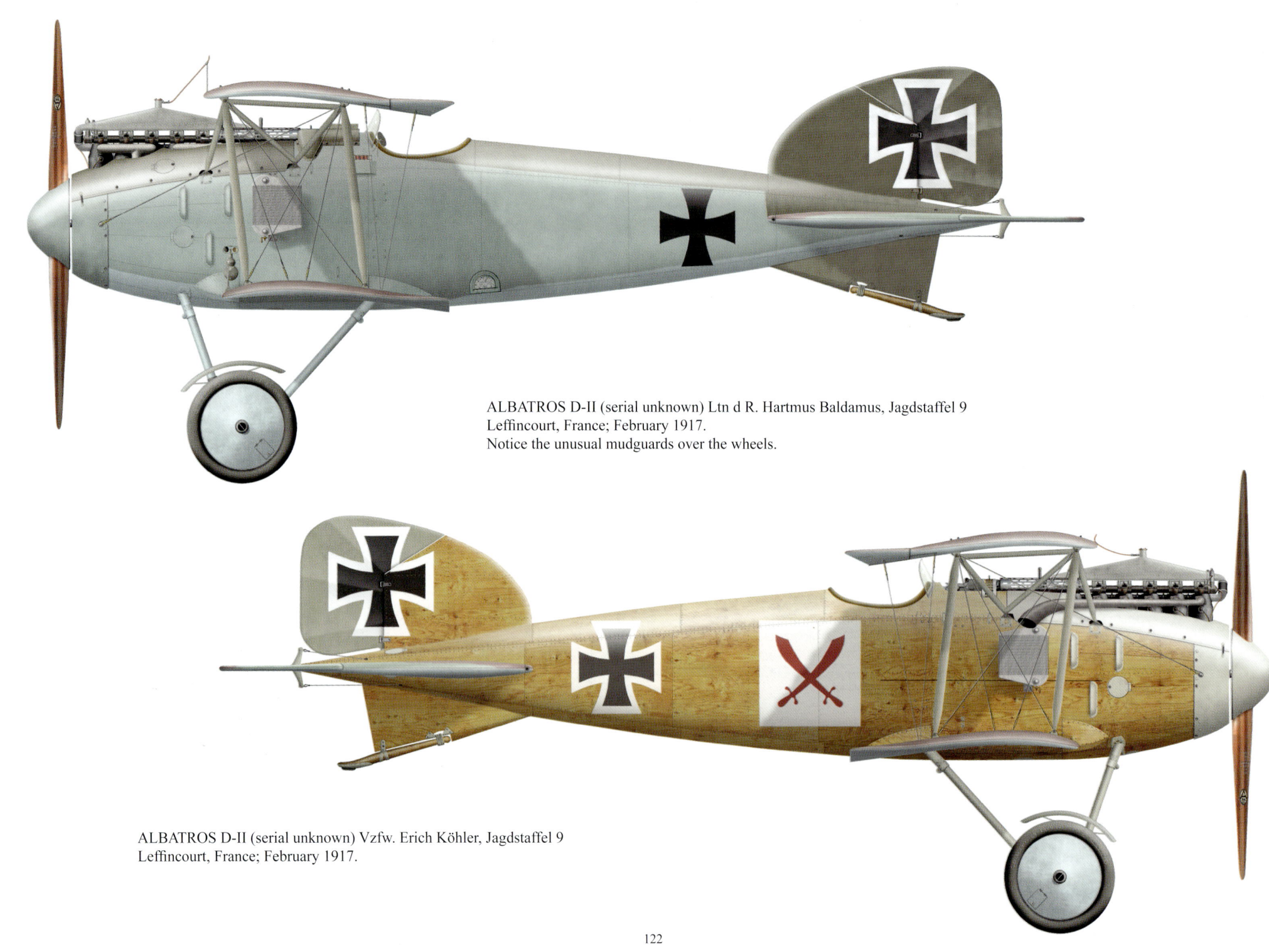

ALBATROS D-II (serial unknown) Ltn d R. Hartmus Baldamus, Jagdstaffel 9
Leffincourt, France; February 1917.
Notice the unusual mudguards over the wheels.

ALBATROS D-II (serial unknown) Vzfw. Erich Köhler, Jagdstaffel 9
Leffincourt, France; February 1917.

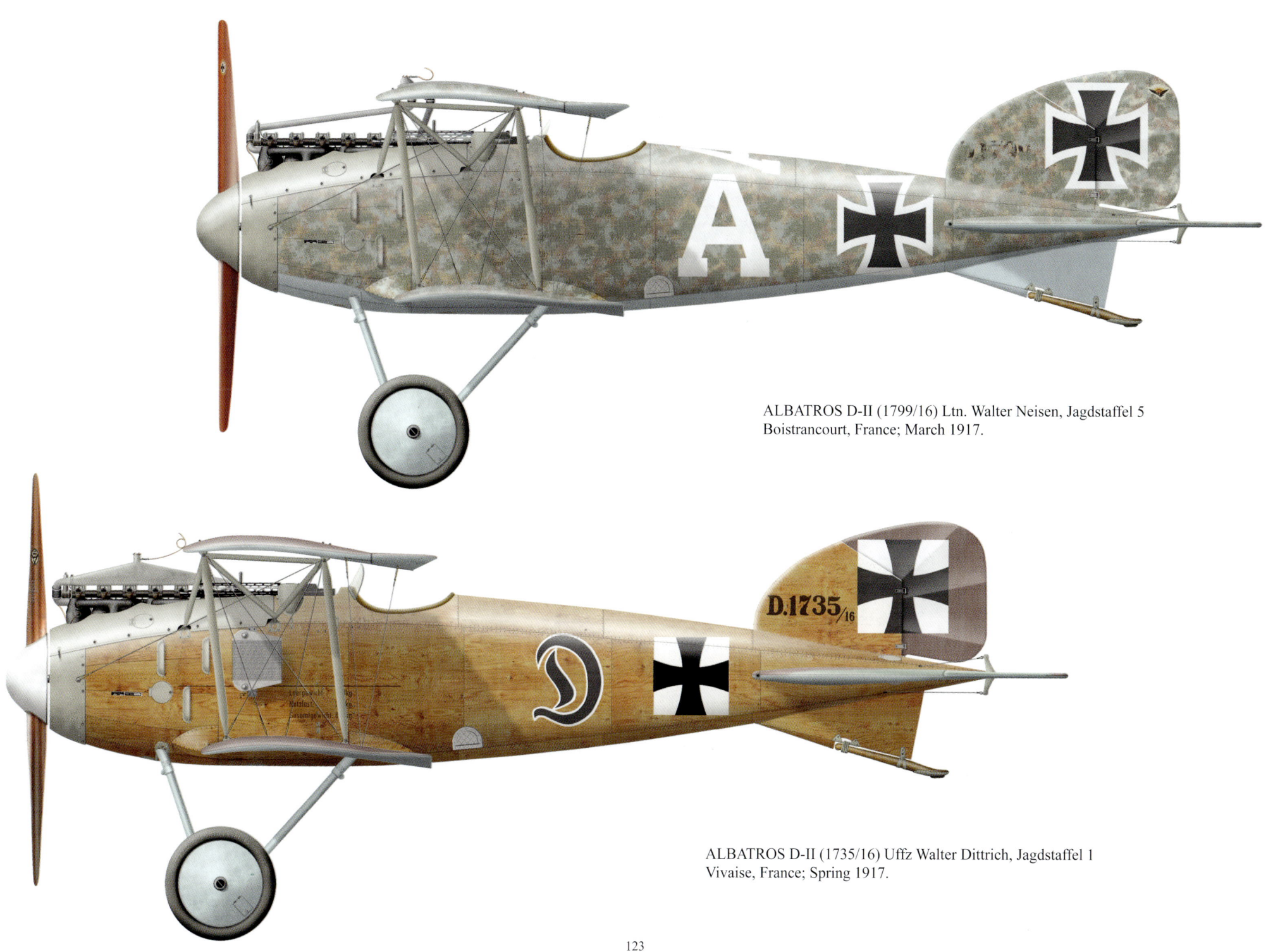

ALBATROS D-II (1799/16) Ltn. Walter Neisen, Jagdstaffel 5
Boistrancourt, France; March 1917.

ALBATROS D-II (1735/16) Uffz Walter Dittrich, Jagdstaffel 1
Vivaise, France; Spring 1917.

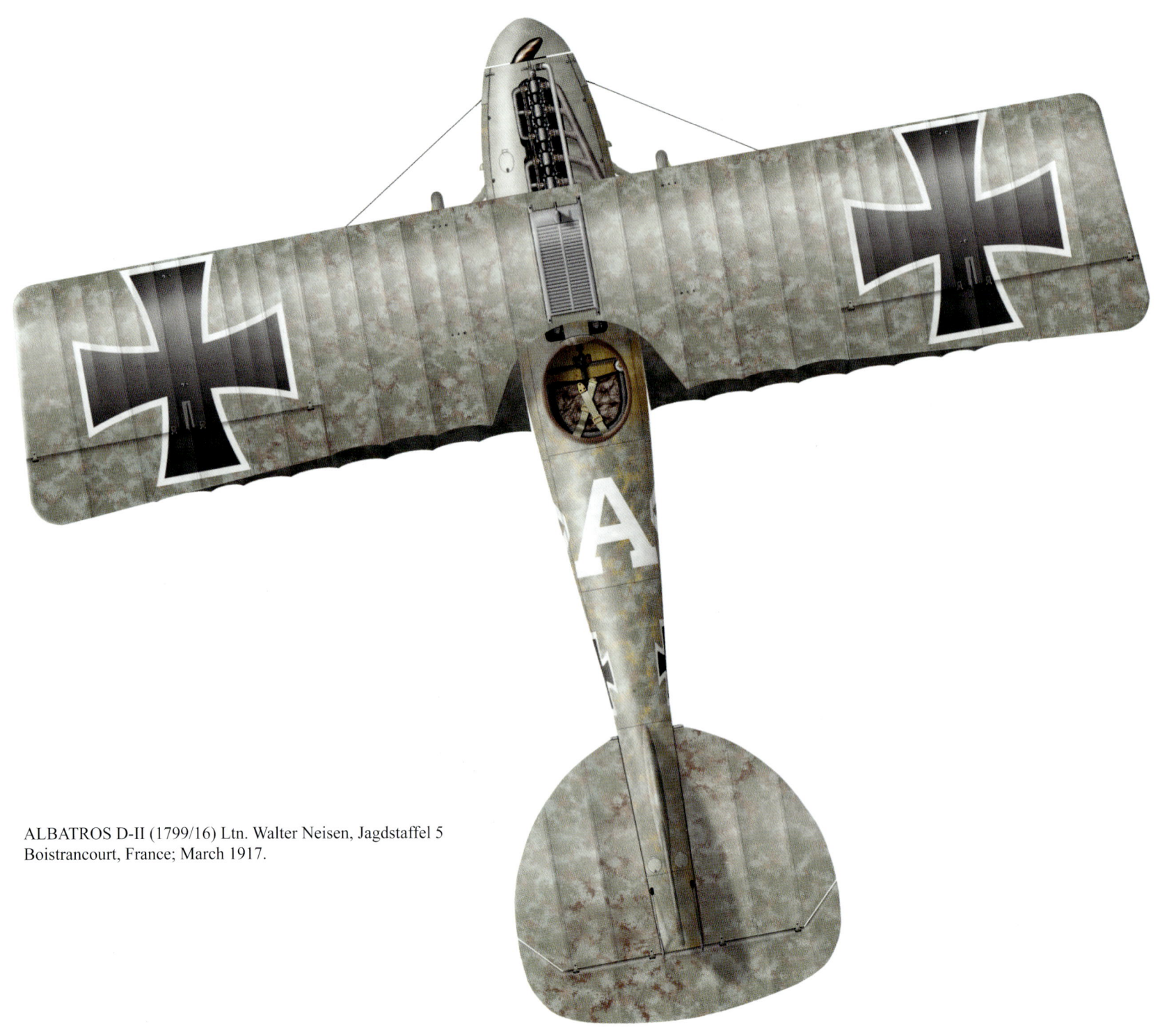

ALBATROS D-II (1799/16) Ltn. Walter Neisen, Jagdstaffel 5
Boistrancourt, France; March 1917.

ALBATROS D-II (1799/16) Ltn. Walter Neisen, Jagdstaffel 5
Boistrancourt, France; March 1917.

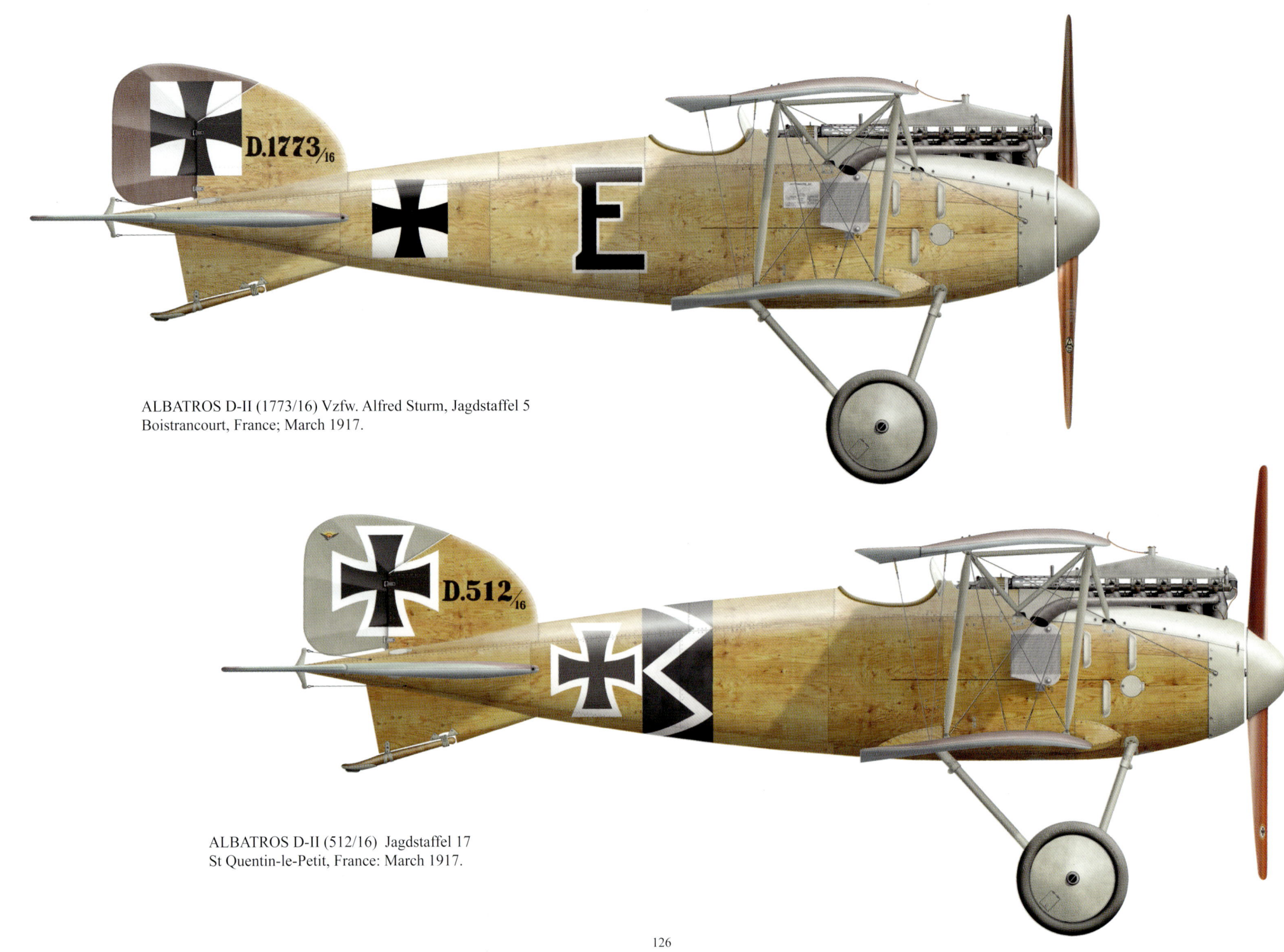

ALBATROS D-II (1773/16) Vzfw. Alfred Sturm, Jagdstaffel 5
Boistrancourt, France; March 1917.

ALBATROS D-II (512/16) Jagdstaffel 17
St Quentin-le-Petit, France; March 1917.

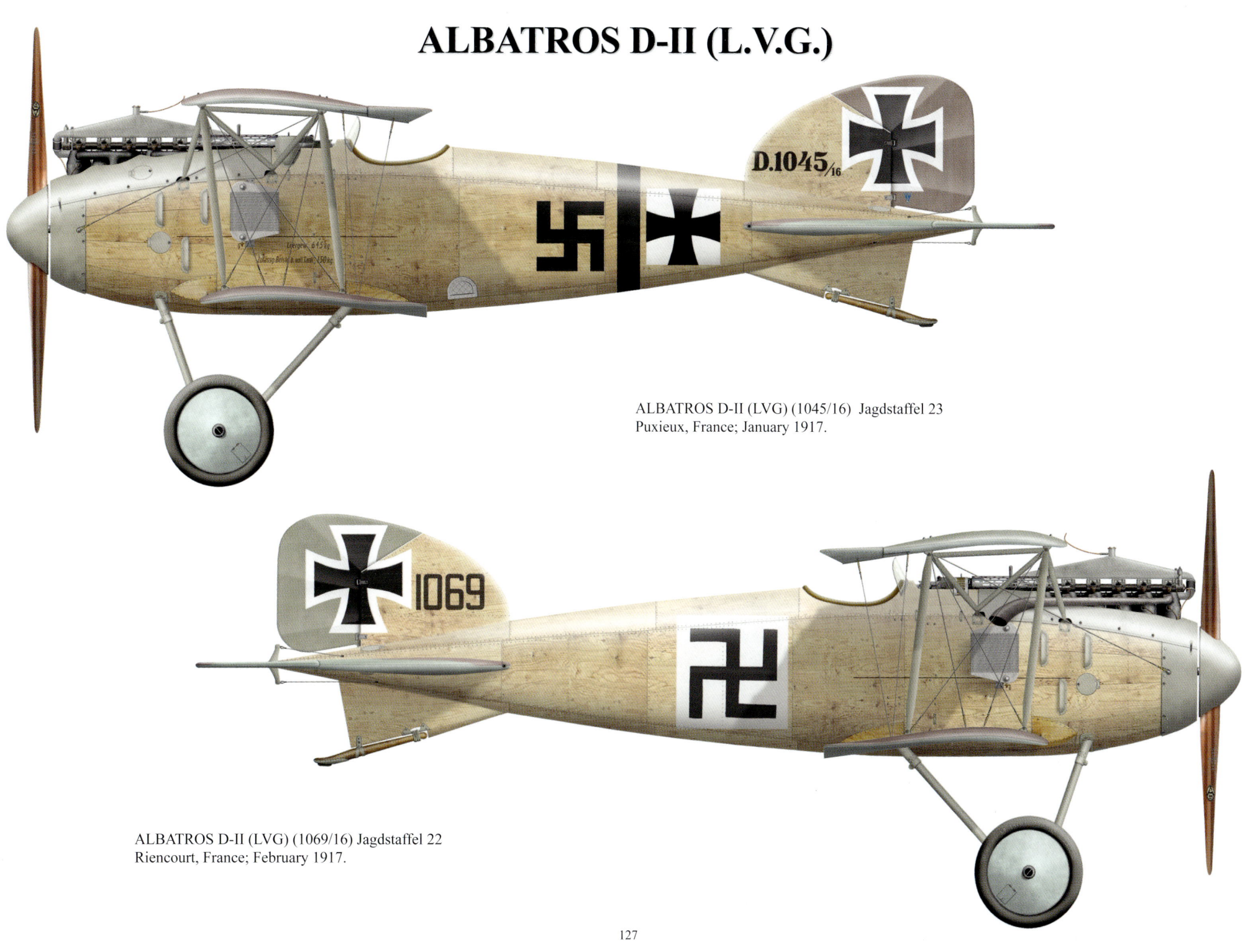

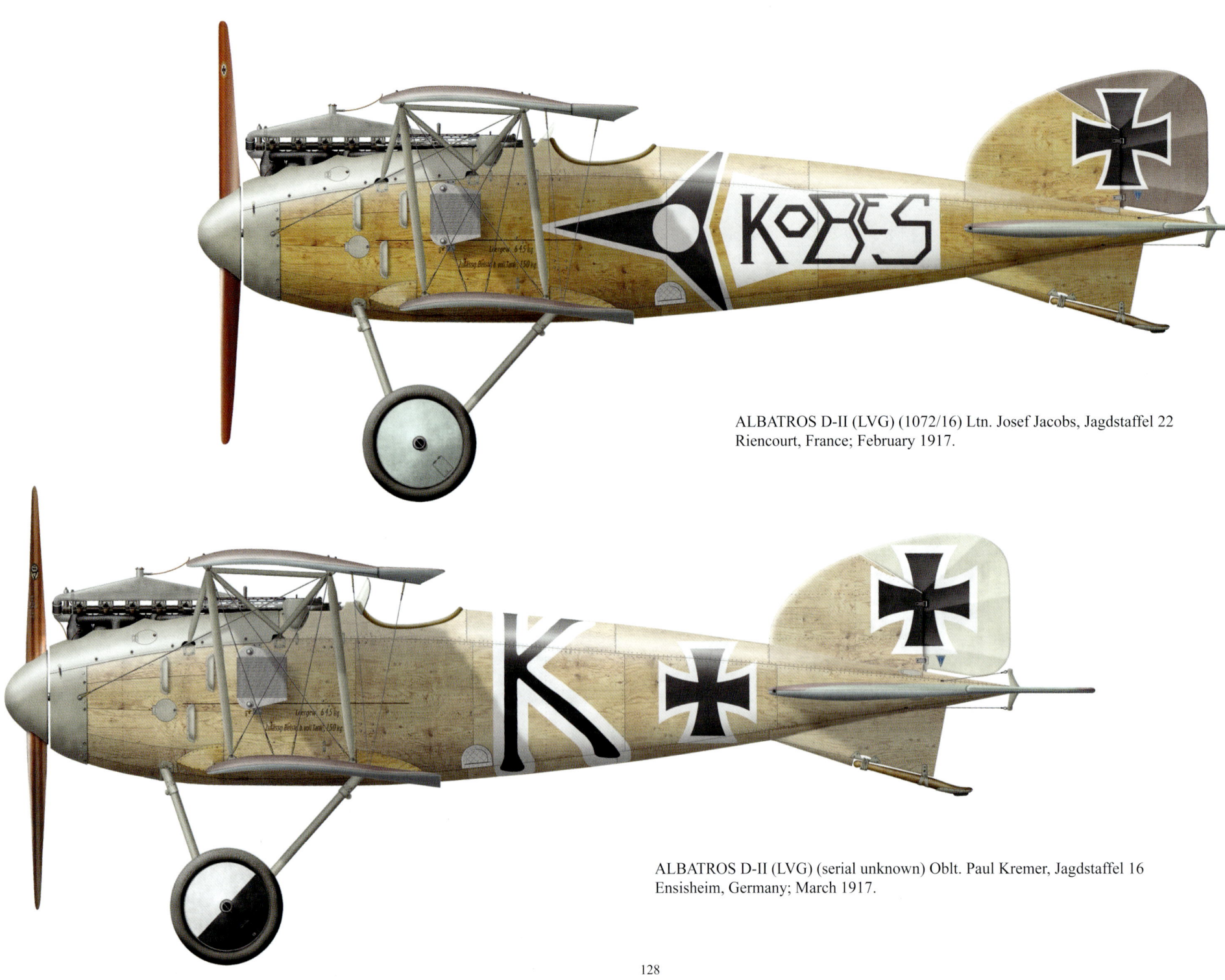

ALBATROS D-II (LVG) (1072/16) Ltn. Josef Jacobs, Jagdstaffel 22
Riencourt, France; February 1917.

ALBATROS D-II (LVG) (serial unknown) Oblt. Paul Kremer, Jagdstaffel 16
Ensisheim, Germany; March 1917.

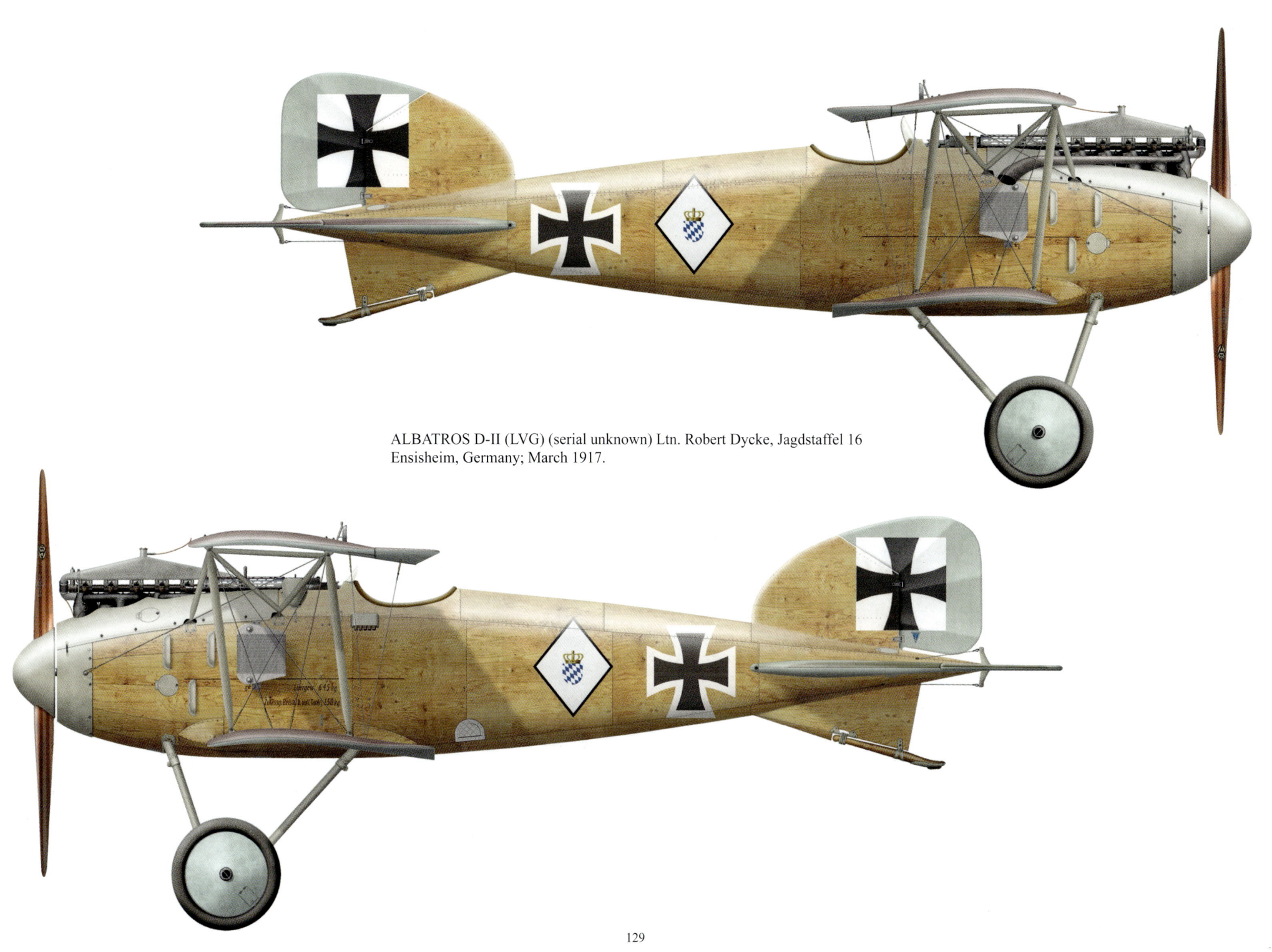

ALBATROS D-II (LVG) (serial unknown) Ltn. Robert Dycke, Jagdstaffel 16
Ensisheim, Germany; March 1917.

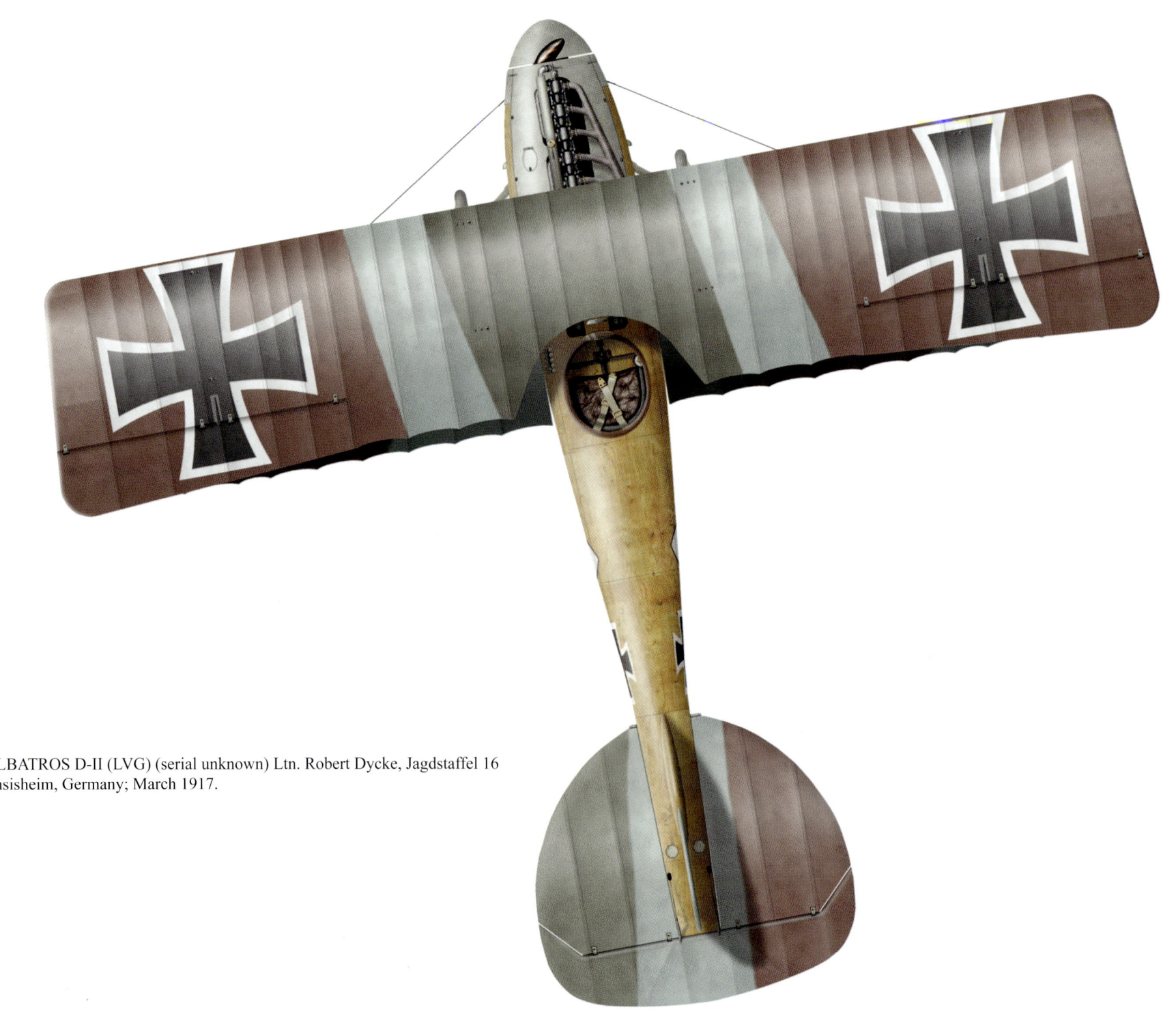

ALBATROS D-II (LVG) (serial unknown) Ltn. Robert Dycke, Jagdstaffel 16 Ensisheim, Germany; March 1917.

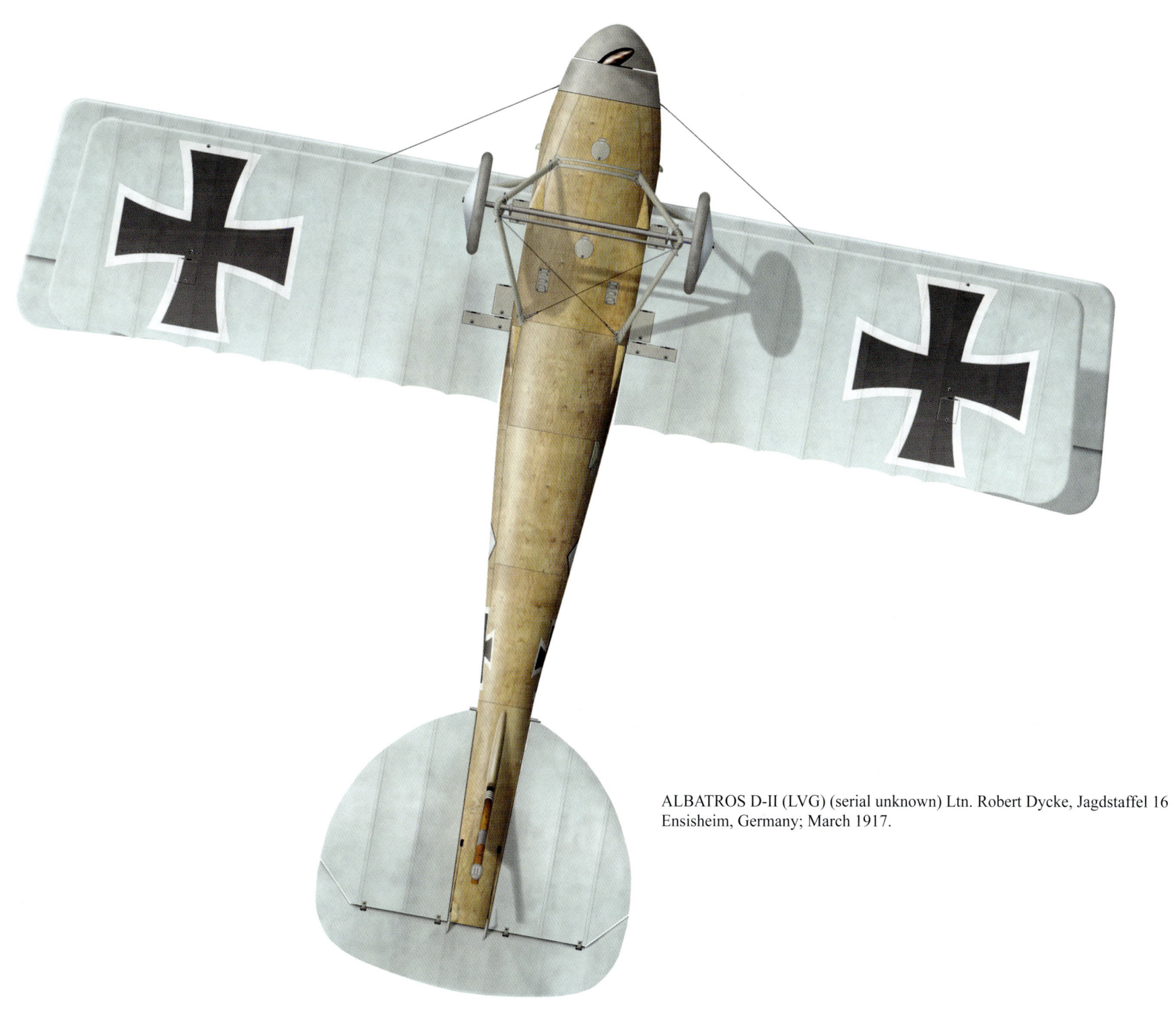

ALBATROS D-II (LVG) (serial unknown) Ltn. Robert Dycke, Jagdstaffel 16
Ensisheim, Germany; March 1917.

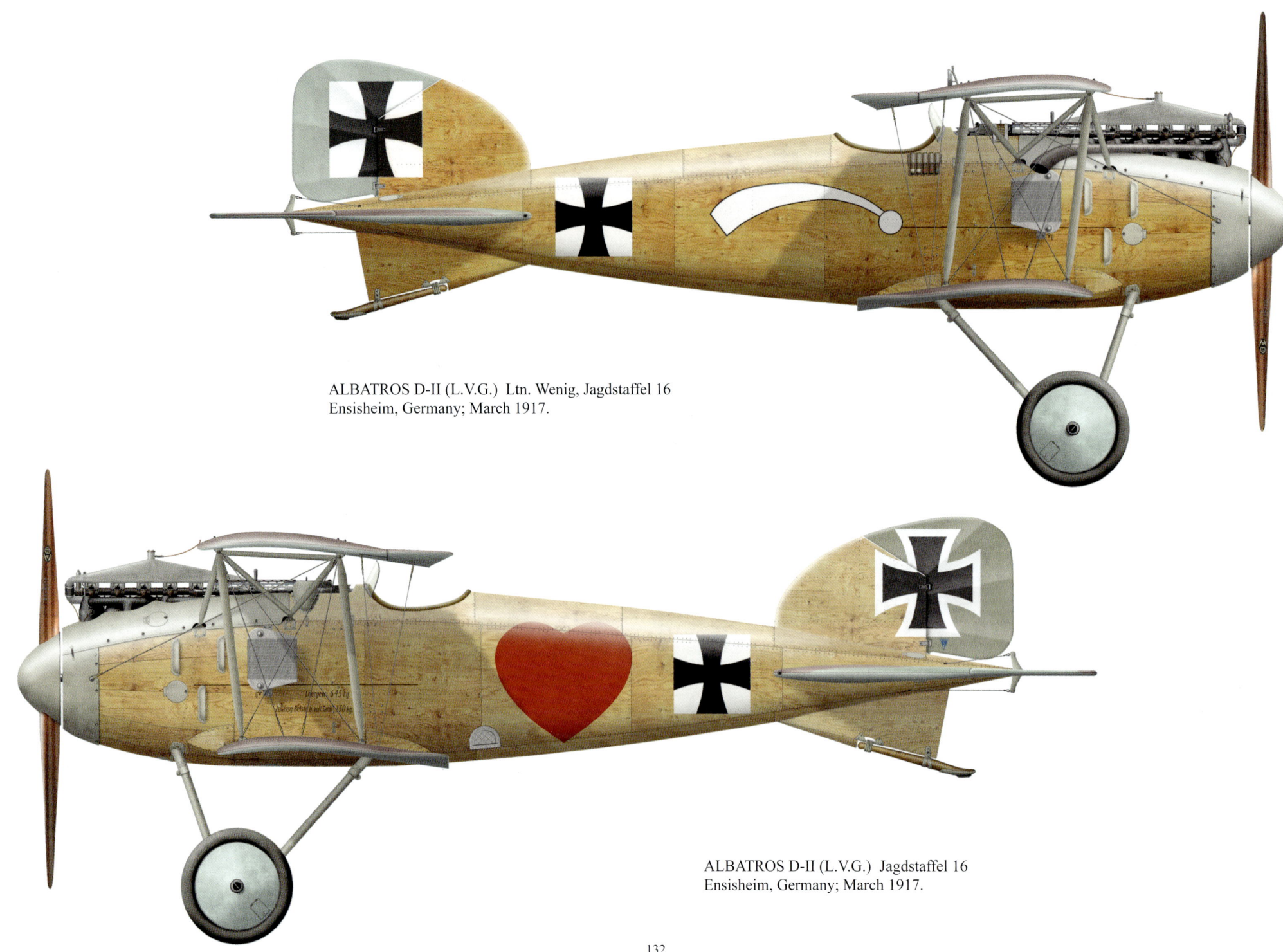

ALBATROS D-II (L.V.G.) Ltn. Wenig, Jagdstaffel 16
Ensisheim, Germany; March 1917.

ALBATROS D-II (L.V.G.) Jagdstaffel 16
Ensisheim, Germany; March 1917.

ALBATROS D-II (O.A.W.)

ALBATROS D-II (O.A.W.) (910/16) Ltn. Max Böhme, Jagdstaffel 5
Gonnelieu, France; March 1917.
Notice overpainted number 4 under the number 8 on the fuselage sides.

ALBATROS D-II (O.A.W.) (910/16) Ltn. Max Böhme, Jagdstaffel 5
Gonnelieu, France; March 1917.

ALBATROS D-II (O.A.W.) (910/16) Ltn. Max Böhme, Jagdstaffel 5
Gonnelieu, France; March 1917.

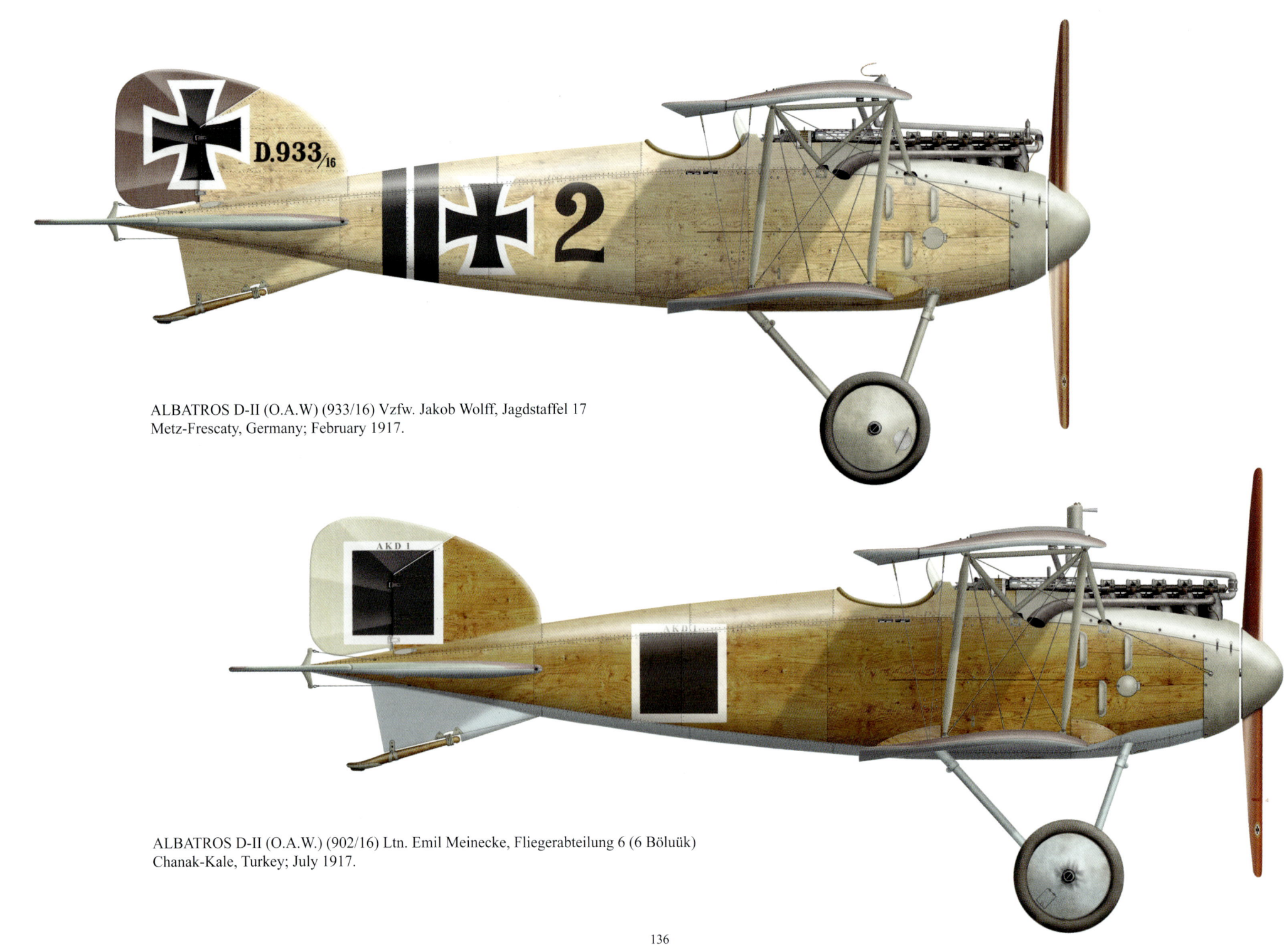

ALBATROS D-II (O.A.W) (933/16) Vzfw. Jakob Wolff, Jagdstaffel 17
Metz-Frescaty, Germany; February 1917.

ALBATROS D-II (O.A.W.) (902/16) Ltn. Emil Meinecke, Fliegerabteilung 6 (6 Bölüük)
Chanak-Kale, Turkey; July 1917.

ALBATROS D-III

ALBATROS D-III (2171/16) Ltn d R. Josef Veltjens, Jagdstaffel 14
Marchais, France; March 1917.

ALBATROS D-III (2182/16) Oblt. Rudolph Berthold, Jagdstaffel 14
Marchais, France; March 1917.

ALBATROS D-III (2171/16) Ltn d R. Josef Veltjens, Jagdstaffel 14
Marchais, France; March 1917.

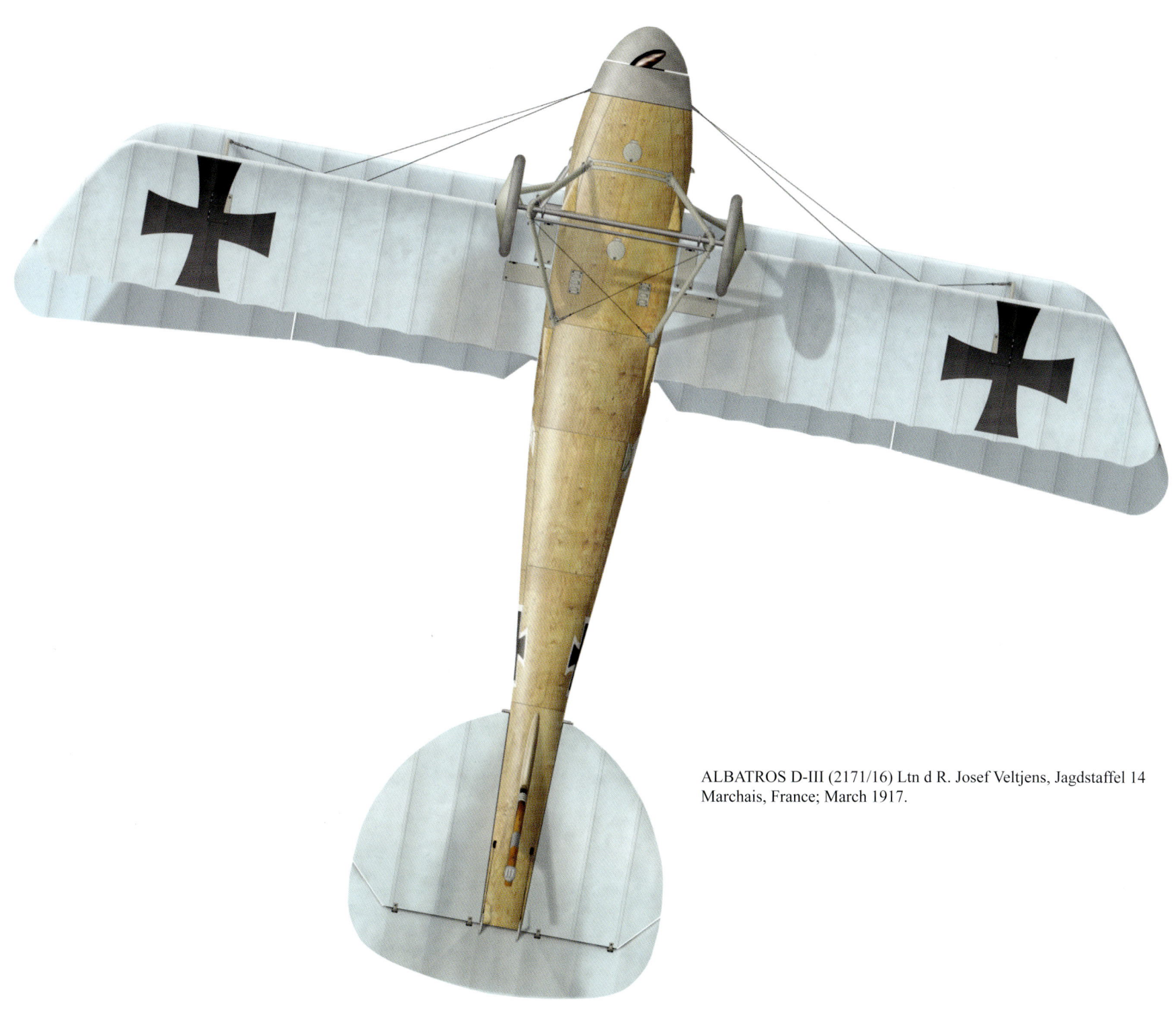

ALBATROS D-III (2171/16) Ltn d R. Josef Veltjens, Jagdstaffel 14
Marchais, France; March 1917.

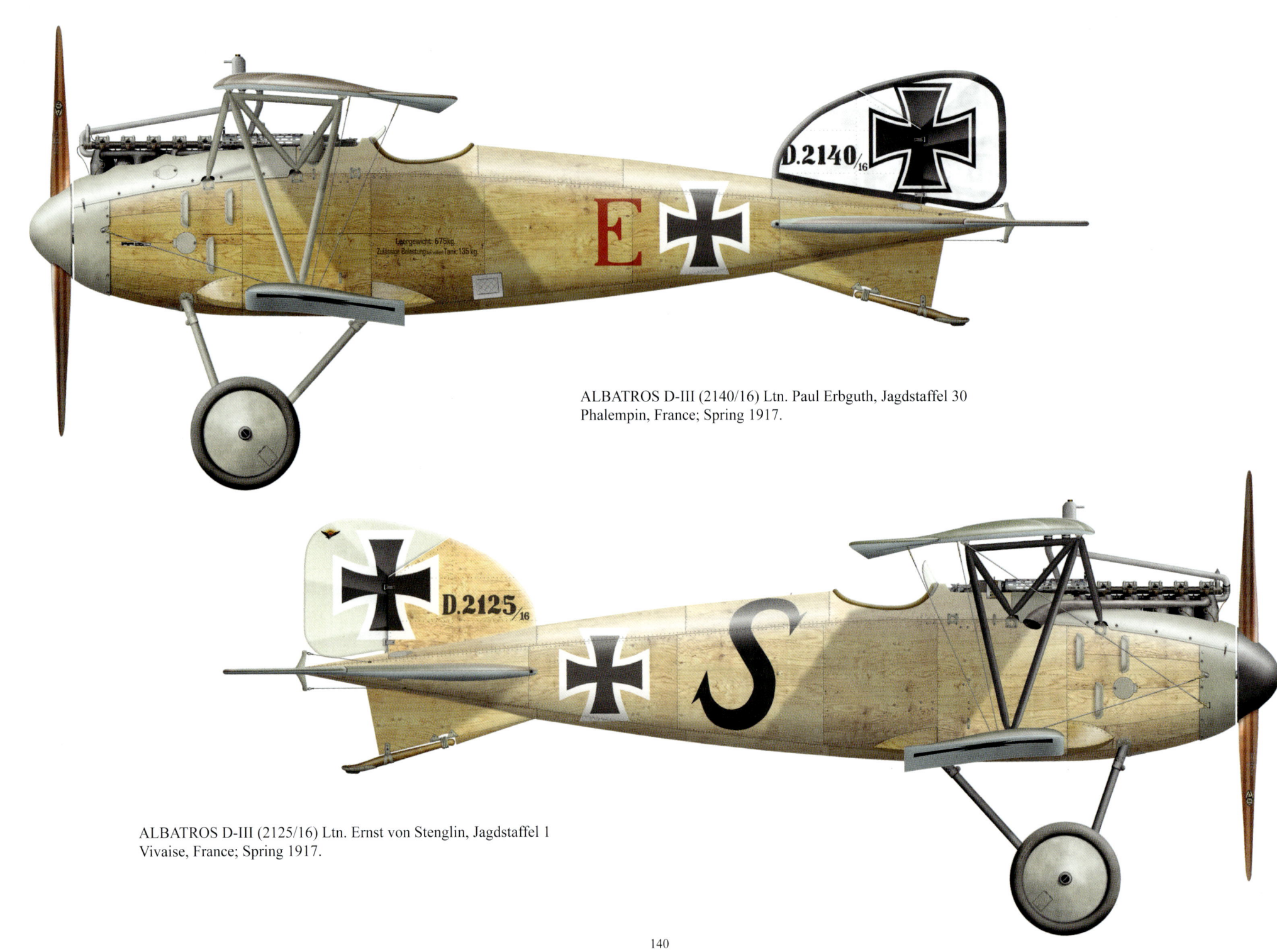

ALBATROS D-III (2140/16) Ltn. Paul Erbguth, Jagdstaffel 30
Phalempin, France; Spring 1917.

ALBATROS D-III (2125/16) Ltn. Ernst von Stenglin, Jagdstaffel 1
Vivaise, France; Spring 1917.

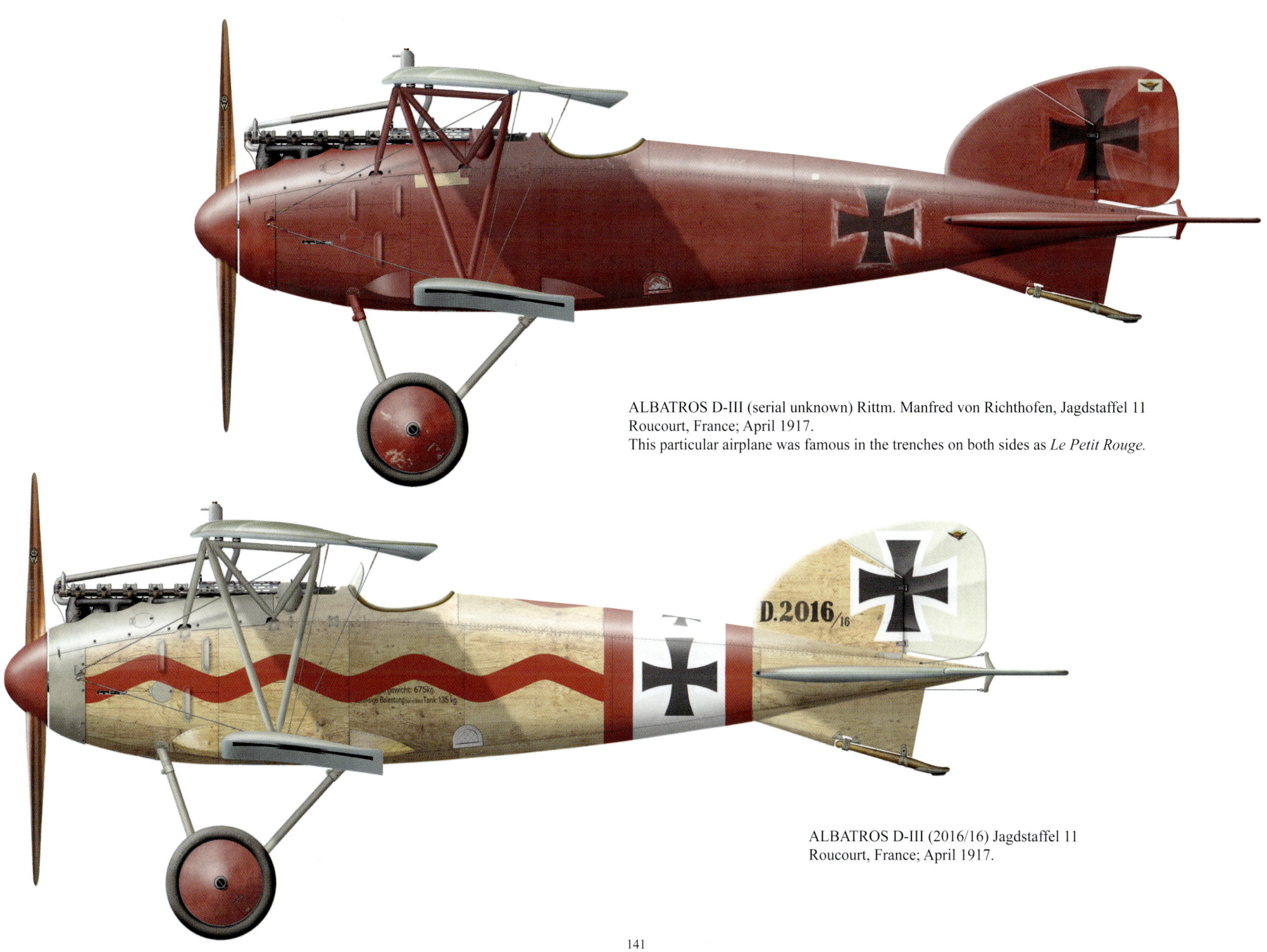

ALBATROS D-III (serial unknown) Rittm. Manfred von Richthofen, Jagdstaffel 11
Roucourt, France; April 1917.
This particular airplane was famous in the trenches on both sides as *Le Petit Rouge*.

ALBATROS D-III (2016/16) Jagdstaffel 11
Roucourt, France; April 1917.

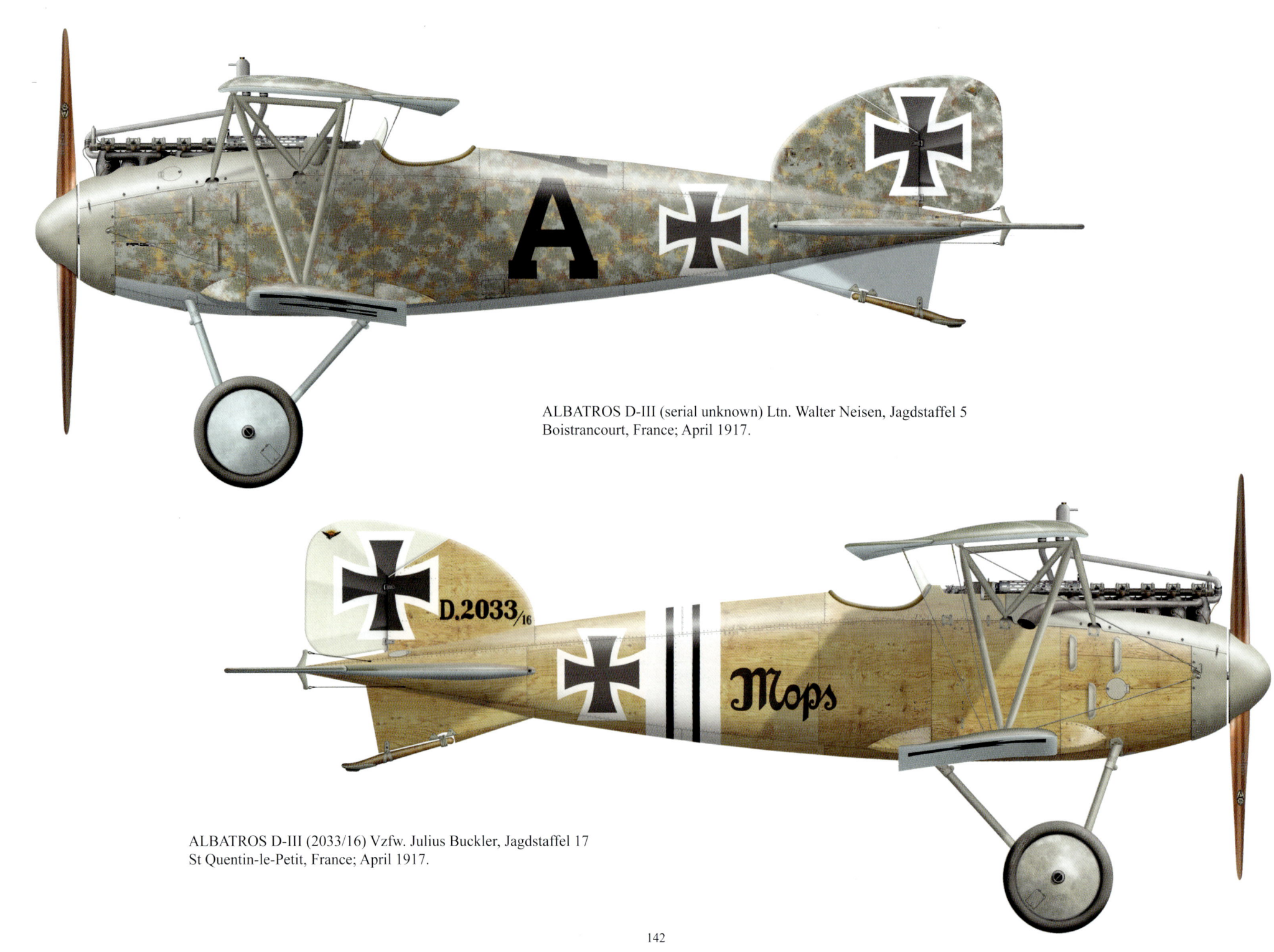

ALBATROS D-III (serial unknown) Ltn. Walter Neisen, Jagdstaffel 5
Boistrancourt, France; April 1917.

ALBATROS D-III (2033/16) Vzfw. Julius Buckler, Jagdstaffel 17
St Quentin-le-Petit, France; April 1917.

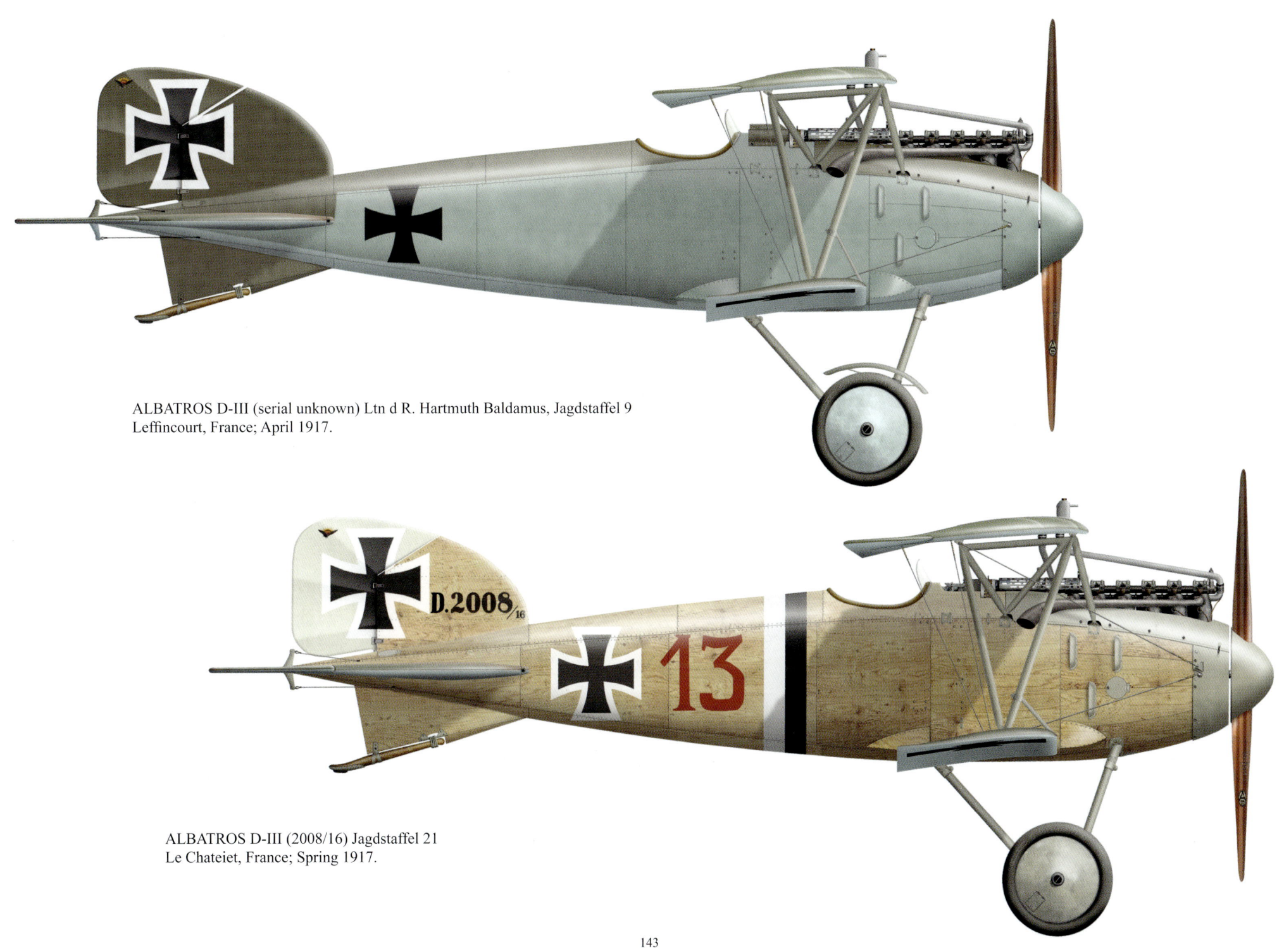

ALBATROS D-III (serial unknown) Ltn d R. Hartmuth Baldamus, Jagdstaffel 9 Leffincourt, France; April 1917.

ALBATROS D-III (2008/16) Jagdstaffel 21 Le Chateiet, France; Spring 1917.

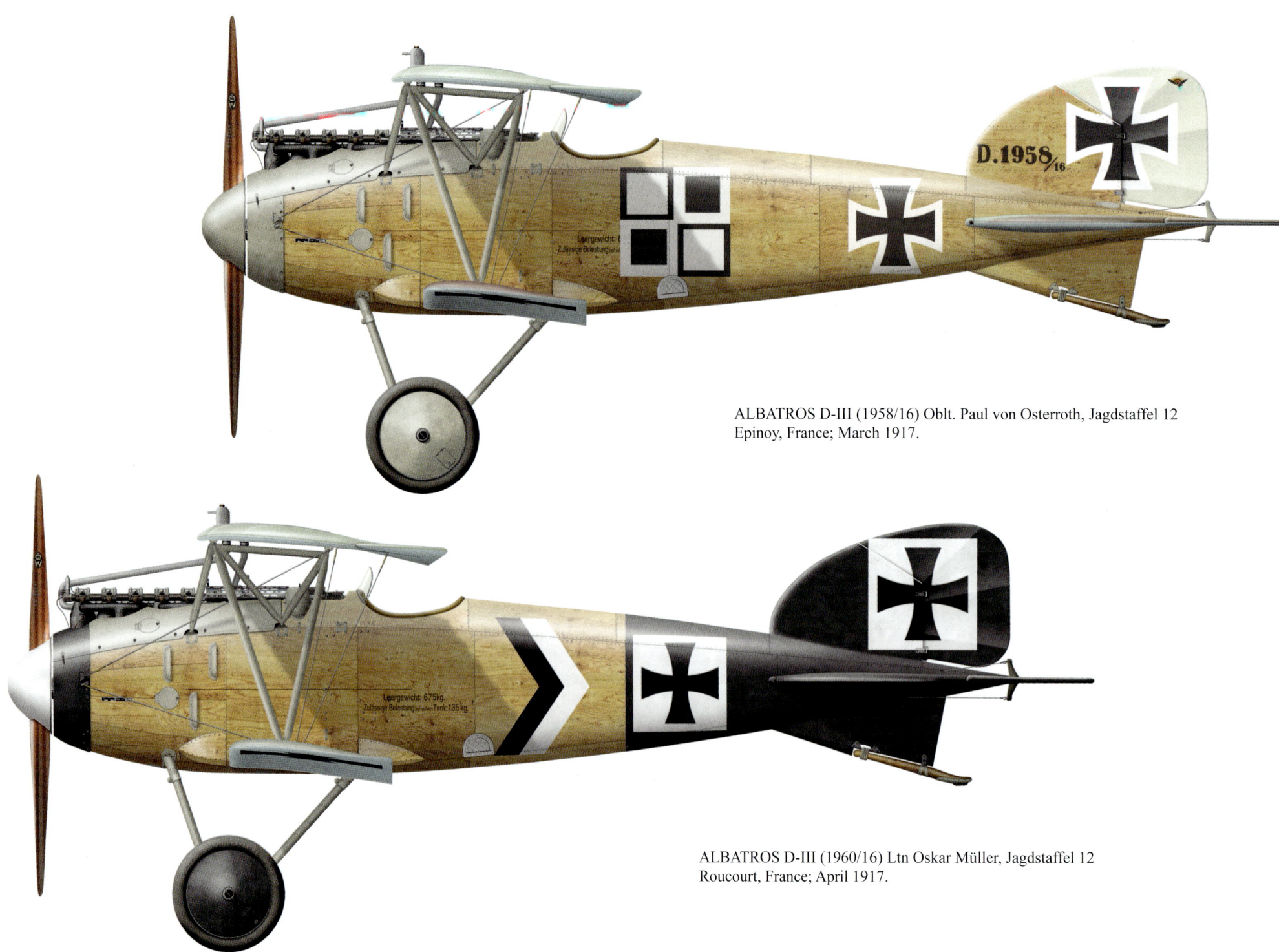

ALBATROS D-III (1958/16) Oblt. Paul von Osterroth, Jagdstaffel 12
Epinoy, France; March 1917.

ALBATROS D-III (1960/16) Ltn Oskar Müller, Jagdstaffel 12
Roucourt, France; April 1917.

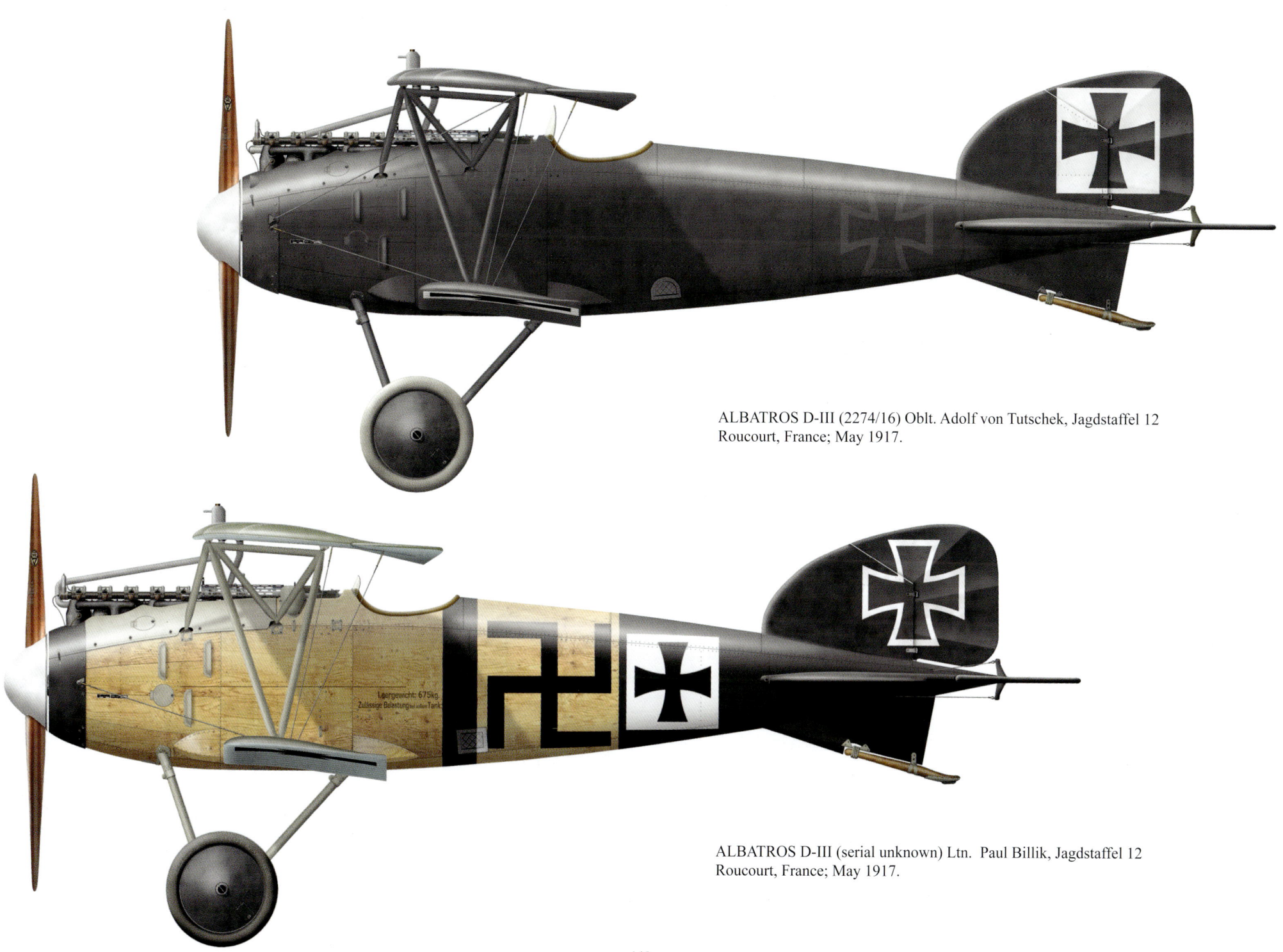

ALBATROS D-III (2274/16) Oblt. Adolf von Tutschek, Jagdstaffel 12
Roucourt, France; May 1917.

ALBATROS D-III (serial unknown) Ltn. Paul Billik, Jagdstaffel 12
Roucourt, France; May 1917.

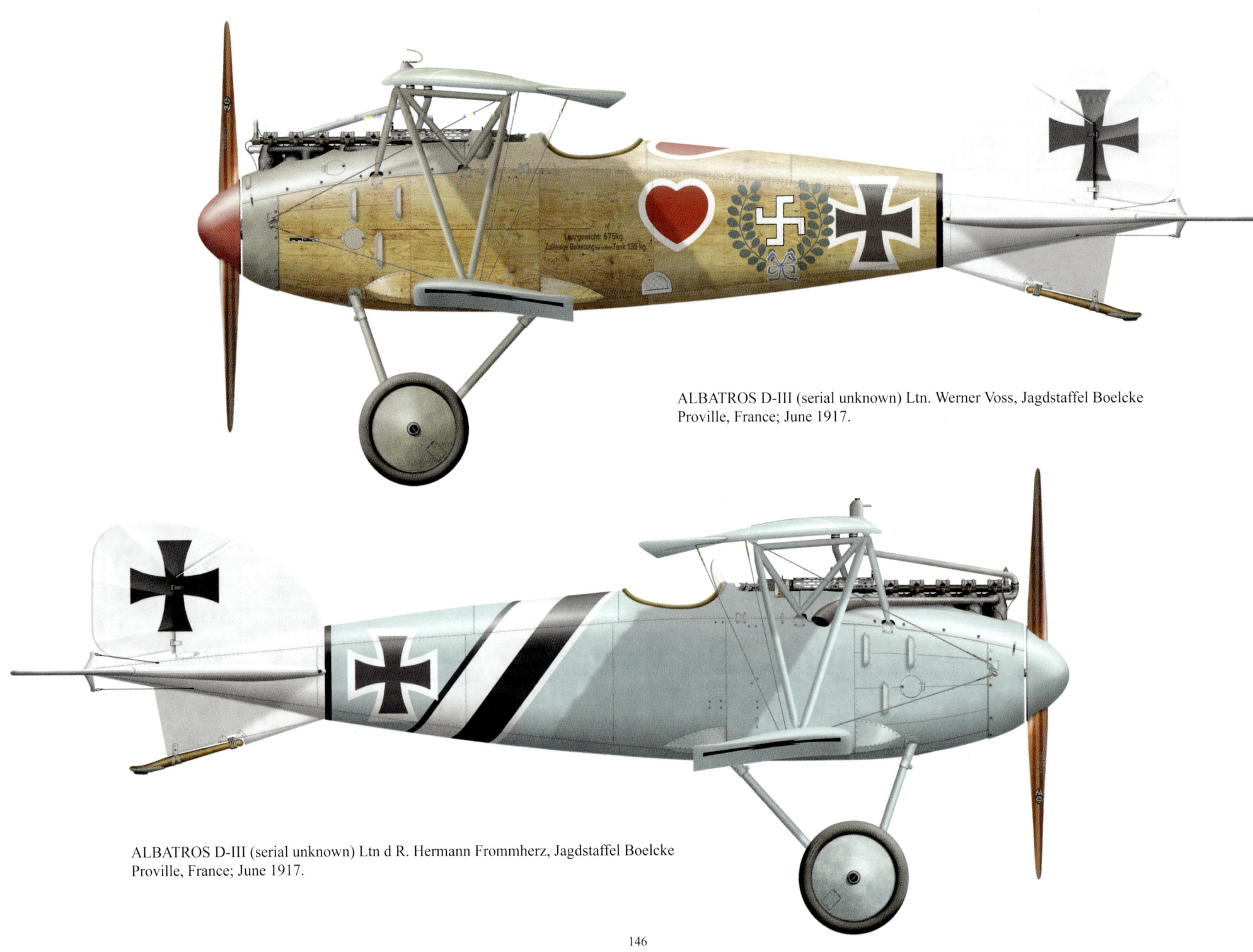

ALBATROS D-III (serial unknown) Ltn. Werner Voss, Jagdstaffel Boelcke
Proville, France; June 1917.

ALBATROS D-III (serial unknown) Ltn d R. Hermann Frommherz, Jagdstaffel Boelcke
Proville, France; June 1917.

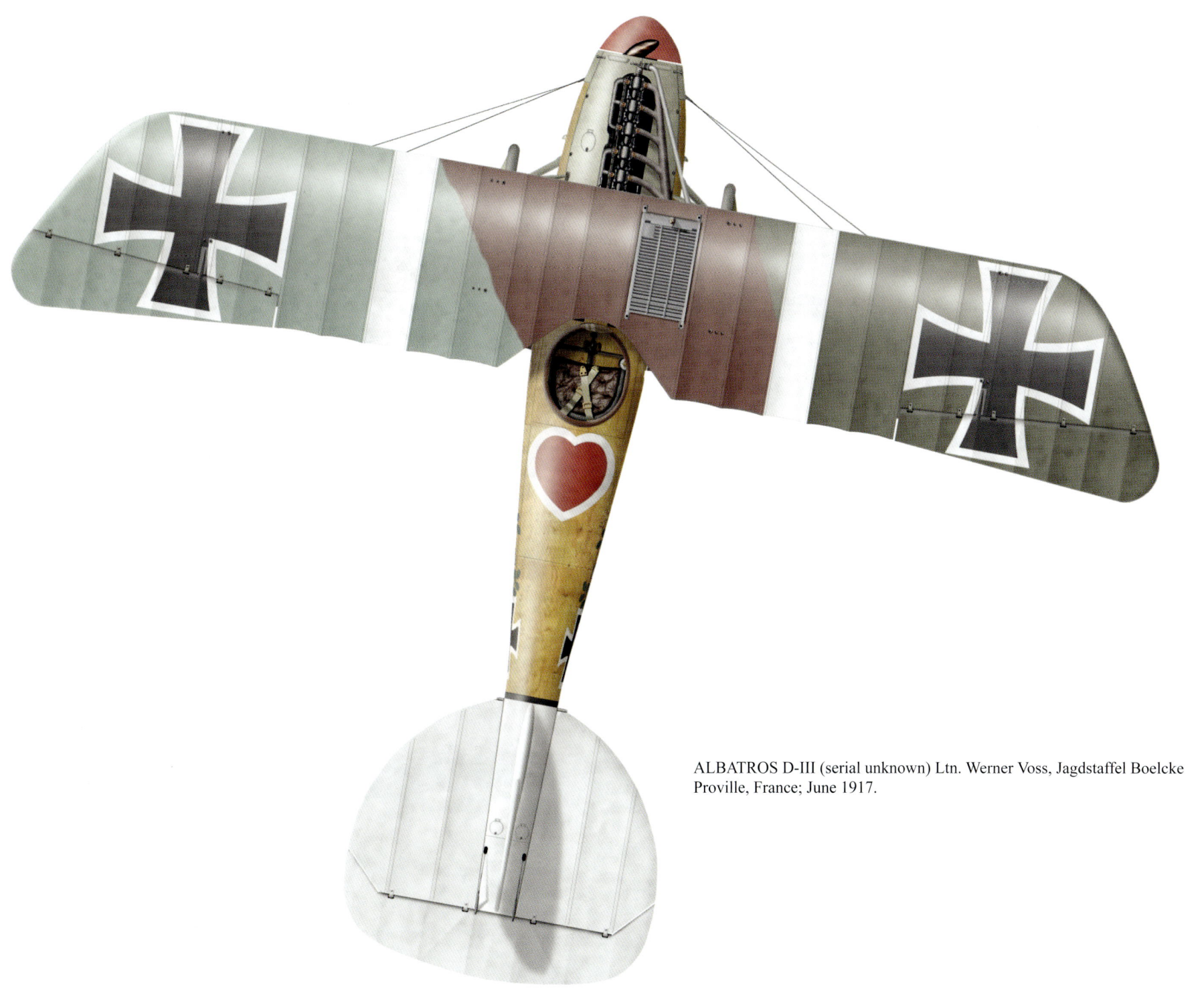

ALBATROS D-III (serial unknown) Ltn. Werner Voss, Jagdstaffel Boelcke Proville, France; June 1917.

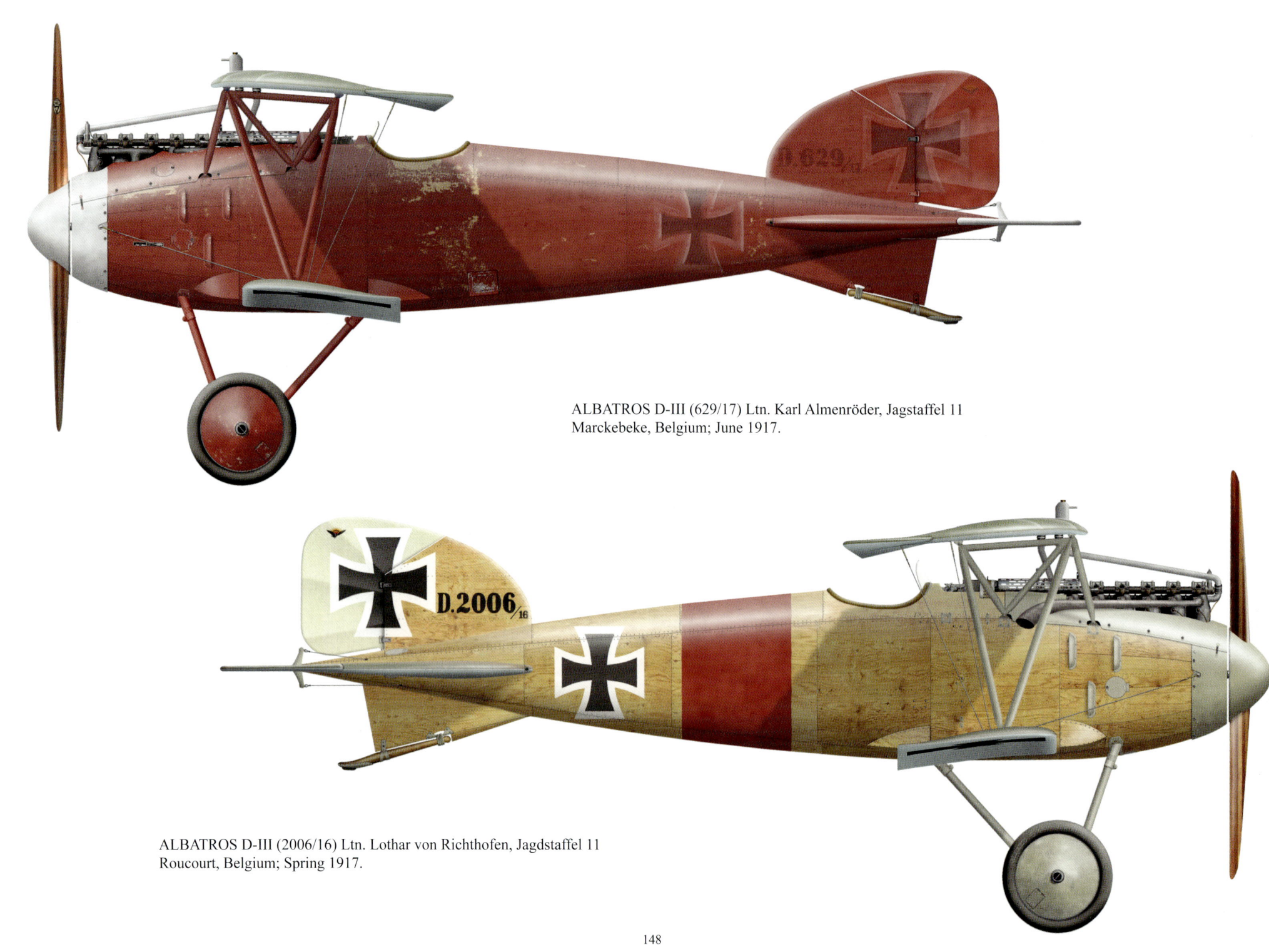

ALBATROS D-III (629/17) Ltn. Karl Almenröder, Jagstaffel 11
Marckebeke, Belgium; June 1917.

ALBATROS D-III (2006/16) Ltn. Lothar von Richthofen, Jagdstaffel 11
Roucourt, Belgium; Spring 1917.

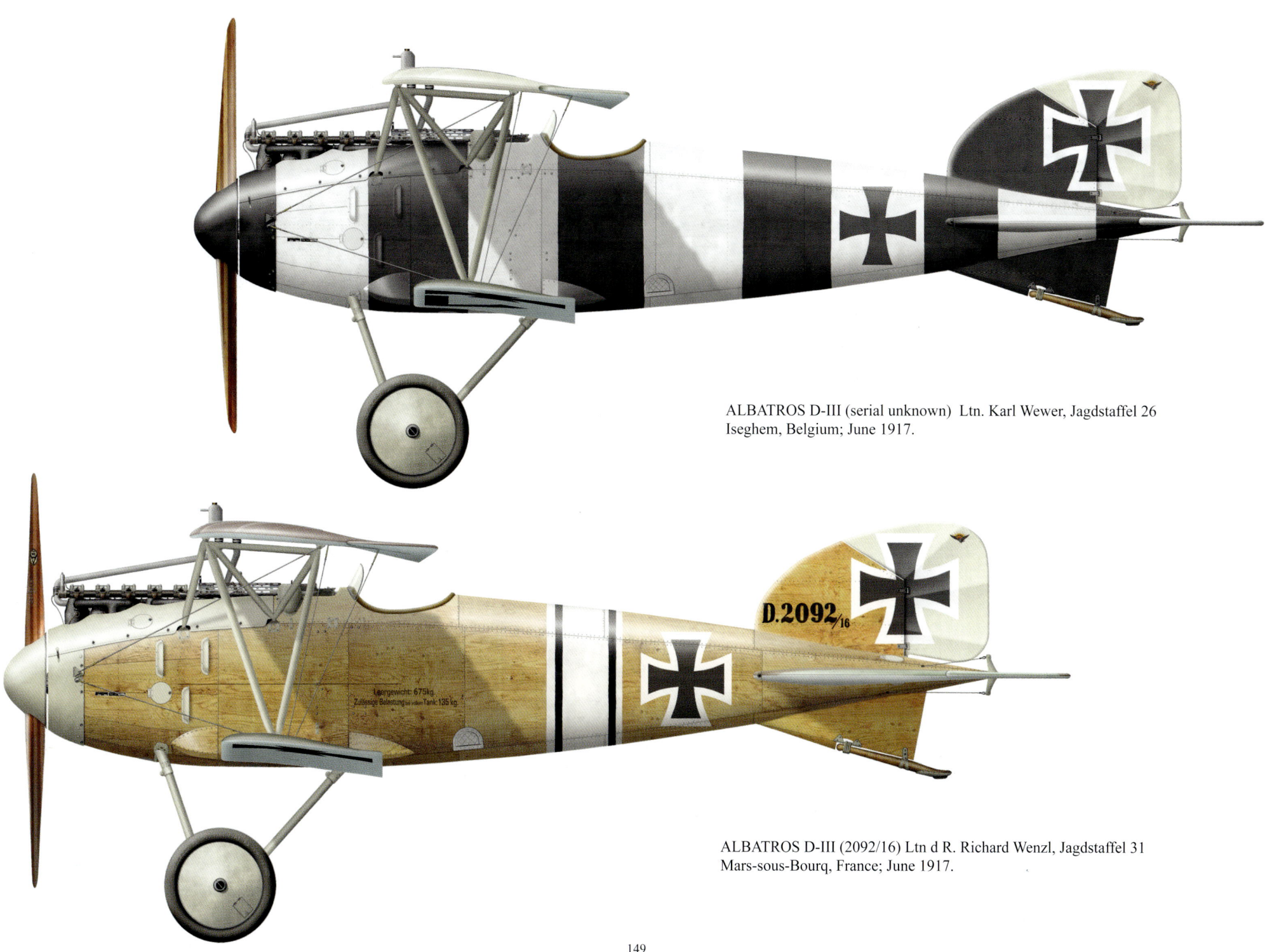

ALBATROS D-III (serial unknown) Ltn. Karl Wewer, Jagdstaffel 26
Iseghem, Belgium; June 1917.

ALBATROS D-III (2092/16) Ltn d R. Richard Wenzl, Jagdstaffel 31
Mars-sous-Bourq, France; June 1917.

ALBATROS D-III (serial unknown) Ltn. Karl Wewer, Jagdstaffel 26
Iseghem, Belgium; June 1917.
Note repair panel of unpainted fabric on left wingtip.

ALBATROS D-III (serial unknown) Ltn. Karl Wewer, Jagdstaffel 26
Iseghem, Belgium; June 1917.

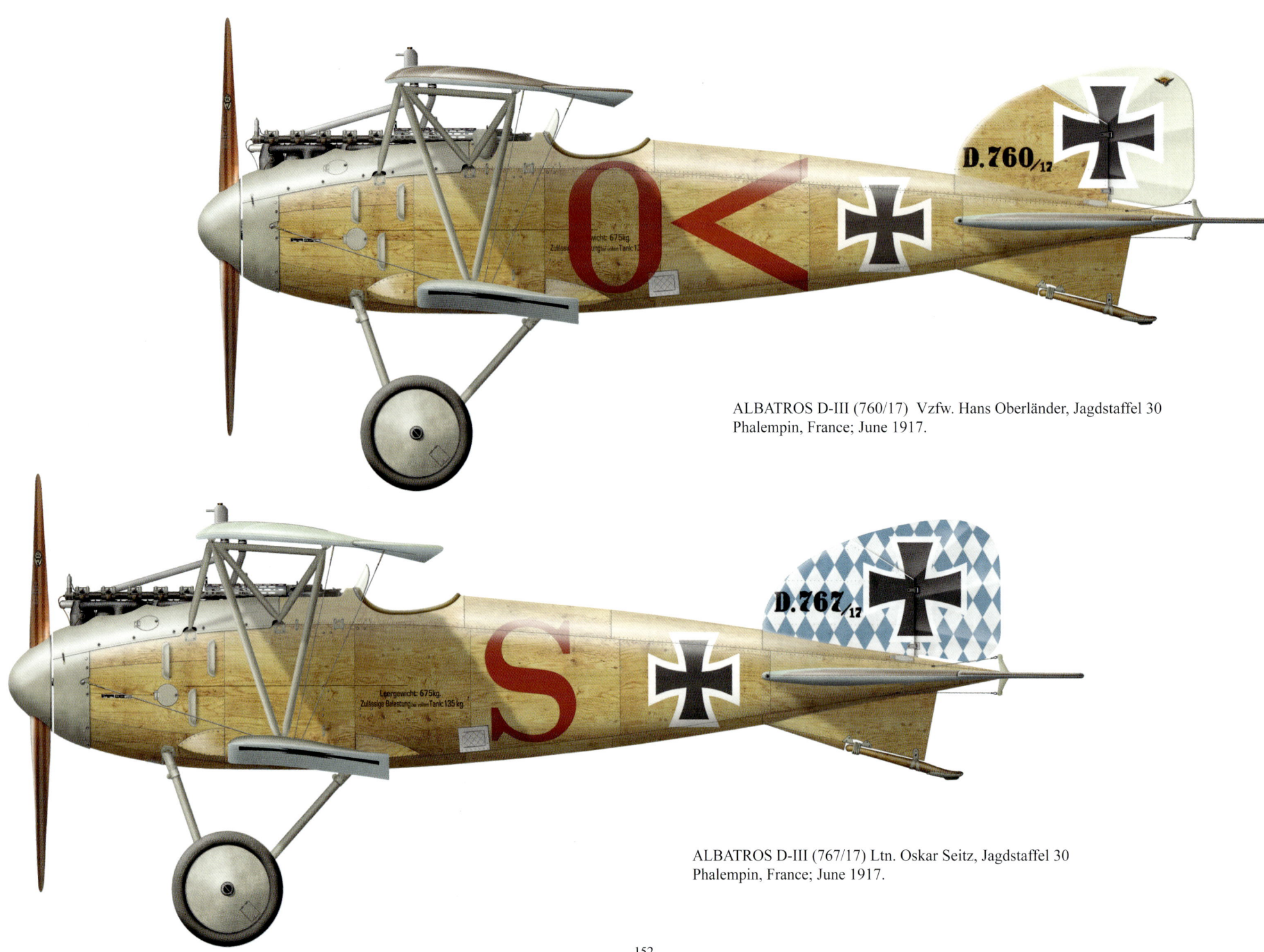

ALBATROS D-III (760/17) Vzfw. Hans Oberländer, Jagdstaffel 30
Phalempin, France; June 1917.

ALBATROS D-III (767/17) Ltn. Oskar Seitz, Jagdstaffel 30
Phalempin, France; June 1917.

ALBATROS D-III (760/17) Ltn, d R Hans Oberländer, Jagdstaffel 30
Phalempin, France; June 1917.

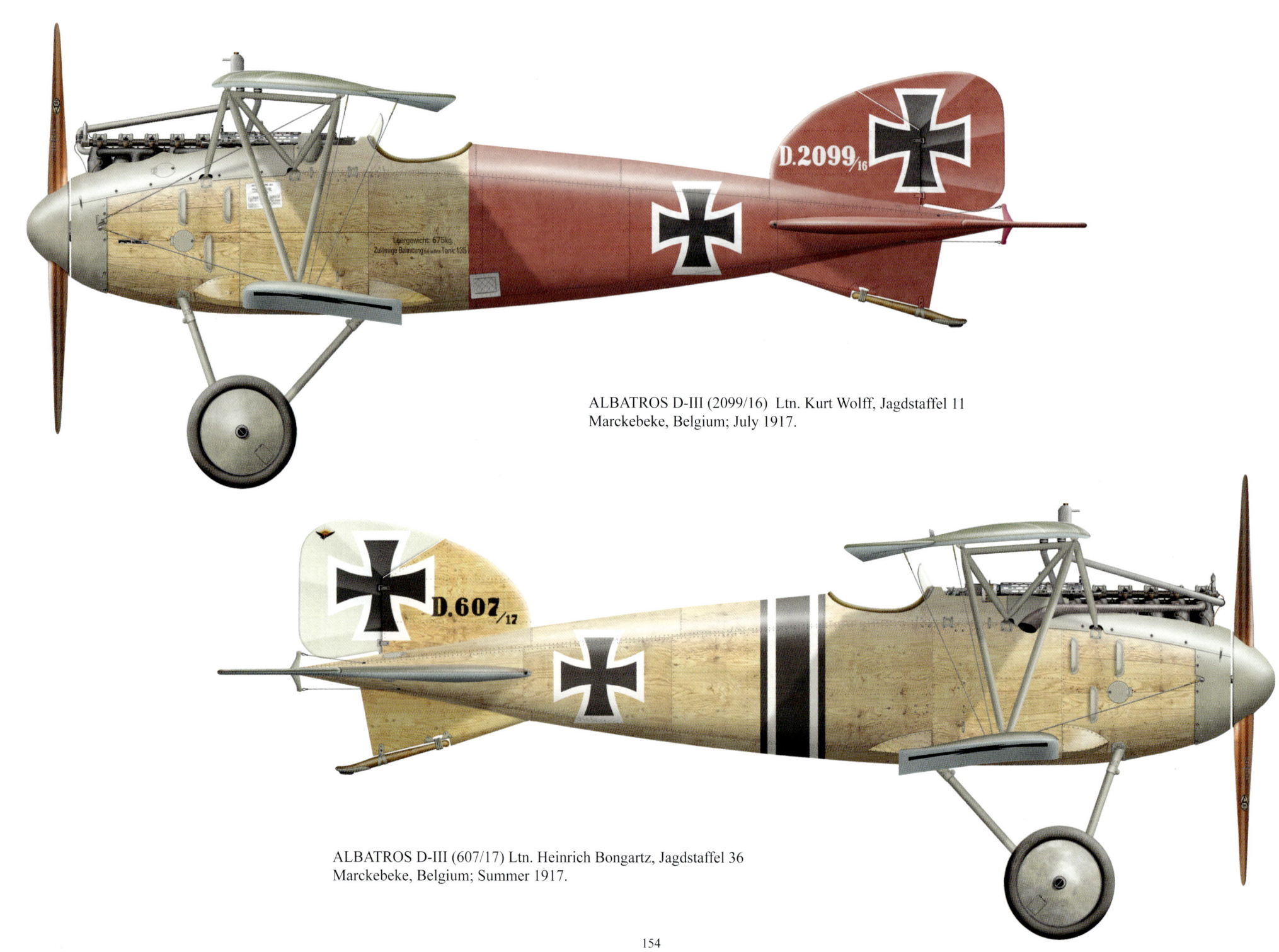

ALBATROS D-III (2099/16) Ltn. Kurt Wolff, Jagdstaffel 11
Marckebeke, Belgium; July 1917.

ALBATROS D-III (607/17) Ltn. Heinrich Bongartz, Jagdstaffel 36
Marckebeke, Belgium; Summer 1917.

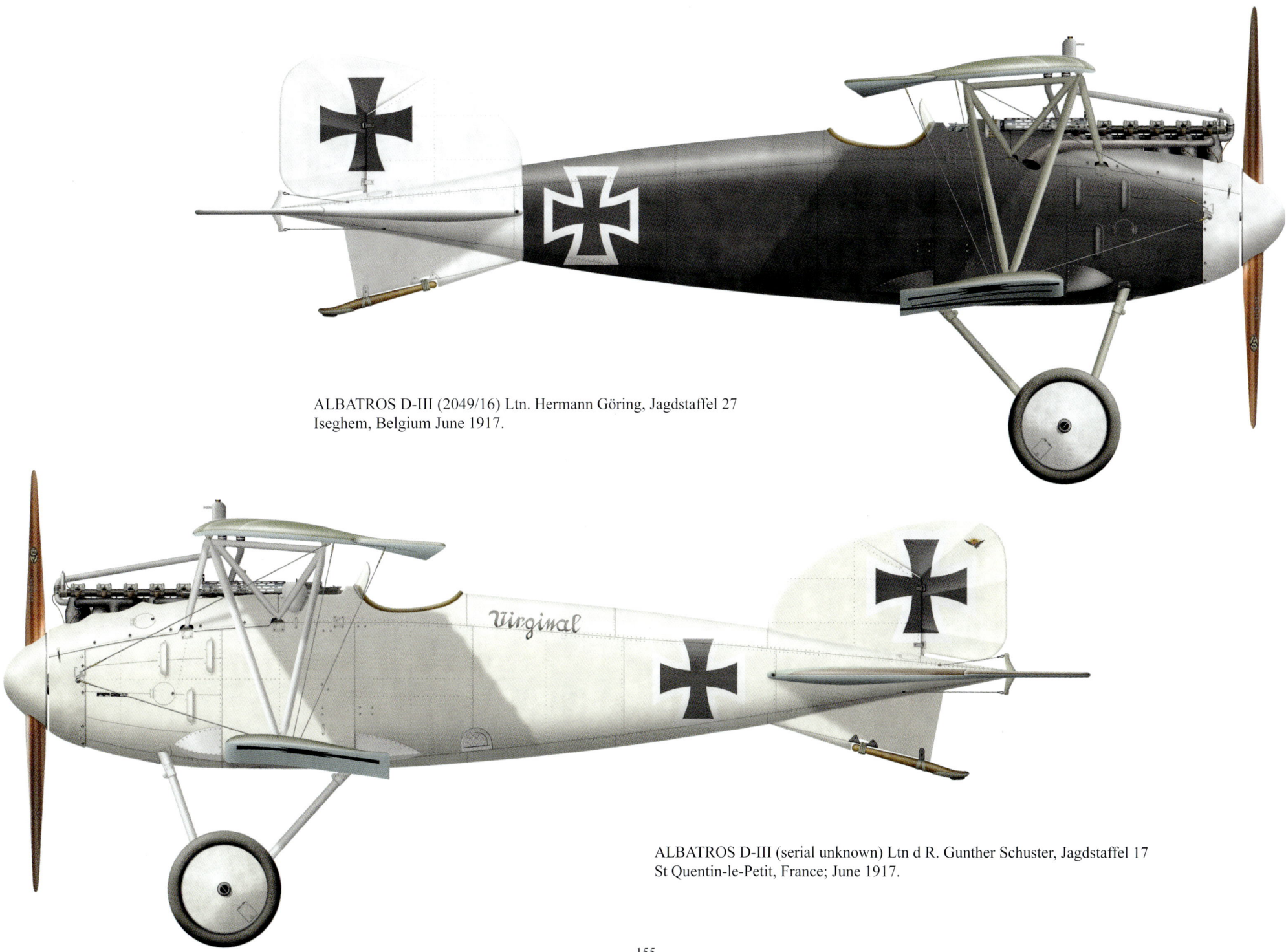

ALBATROS D-III (2049/16) Ltn. Hermann Göring, Jagdstaffel 27
Iseghem, Belgium June 1917.

ALBATROS D-III (serial unknown) Ltn d R. Gunther Schuster, Jagdstaffel 17
St Quentin-le-Petit, France; June 1917.

ALBATROS D-III (2049/16) Ltn. Hermann Göring, Jagdstaffel 27
Iseghem, Belgium June 1917.

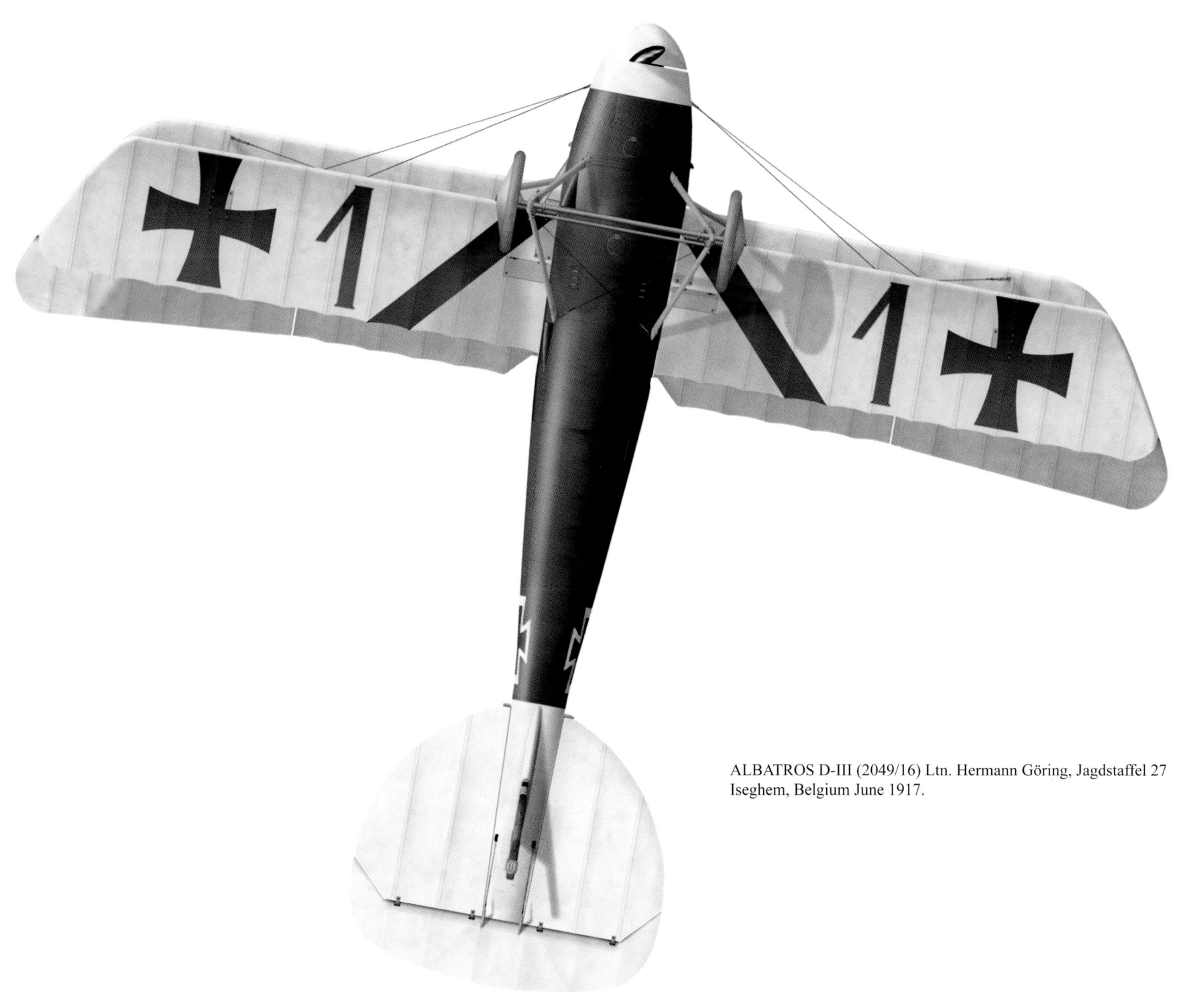

ALBATROS D-III (2049/16) Ltn. Hermann Göring, Jagdstaffel 27 Iseghem, Belgium June 1917.

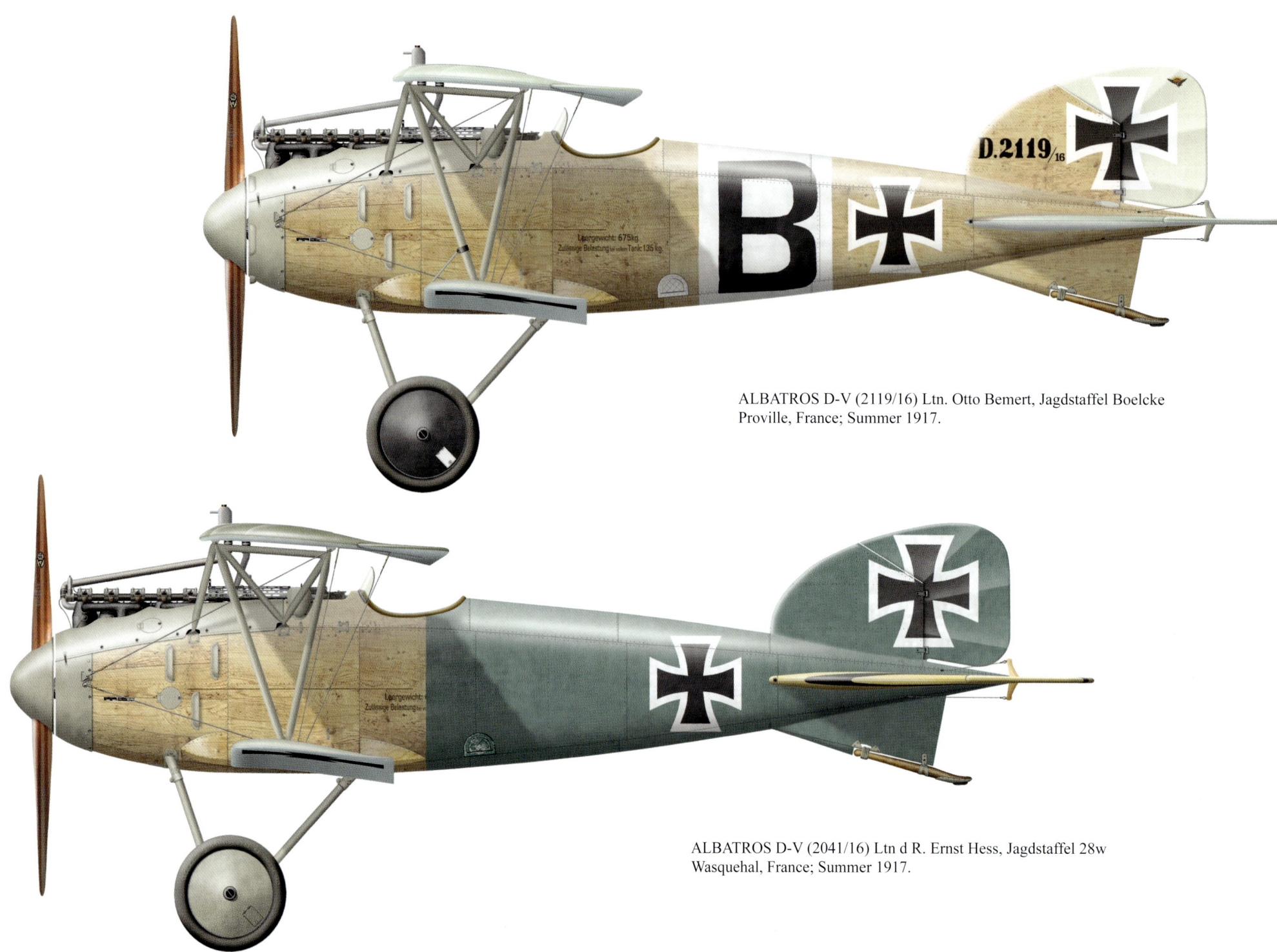

ALBATROS D-V (2119/16) Ltn. Otto Bemert, Jagdstaffel Boelcke
Proville, France; Summer 1917.

ALBATROS D-V (2041/16) Ltn d R. Ernst Hess, Jagdstaffel 28w
Wasquehal, France; Summer 1917.

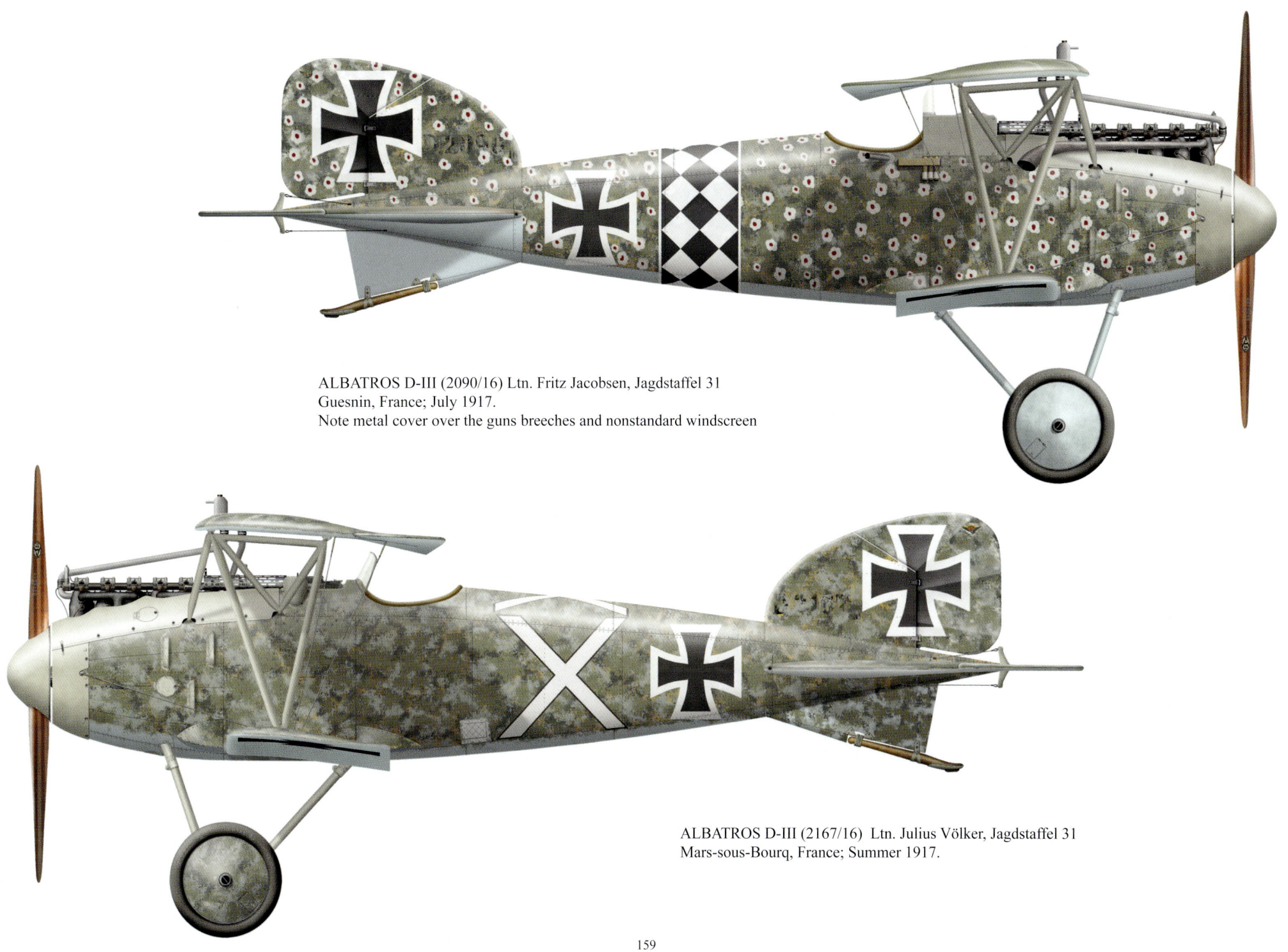

ALBATROS D-III (2090/16) Ltn. Fritz Jacobsen, Jagdstaffel 31
Guesnin, France; July 1917.
Note metal cover over the guns breeches and nonstandard windscreen

ALBATROS D-III (2167/16) Ltn. Julius Völker, Jagdstaffel 31
Mars-sous-Bourq, France; Summer 1917.

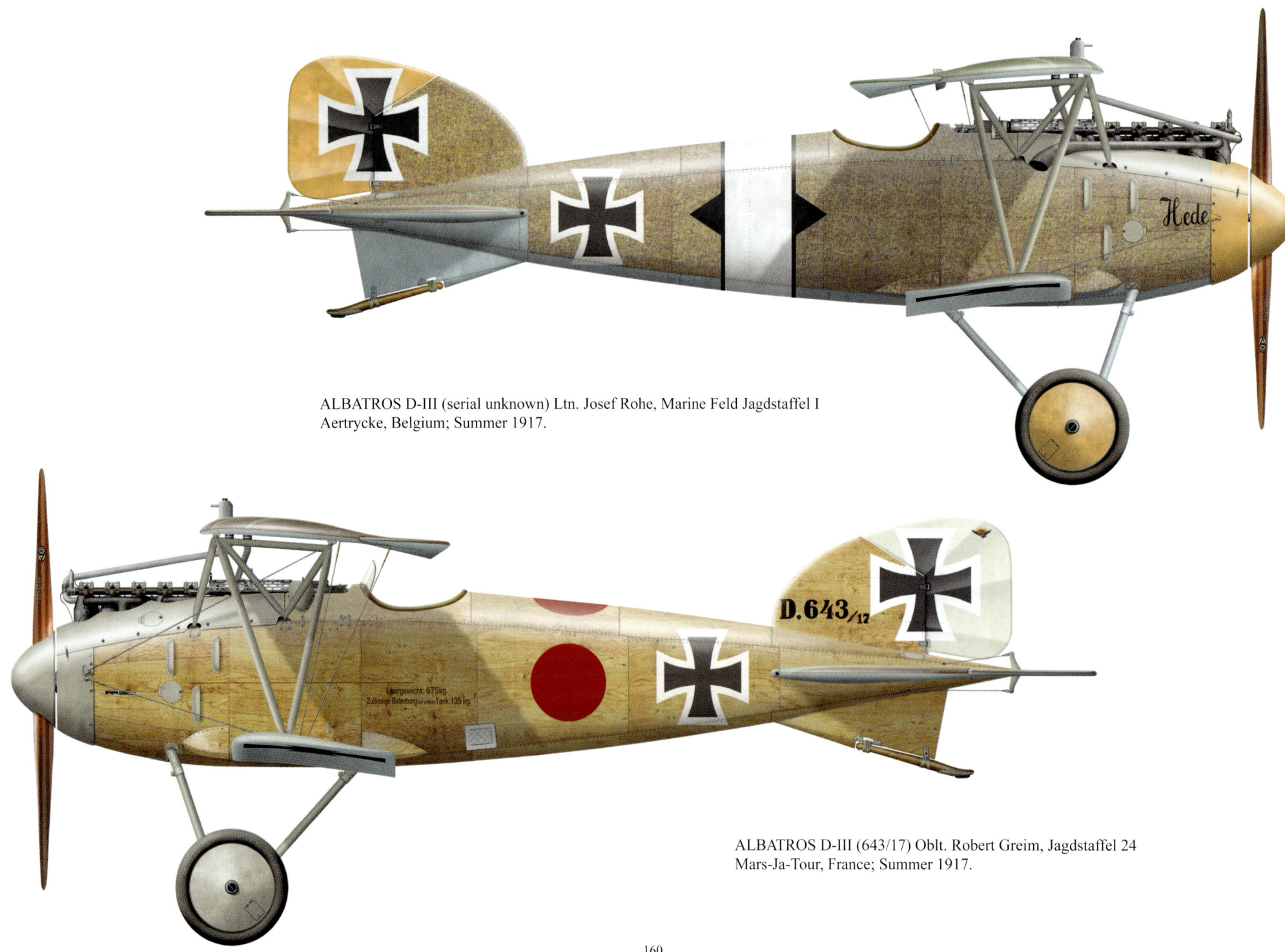

ALBATROS D-III (serial unknown) Ltn. Josef Rohe, Marine Feld Jagdstaffel I
Aertrycke, Belgium; Summer 1917.

ALBATROS D-III (643/17) Oblt. Robert Greim, Jagdstaffel 24
Mars-Ja-Tour, France; Summer 1917.

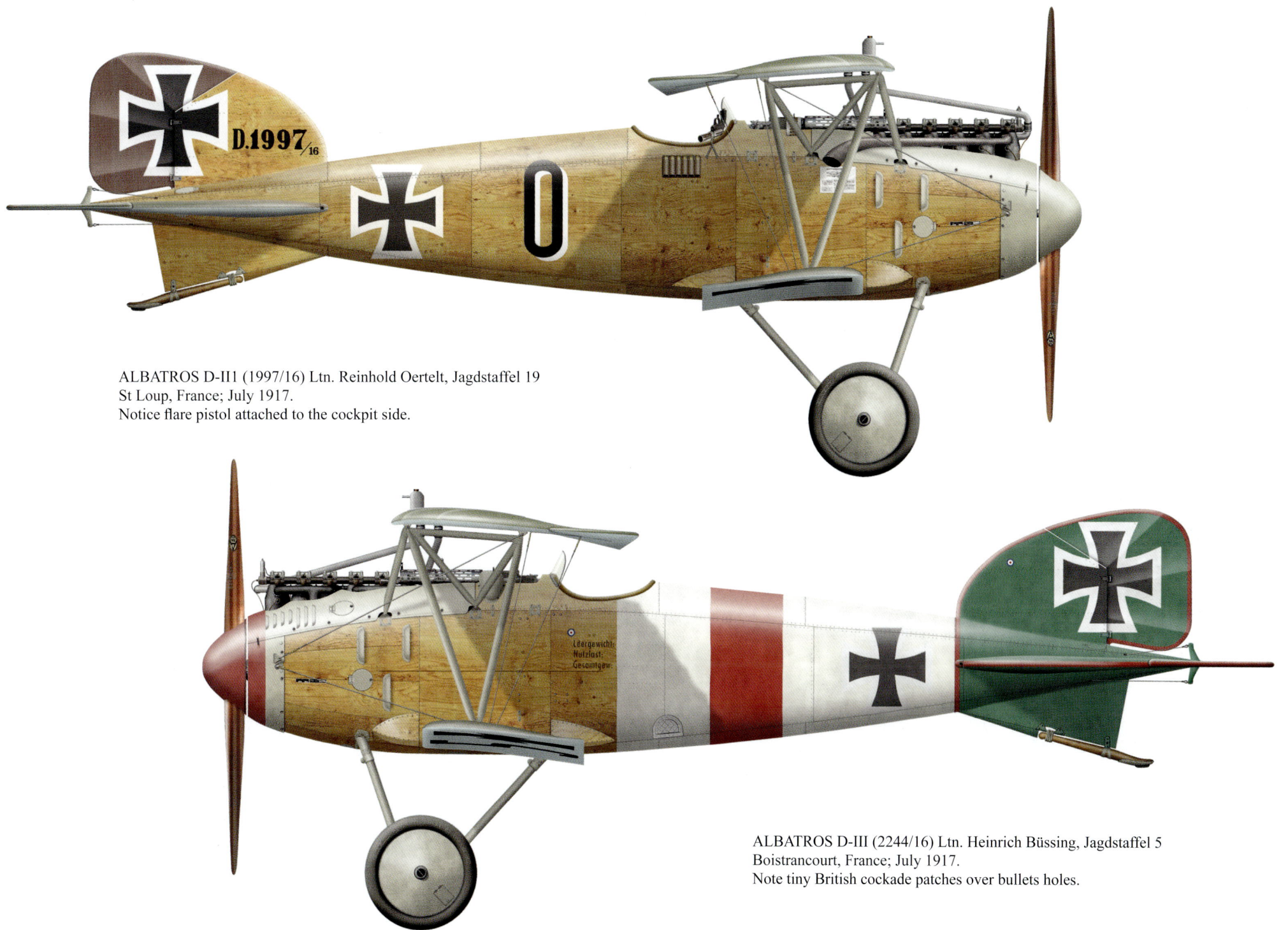

ALBATROS D-III (1997/16) Ltn. Reinhold Oertelt, Jagdstaffel 19
St Loup, France; July 1917.
Notice flare pistol attached to the cockpit side.

ALBATROS D-III (2244/16) Ltn. Heinrich Büssing, Jagdstaffel 5
Boistrancourt, France; July 1917.
Note tiny British cockade patches over bullets holes.

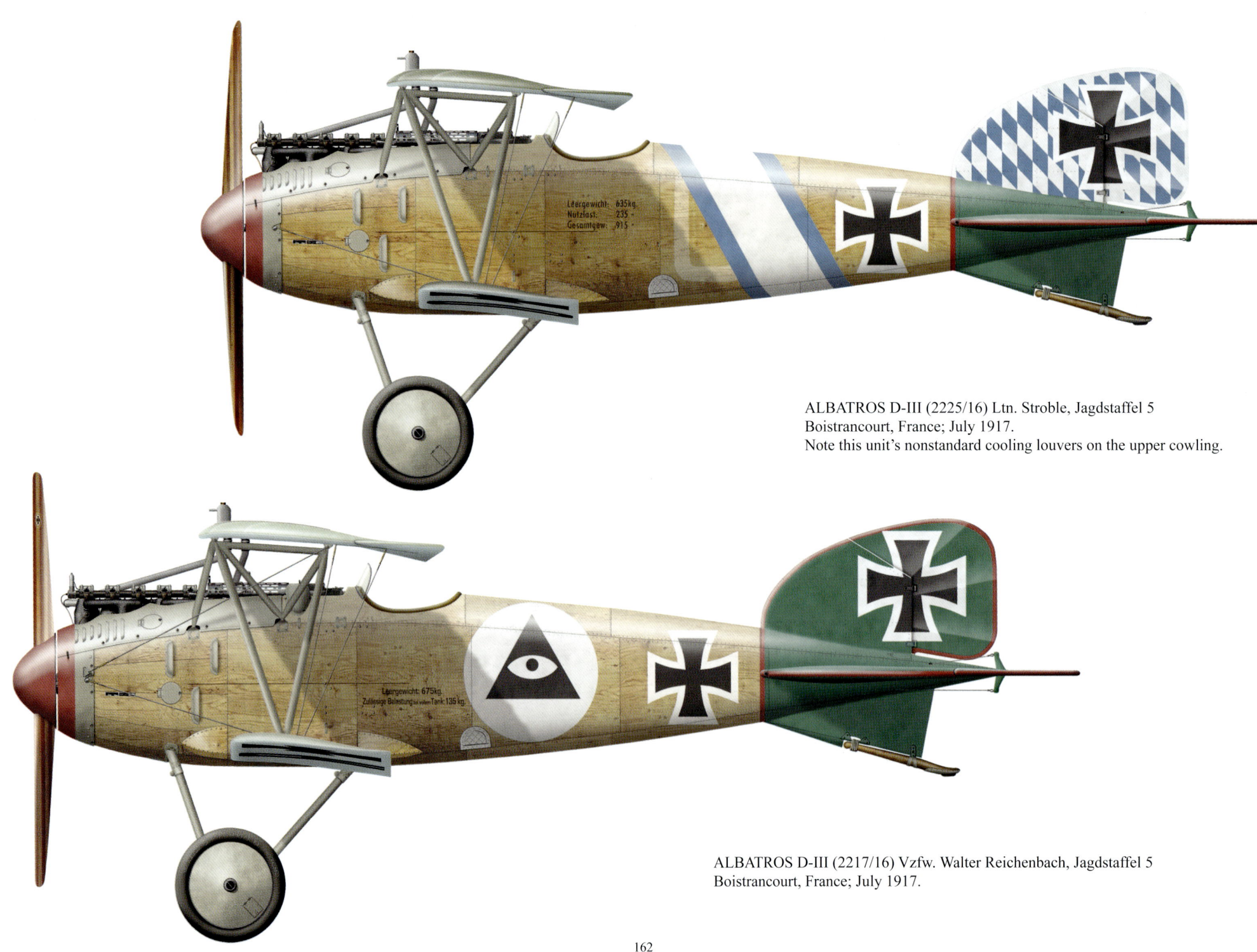

ALBATROS D-III (2225/16) Ltn. Stroble, Jagdstaffel 5
Boistrancourt, France; July 1917.
Note this unit's nonstandard cooling louvers on the upper cowling.

ALBATROS D-III (2217/16) Vzfw. Walter Reichenbach, Jagdstaffel 5
Boistrancourt, France; July 1917.

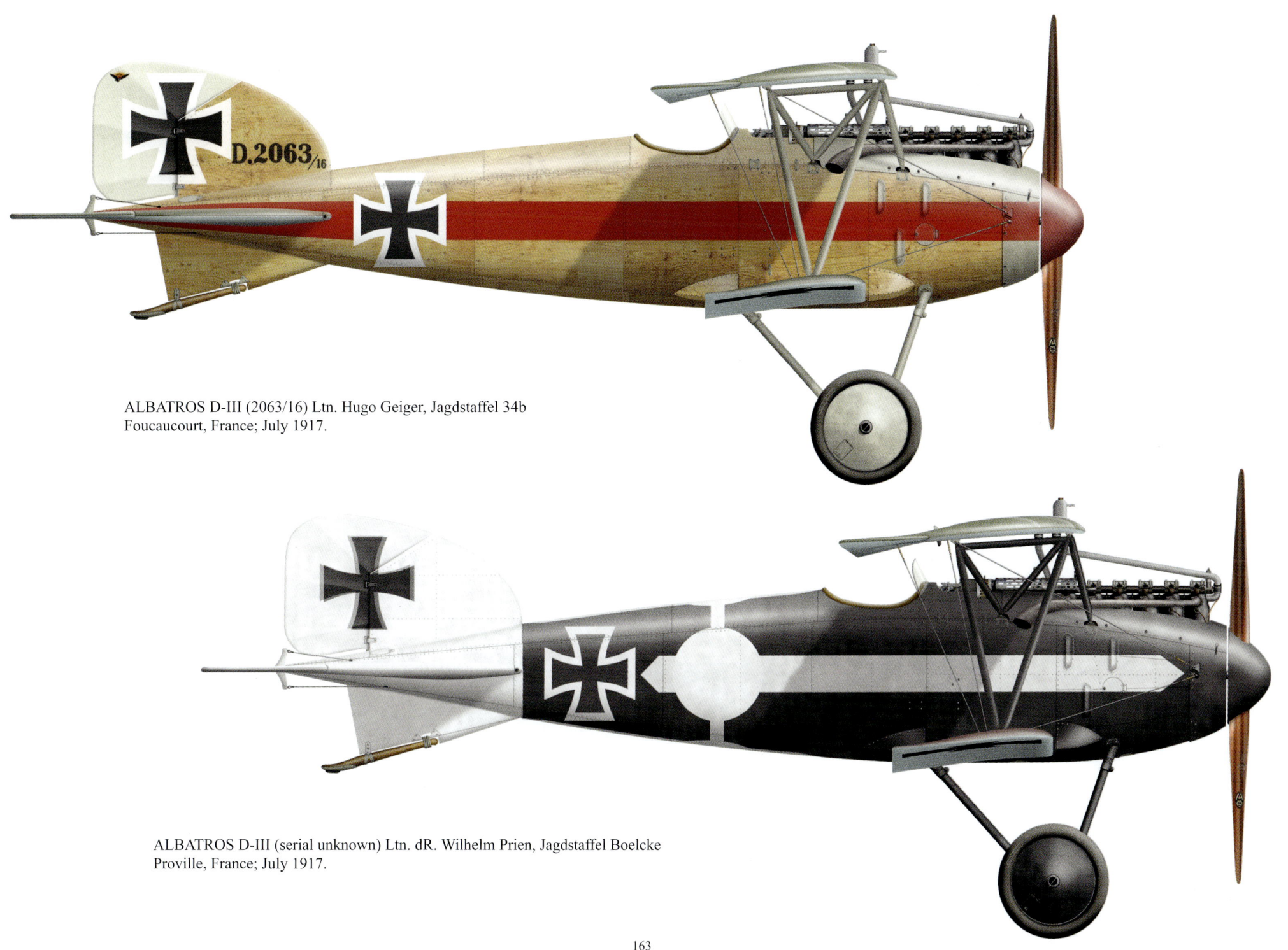

ALBATROS D-III (2063/16) Ltn. Hugo Geiger, Jagdstaffel 34b
Foucaucourt, France; July 1917.

ALBATROS D-III (serial unknown) Ltn. dR. Wilhelm Prien, Jagdstaffel Boelcke
Proville, France; July 1917.

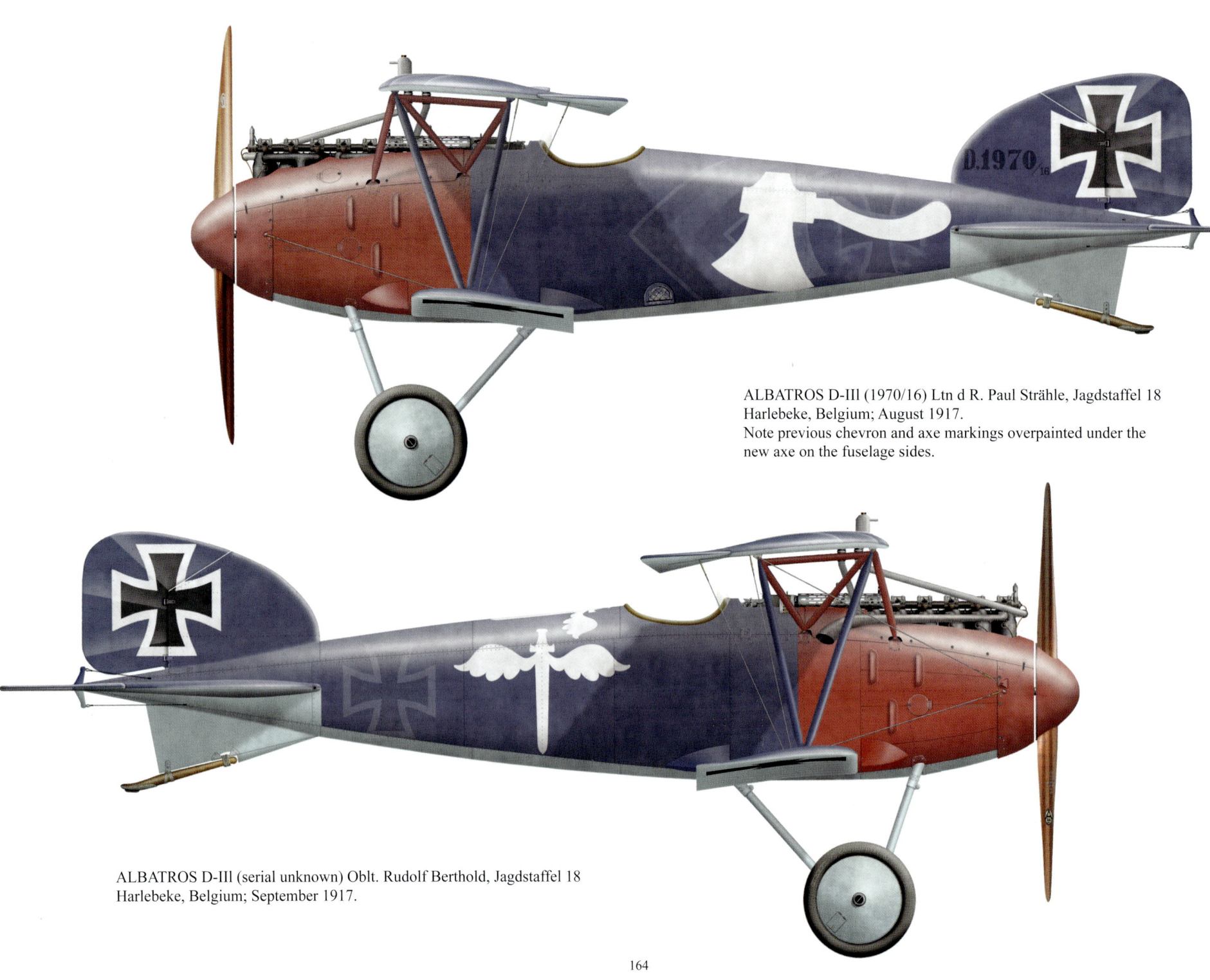

ALBATROS D-III (1970/16) Ltn d R. Paul Strähle, Jagdstaffel 18
Harlebeke, Belgium; August 1917.
Note previous chevron and axe markings overpainted under the
new axe on the fuselage sides.

ALBATROS D-III (serial unknown) Oblt. Rudolf Berthold, Jagdstaffel 18
Harlebeke, Belgium; September 1917.

ALBATROS D-III (O.A.W.)

ALBATROS D-III (O.A.W.) (serial unknown) Ltn. Kurt Wüsthoff, Jagdstaffel 4
Marckebeke, Belgium; August 1917.

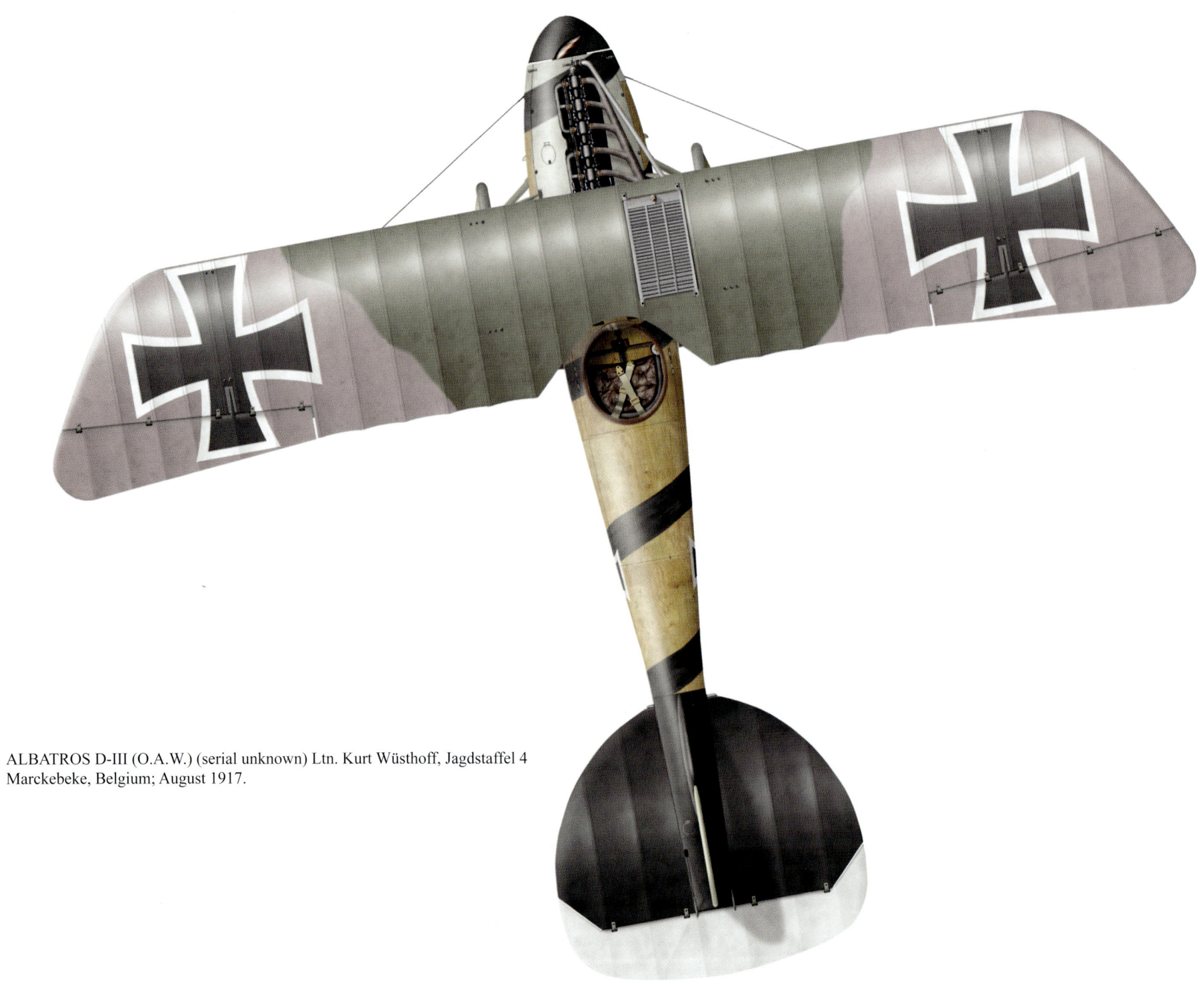

ALBATROS D-III (O.A.W.) (serial unknown) Ltn. Kurt Wüsthoff, Jagdstaffel 4 Marckebeke, Belgium; August 1917.

ALBATROS D-III (O.A.W.) (serial unknown) Ltn. Kurt Wüsthoff, Jagdstaffel 4
Marckebeke, Belgium; August 1917.

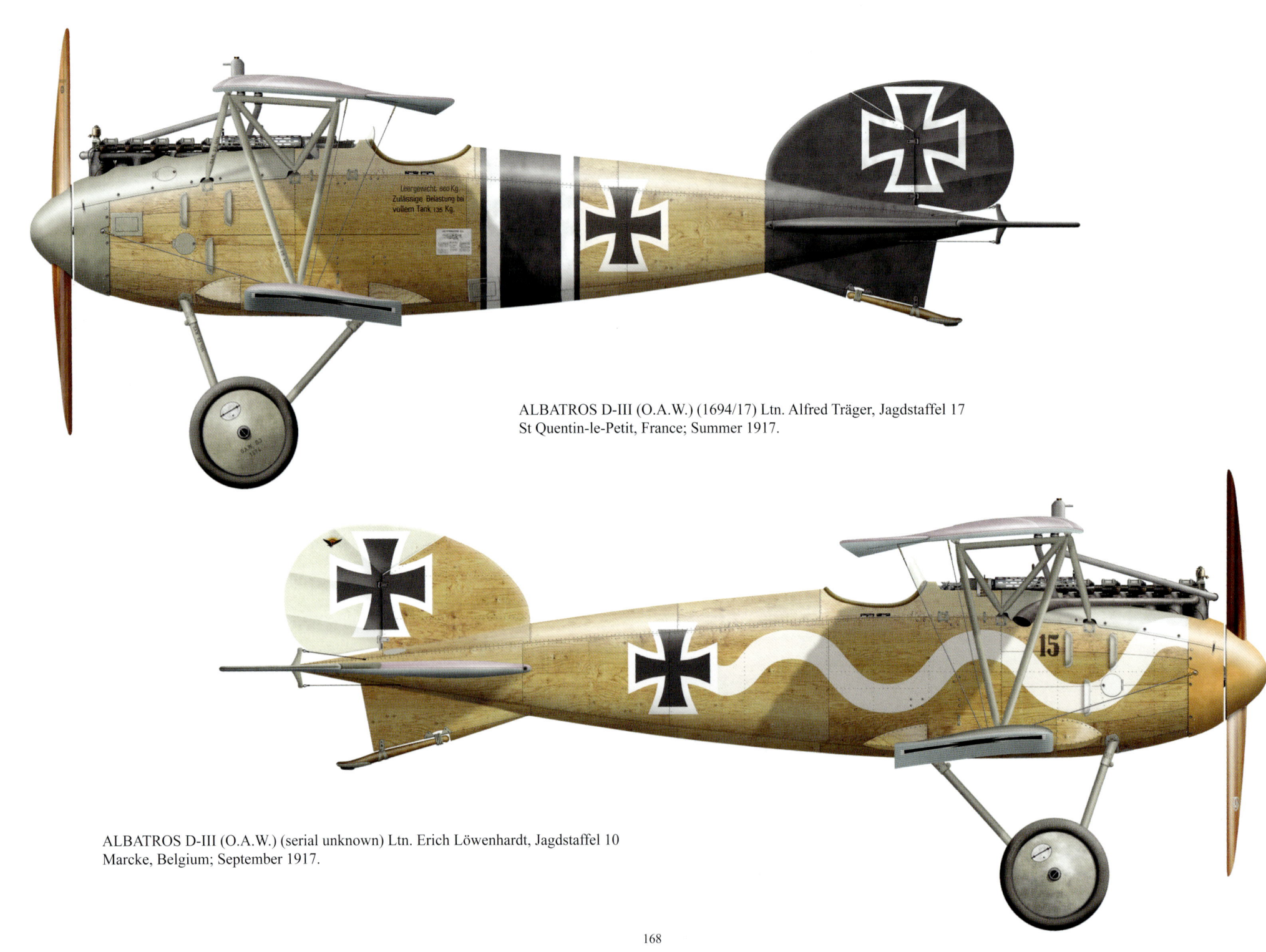

ALBATROS D-III (O.A.W.) (1694/17) Ltn. Alfred Träger, Jagdstaffel 17
St Quentin-le-Petit, France; Summer 1917.

ALBATROS D-III (O.A.W.) (serial unknown) Ltn. Erich Löwenhardt, Jagdstaffel 10
Marcke, Belgium; September 1917.

ALBATROS D-III (O.A.W.) (serial unknown) Ltn. Erich Löwenhardt, Jagdstaffel 10
Marcke, Belgium; September 1917.

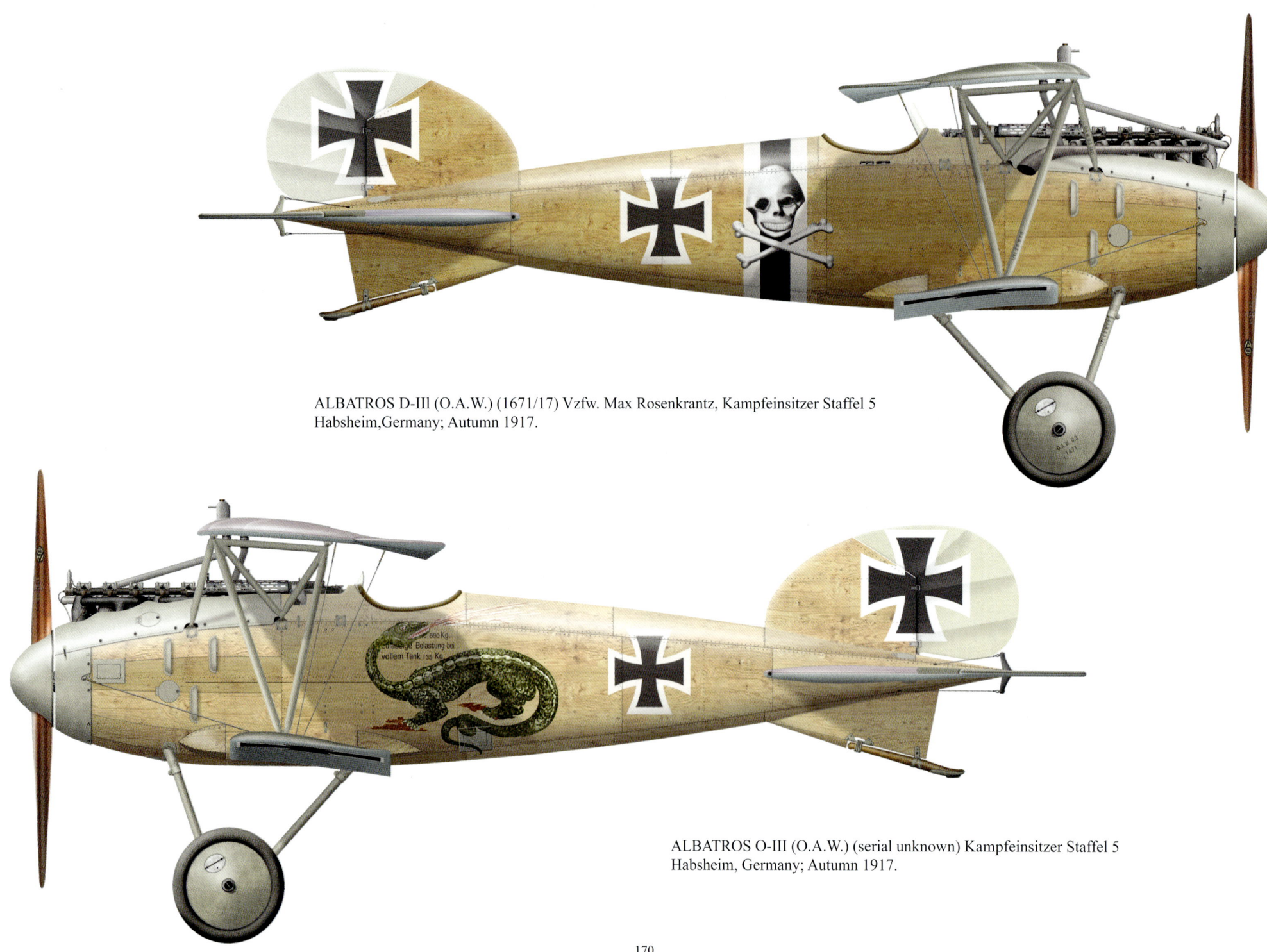

ALBATROS D-III (O.A.W.) (1671/17) Vzfw. Max Rosenkrantz, Kampfeinsitzer Staffel 5
Habsheim, Germany; Autumn 1917.

ALBATROS O-III (O.A.W.) (serial unknown) Kampfeinsitzer Staffel 5
Habsheim, Germany; Autumn 1917.

ALBATROS D-III (O.A.W.) (1671/17) Vzfw. Max Rosenkrantz, Kampfeinsitzer Staffel 5 Habsheim, Germany; Autumn 1917.

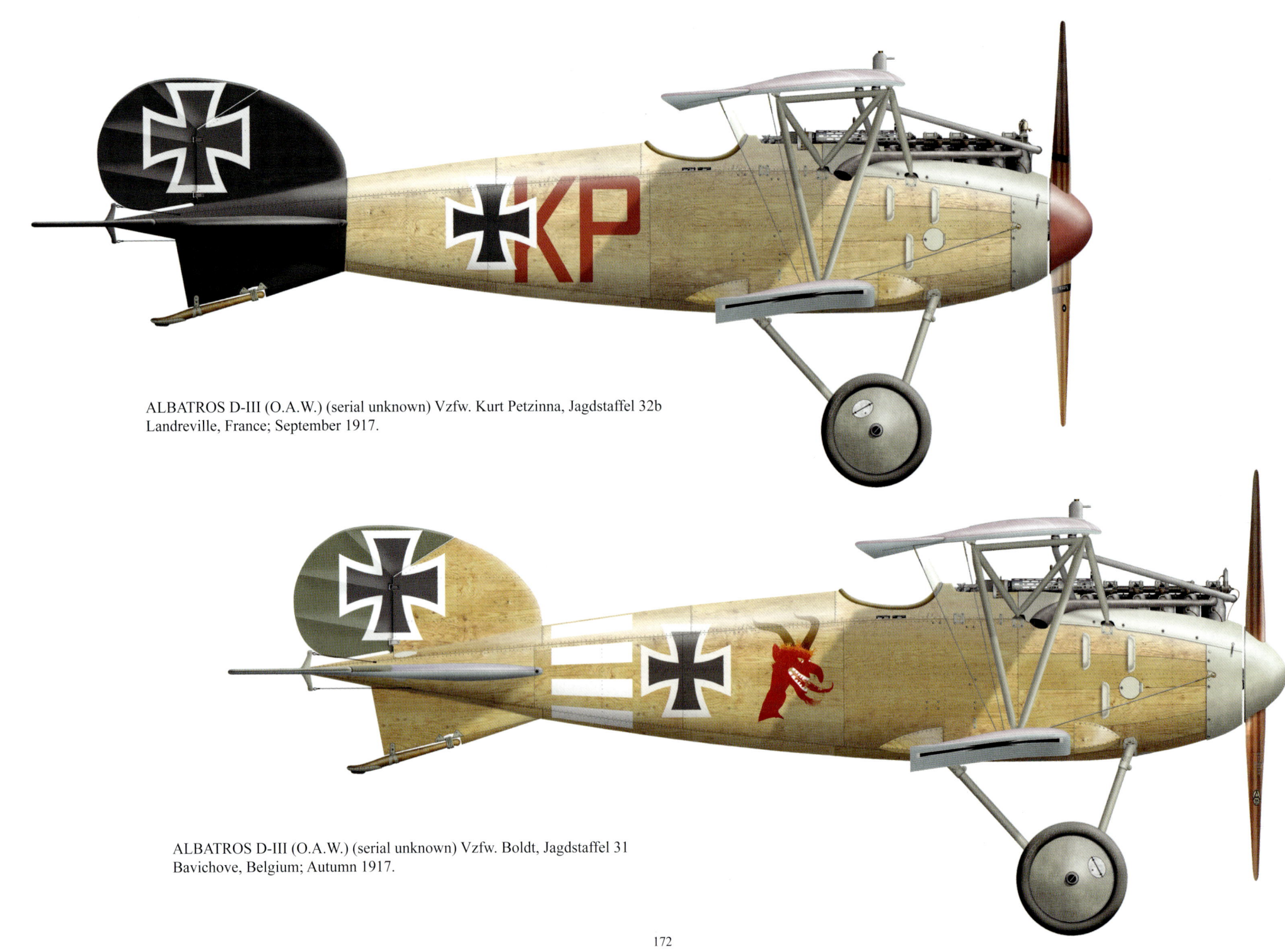

ALBATROS D-III (O.A.W.) (serial unknown) Vzfw. Kurt Petzinna, Jagdstaffel 32b
Landreville, France; September 1917.

ALBATROS D-III (O.A.W.) (serial unknown) Vzfw. Boldt, Jagdstaffel 31
Bavichove, Belgium; Autumn 1917.

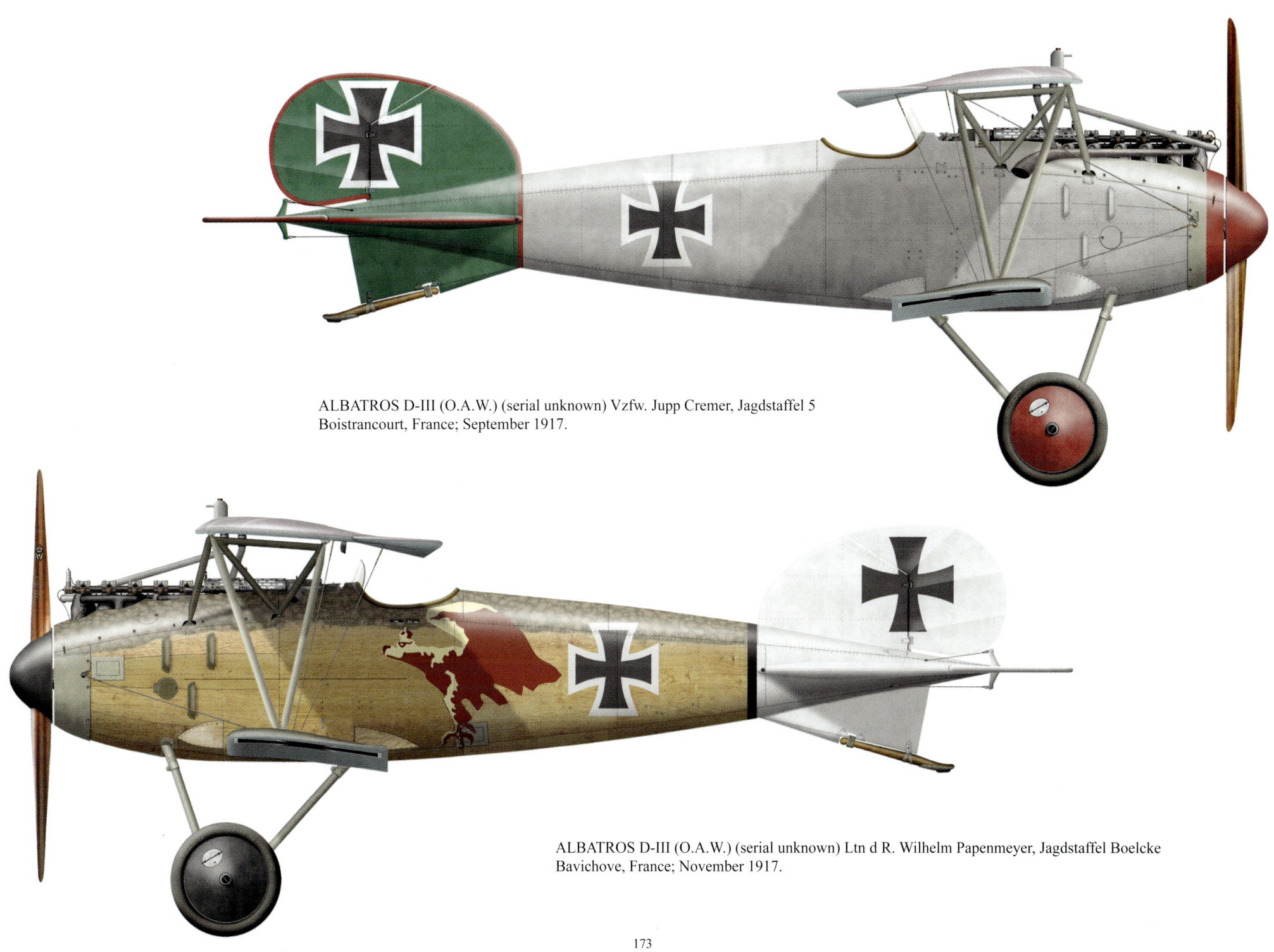

ALBATROS D-III (O.A.W.) (serial unknown) Vzfw. Jupp Cremer, Jagdstaffel 5
Boistrancourt, France; September 1917.

ALBATROS D-III (O.A.W.) (serial unknown) Ltn d R. Wilhelm Papenmeyer, Jagdstaffel Boelcke
Bavichove, France; November 1917.

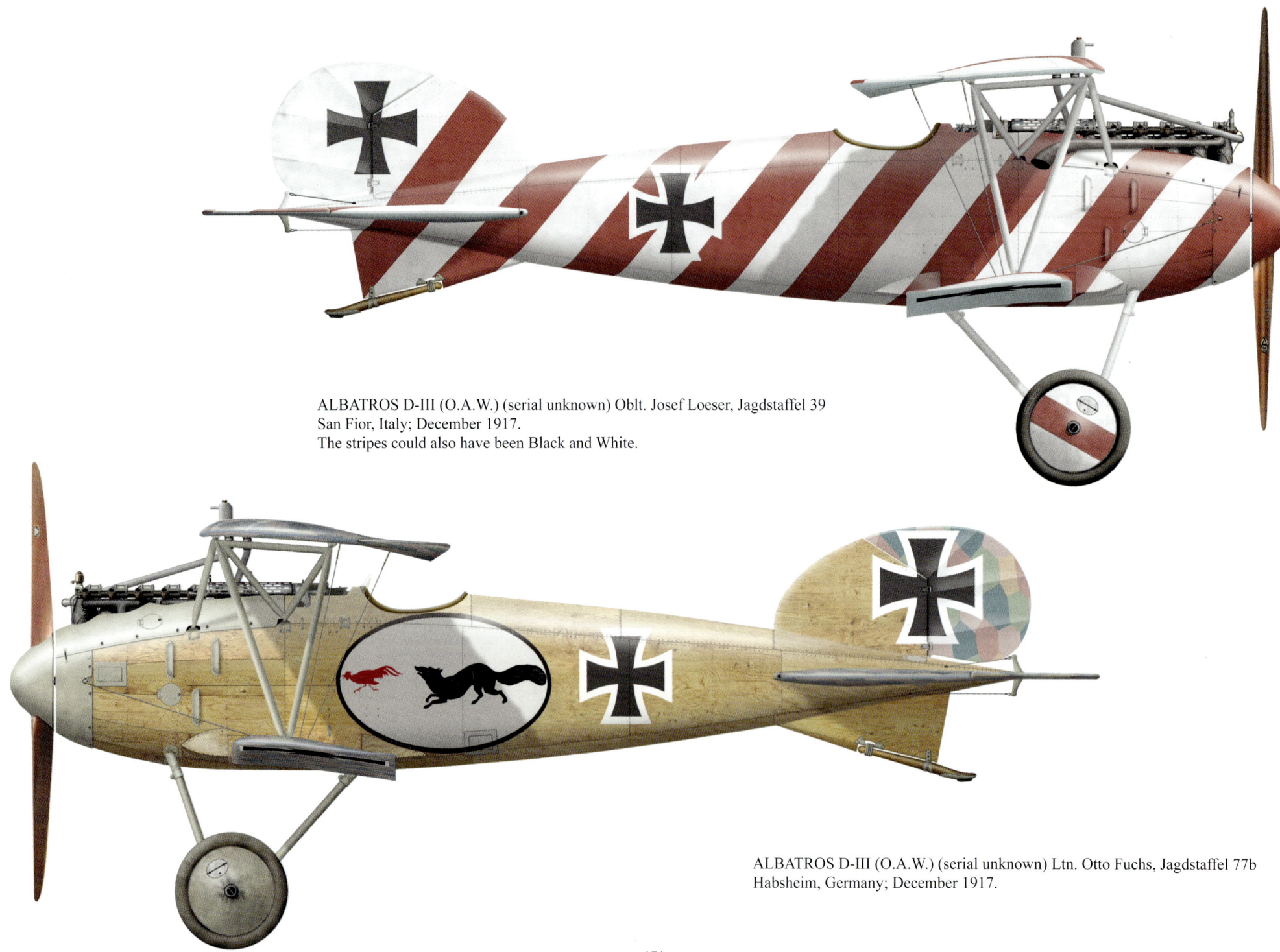

ALBATROS D-III (O.A.W.) (serial unknown) Oblt. Josef Loeser, Jagdstaffel 39
San Fior, Italy; December 1917.
The stripes could also have been Black and White.

ALBATROS D-III (O.A.W.) (serial unknown) Ltn. Otto Fuchs, Jagdstaffel 77b
Habsheim, Germany; December 1917.

ALBATROS D-III (O.A.W.) (serial unknown) Oblt. Josef Loeser, Jagdstaffel 39
San Fior, Italy; December 1917.

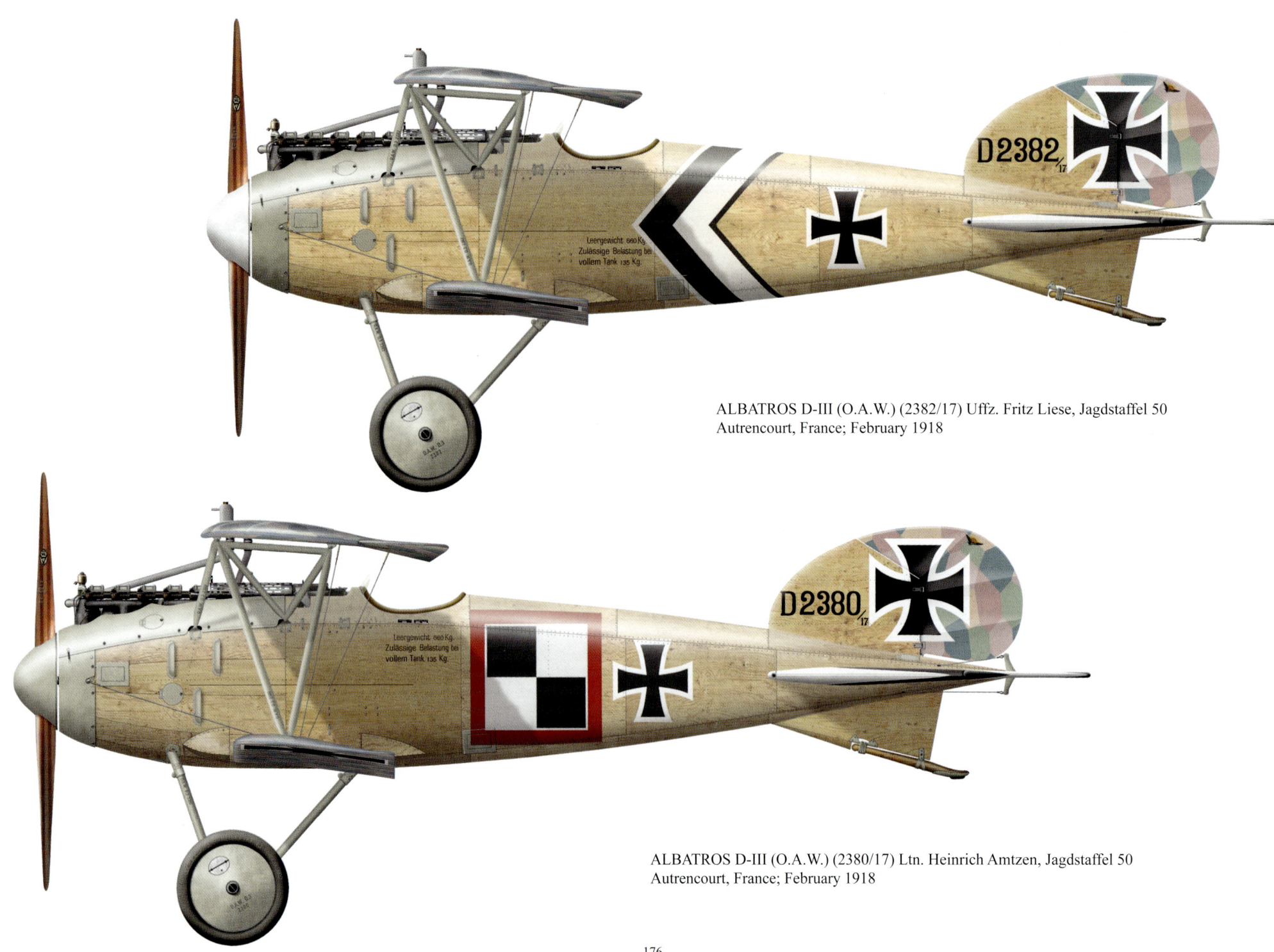

ALBATROS D-III (O.A.W.) (2382/17) Uffz. Fritz Liese, Jagdstaffel 50
Autrencourt, France; February 1918

ALBATROS D-III (O.A.W.) (2380/17) Ltn. Heinrich Amtzen, Jagdstaffel 50
Autrencourt, France; February 1918

ALBATROS D-III (O.A.W.) (2382/17) Uffz. Fritz Liese, Jagdstaffel 50
Autrencourt, France; February 1918

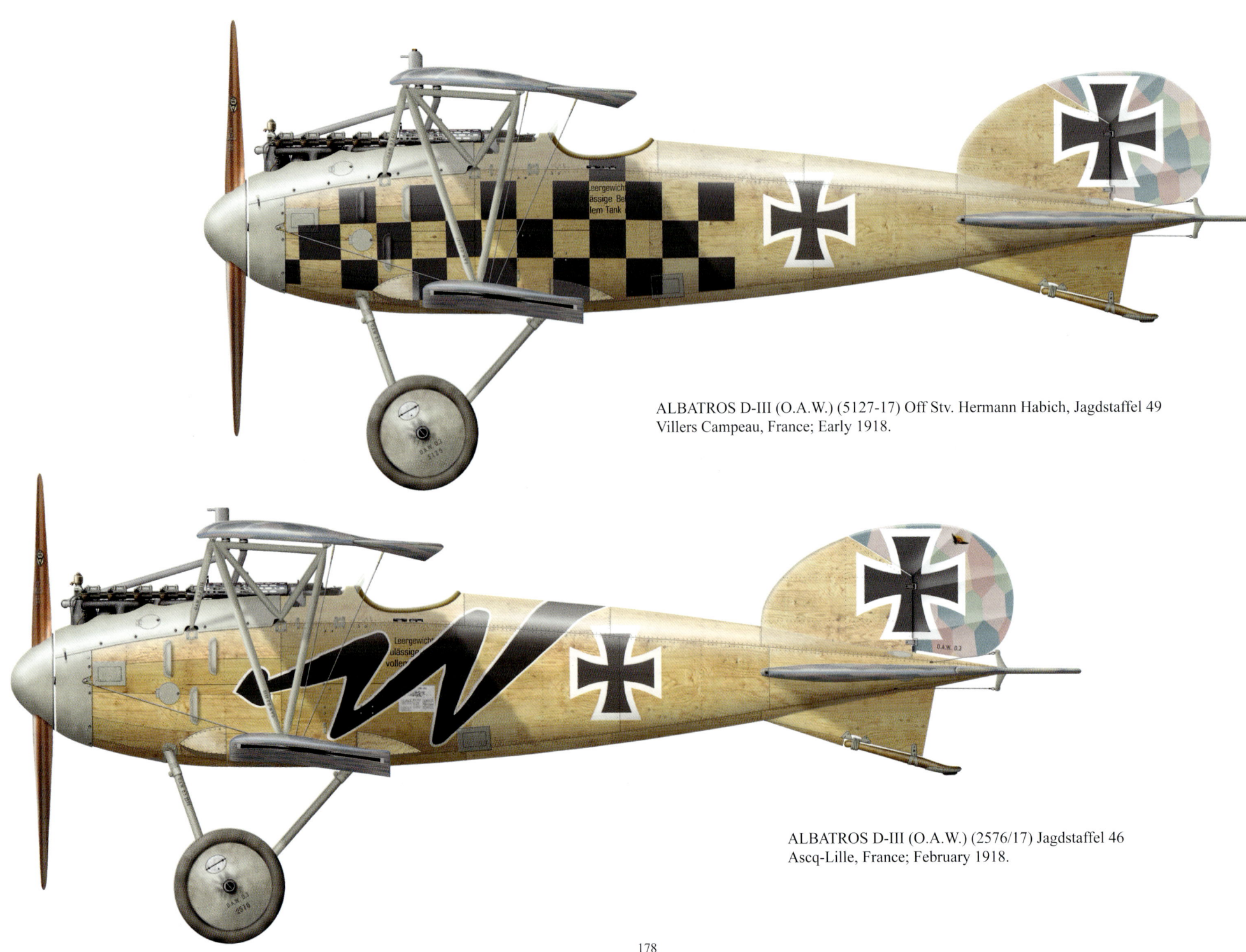

ALBATROS D-III (O.A.W.) (5127-17) Off Stv. Hermann Habich, Jagdstaffel 49
Villers Campeau, France; Early 1918.

ALBATROS D-III (O.A.W.) (2576/17) Jagdstaffel 46
Ascq-Lille, France; February 1918.

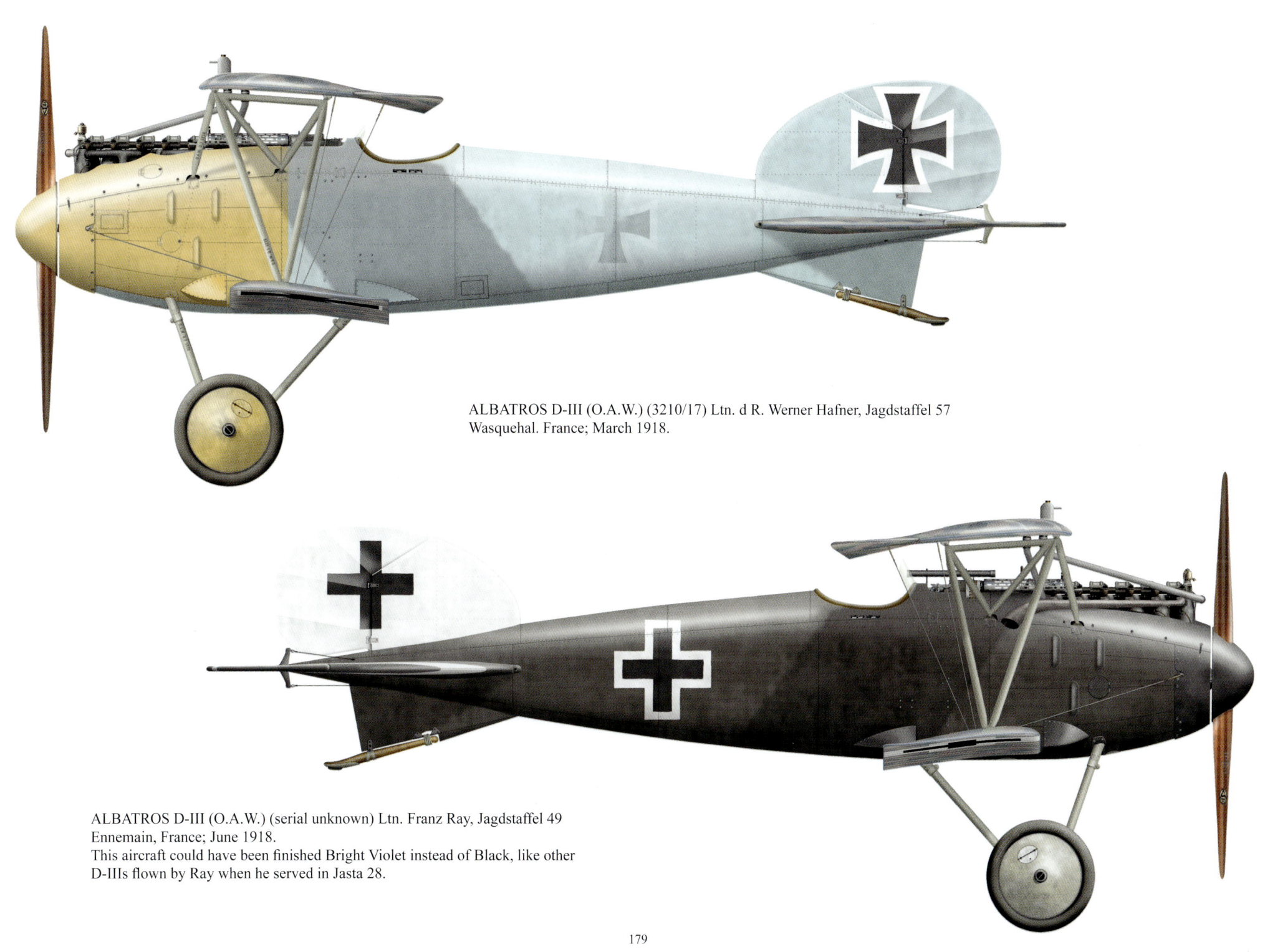

ALBATROS D-III (O.A.W.) (3210/17) Ltn. d R. Werner Hafner, Jagdstaffel 57
Wasquehal, France; March 1918.

ALBATROS D-III (O.A.W.) (serial unknown) Ltn. Franz Ray, Jagdstaffel 49
Ennemain, France; June 1918.
This aircraft could have been finished Bright Violet instead of Black, like other
D-IIIs flown by Ray when he served in Jasta 28.

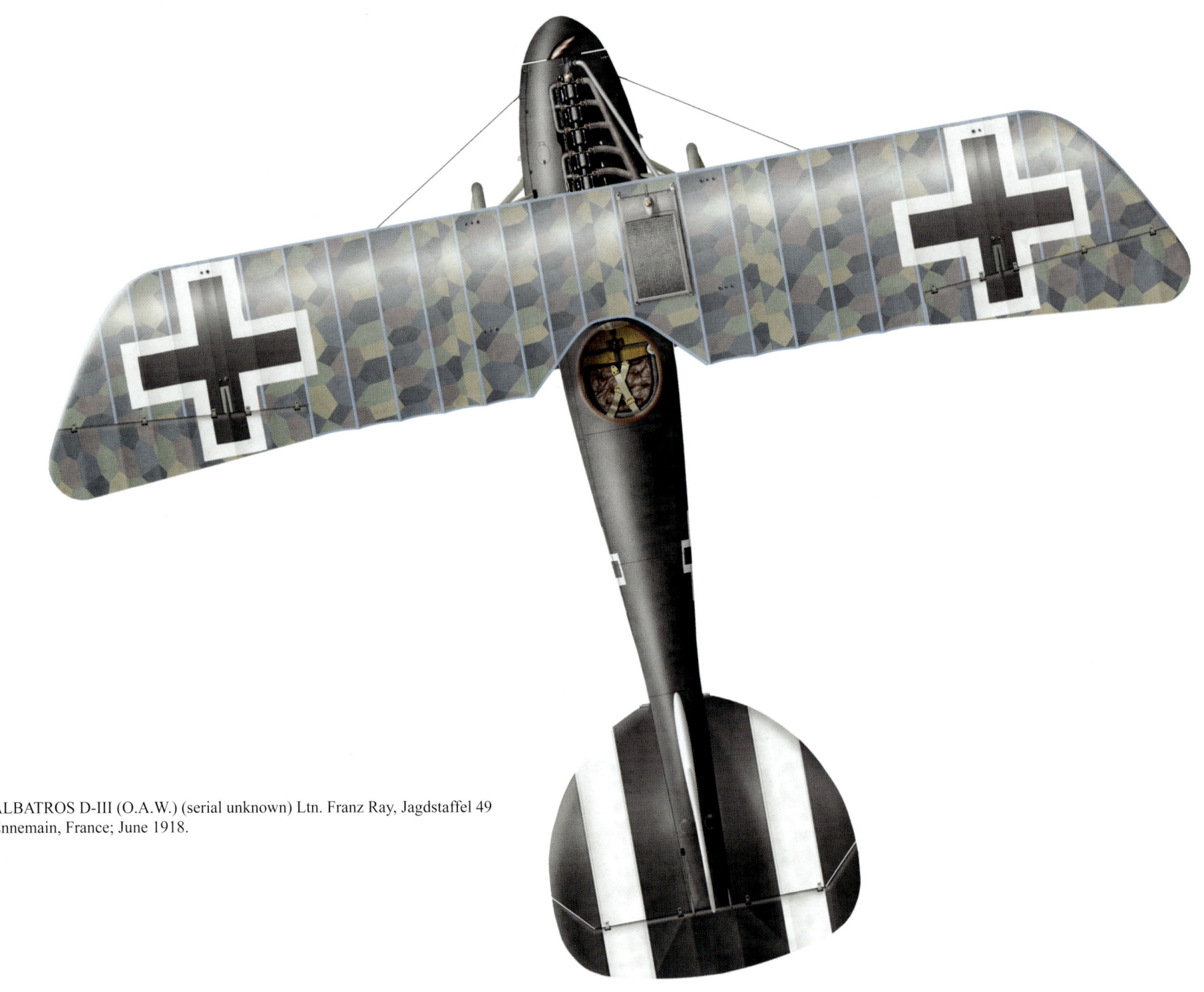

ALBATROS D-III (O.A.W.) (serial unknown) Ltn. Franz Ray, Jagdstaffel 49
Ennemain, France; June 1918.

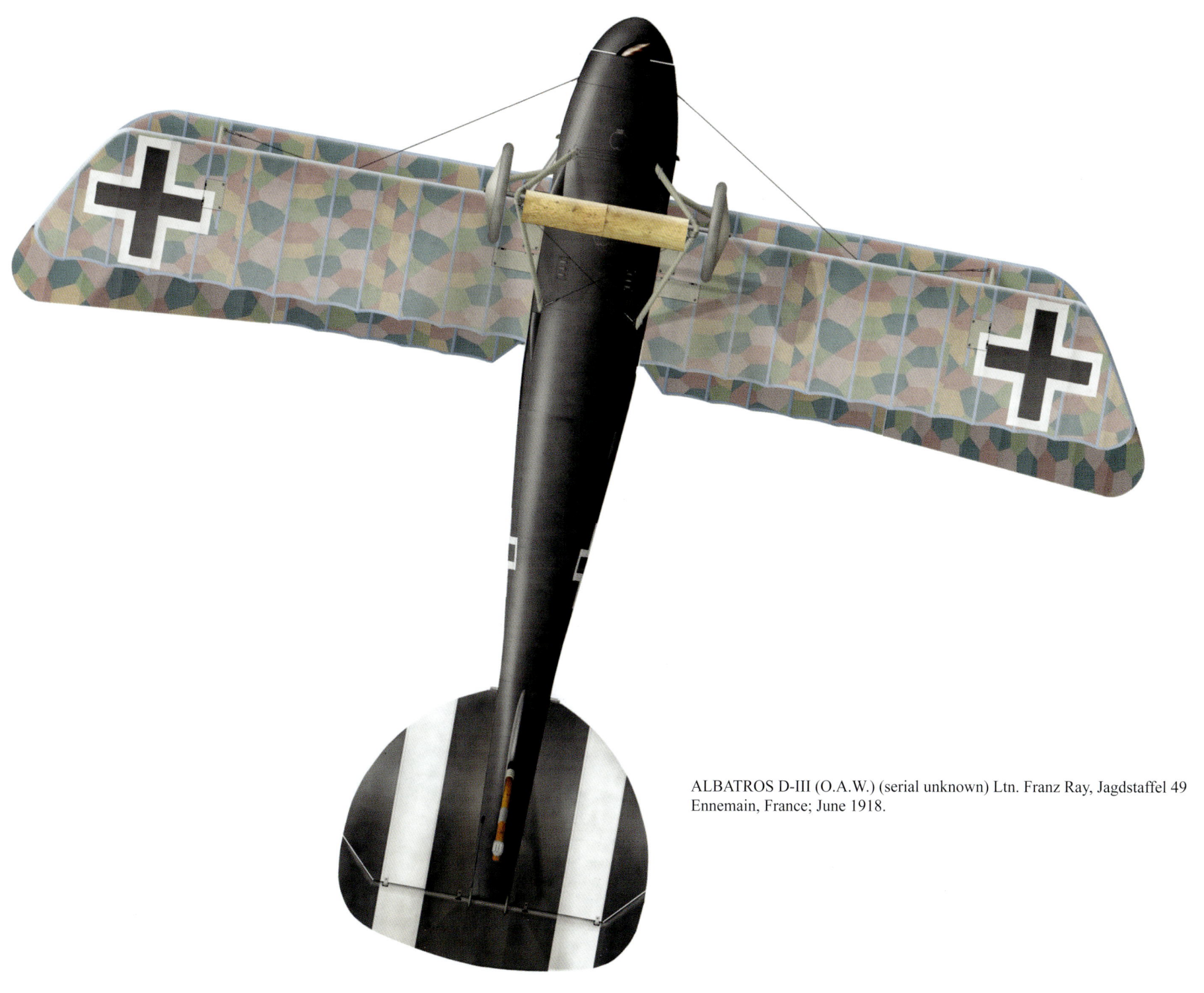

ALBATROS D-III (O.A.W.) (serial unknown) Ltn. Franz Ray, Jagdstaffel 49
Ennemain, France; June 1918.

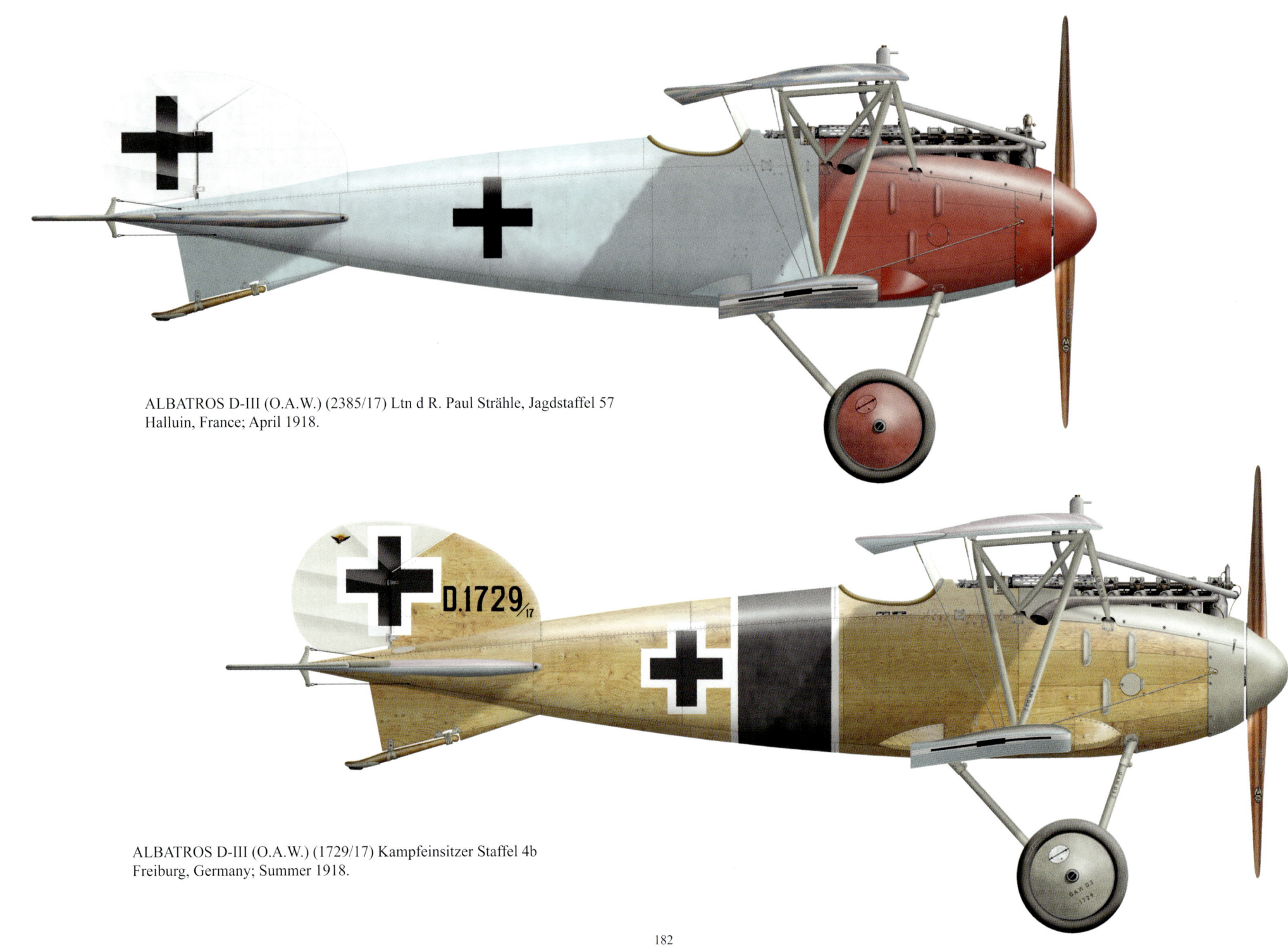

ALBATROS D-III (O.A.W.) (2385/17) Ltn d R. Paul Strähle, Jagdstaffel 57
Halluin, France; April 1918.

ALBATROS D-III (O.A.W.) (1729/17) Kampfeinsitzer Staffel 4b
Freiburg, Germany; Summer 1918.

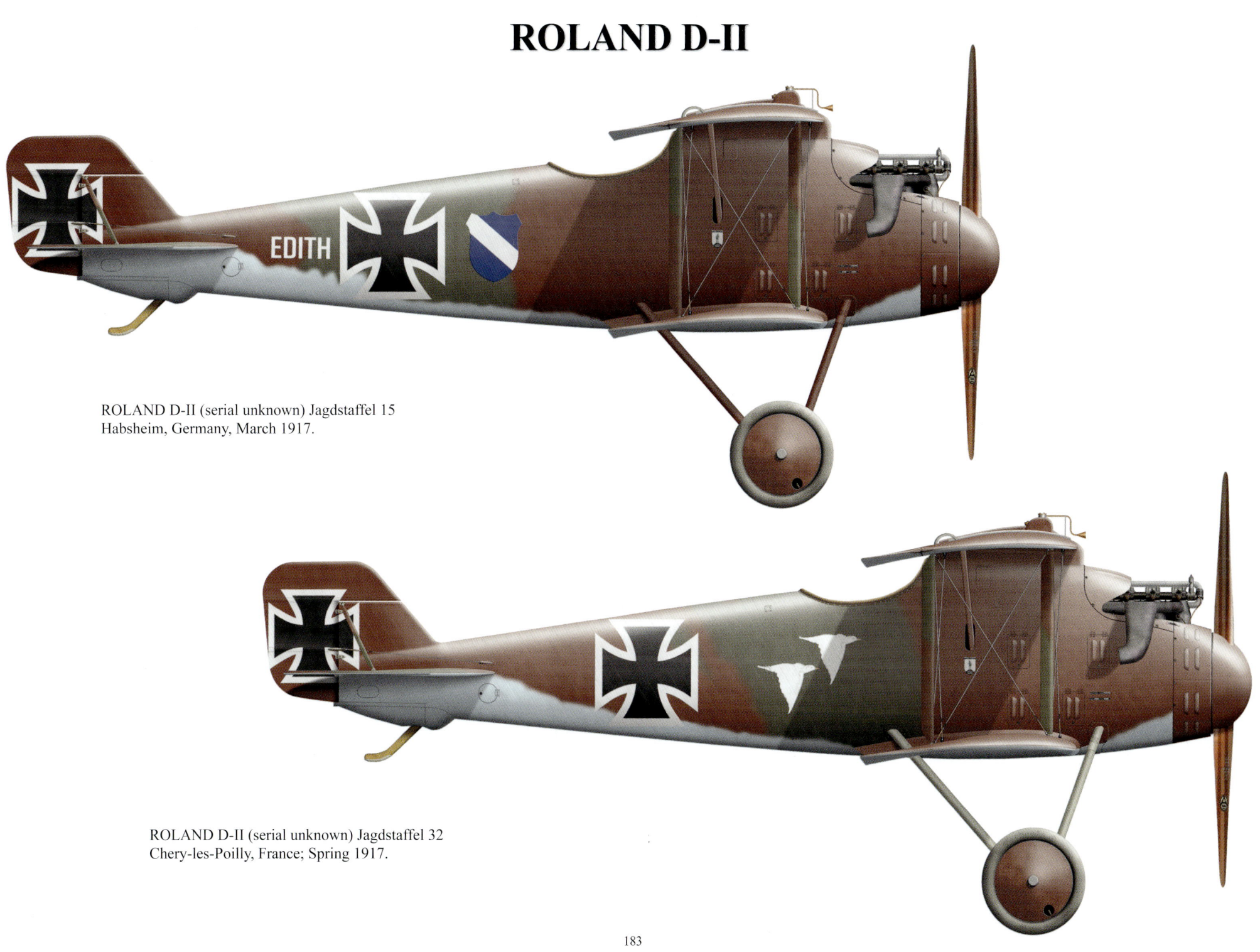

ROLAND D-II (serial unknown) Jagdstaffel 15
Habsheim, Germany, March 1917.

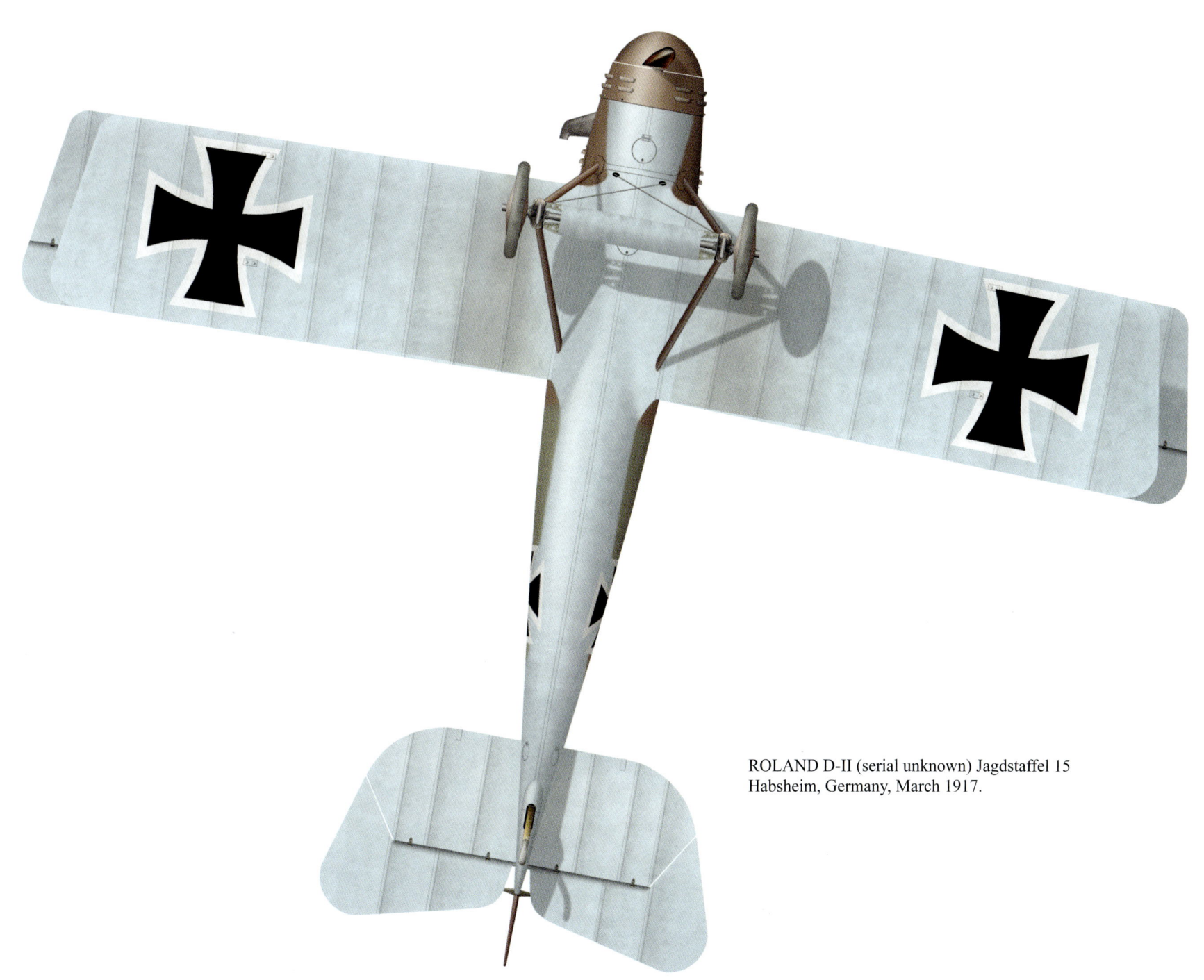

ROLAND D-II (serial unknown) Jagdstaffel 15
Habsheim, Germany, March 1917.

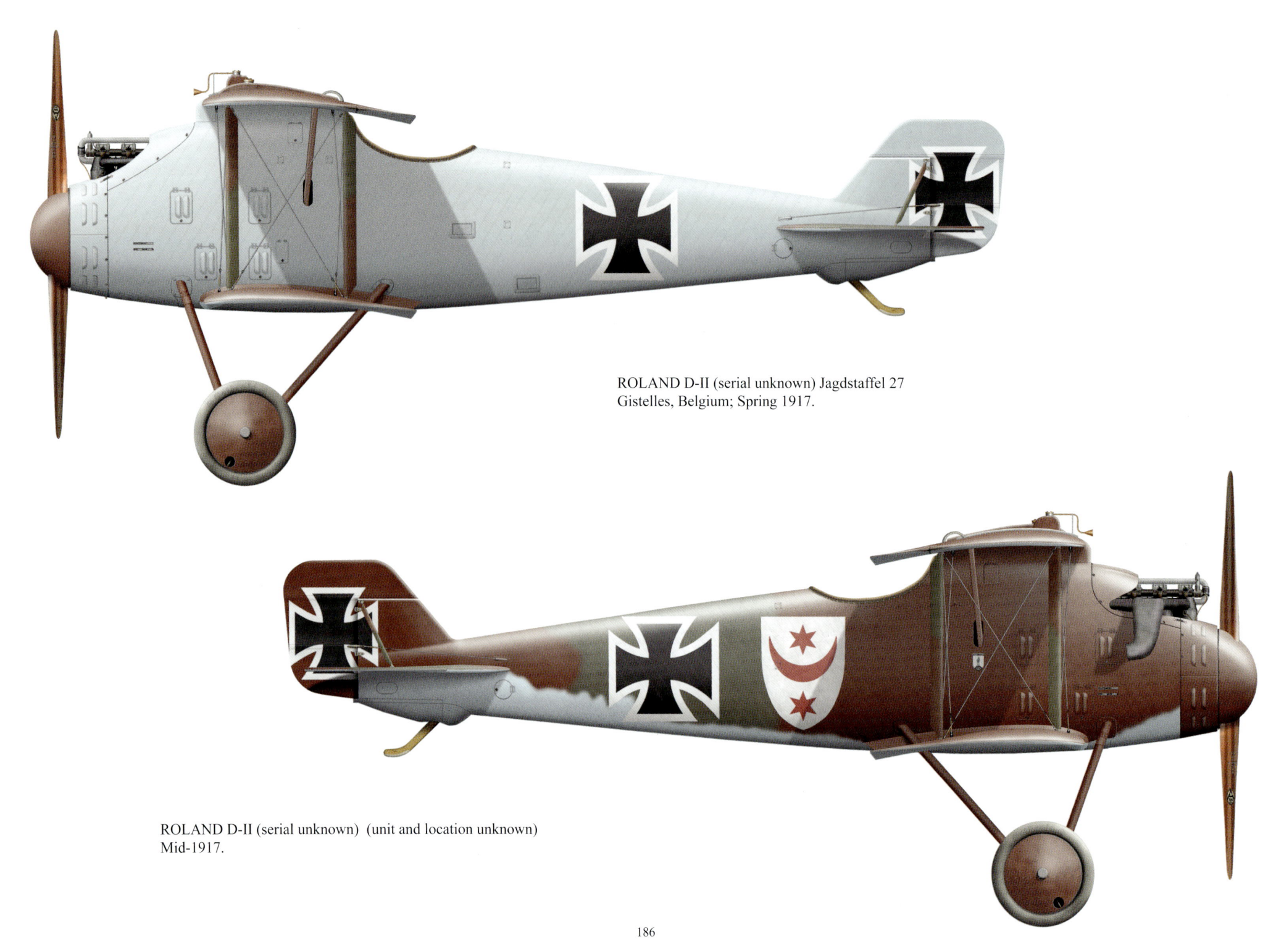

ROLAND D-II (serial unknown) Jagdstaffel 27
Gistelles, Belgium; Spring 1917.

ROLAND D-II (serial unknown) (unit and location unknown)
Mid-1917.

ROLAND D-IIa

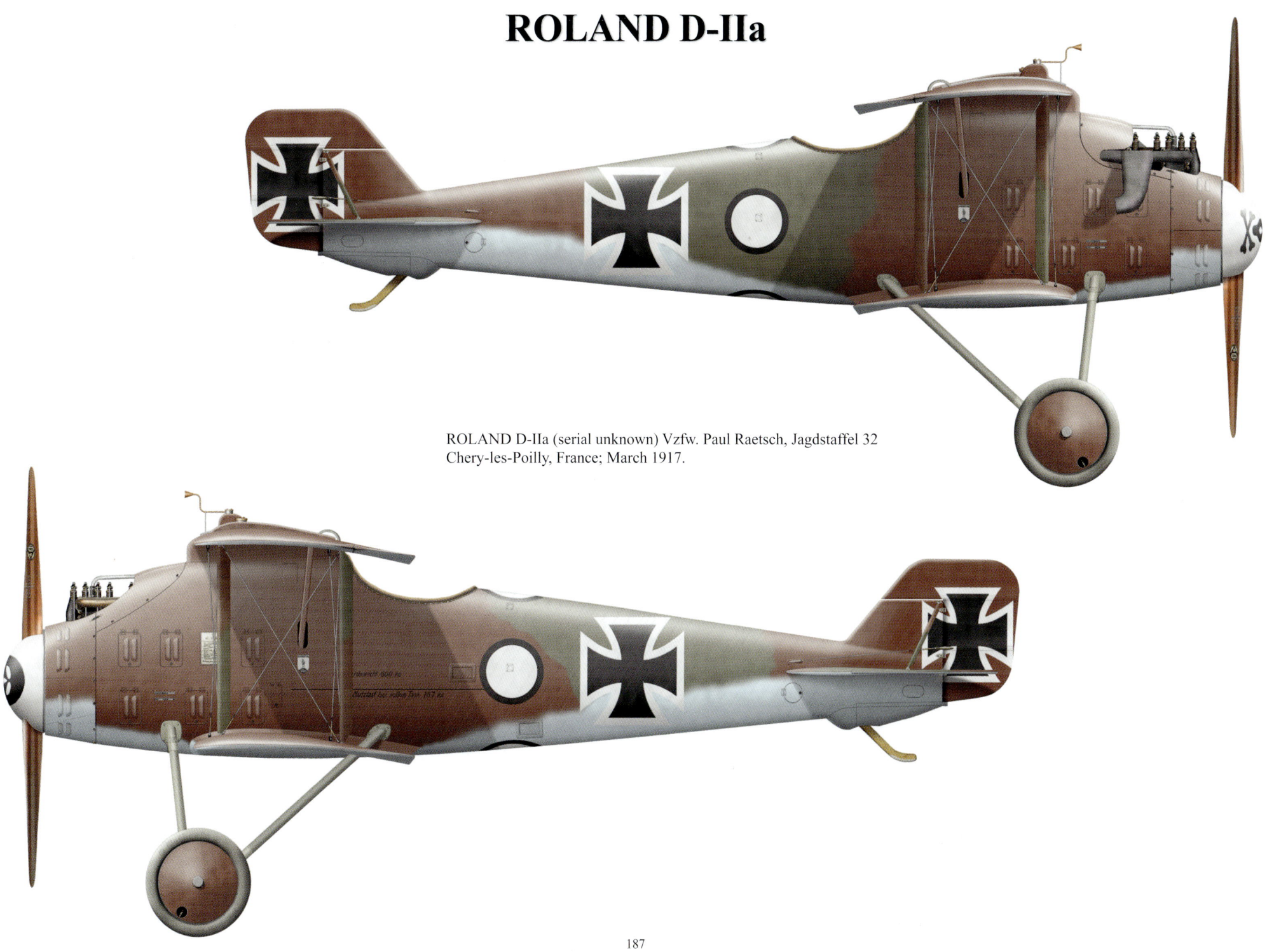

ROLAND D-IIa (serial unknown) Vzfw. Paul Raetsch, Jagdstaffel 32
Chery-les-Poilly, France; March 1917.

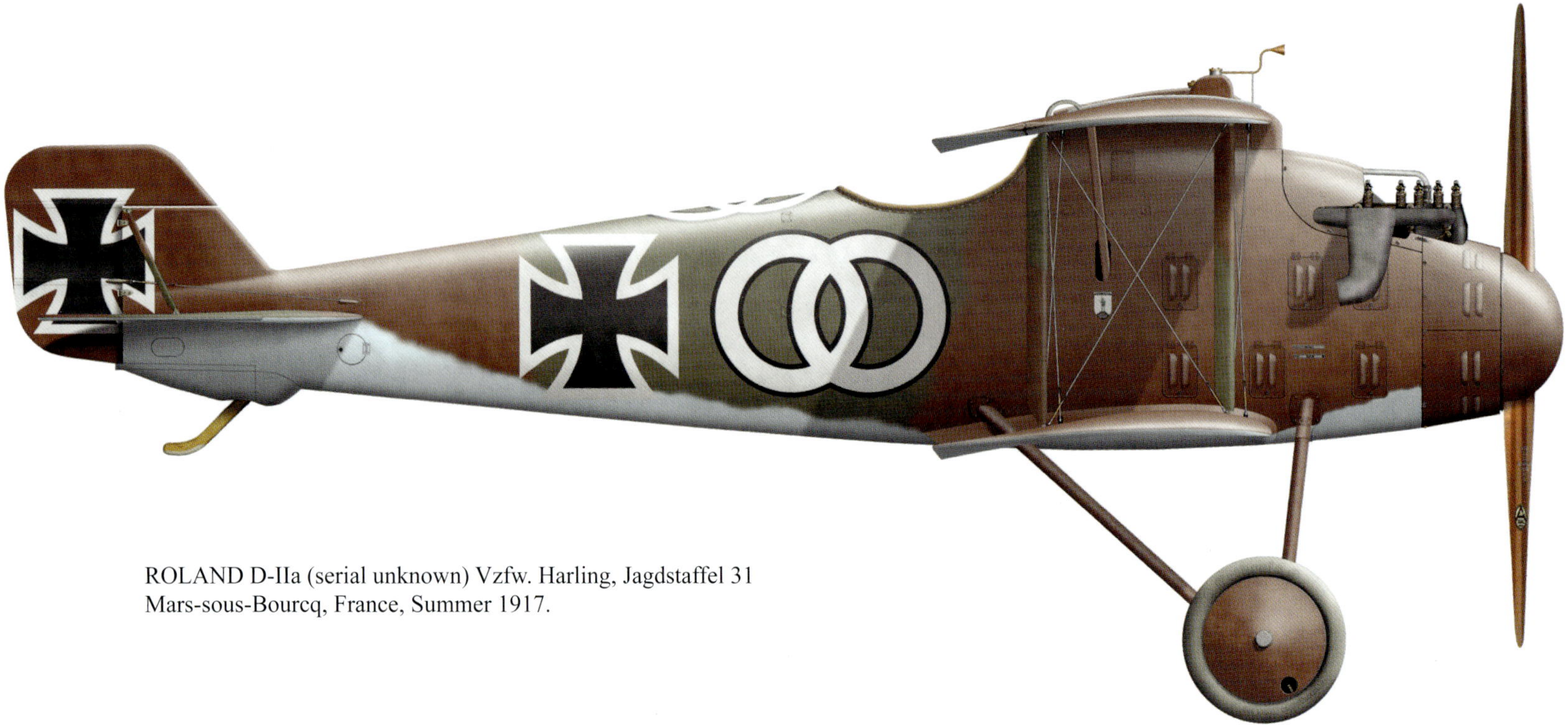

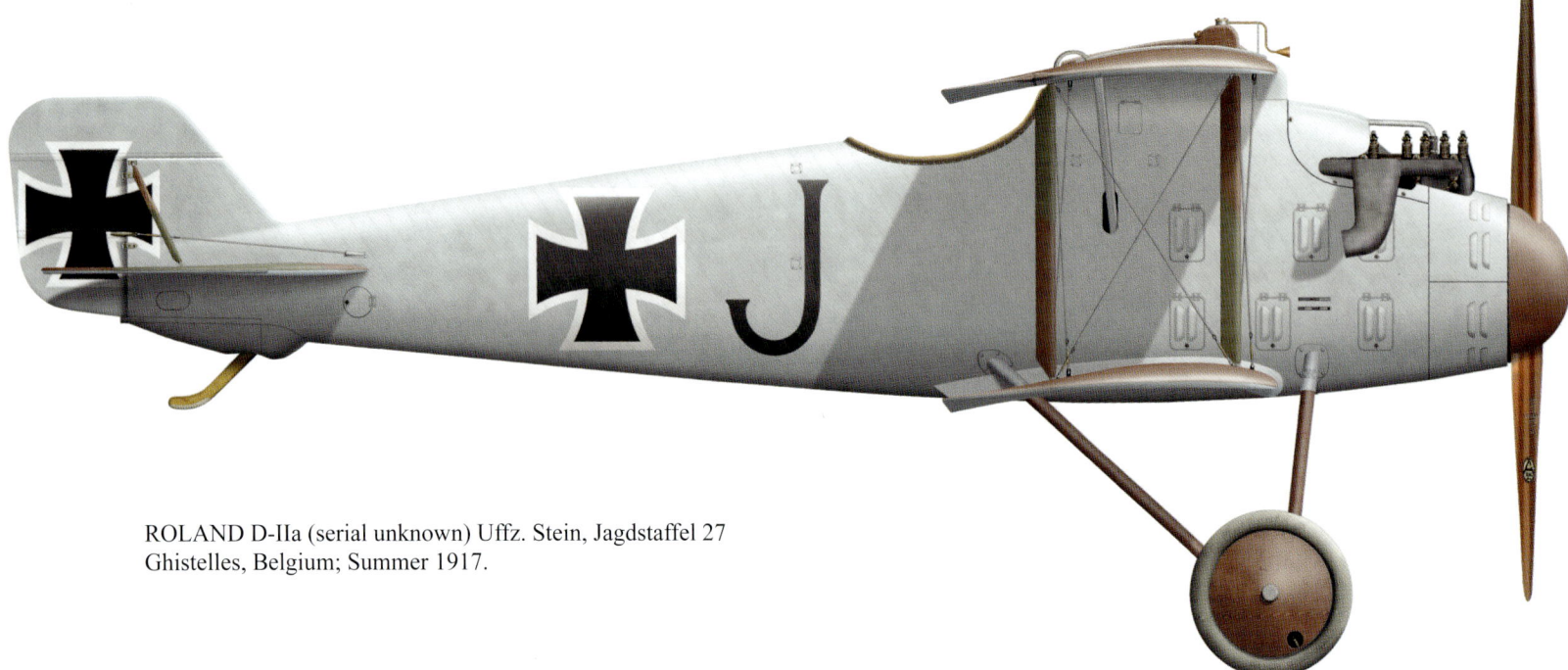

ROLAND D-IIa (serial unknown) Vzfw. Harling, Jagdstaffel 31
Mars-sous-Bourcq, France, Summer 1917.

ROLAND D-IIa (serial unknown) Uffz. Stein, Jagdstaffel 27
Ghistelles, Belgium; Summer 1917.

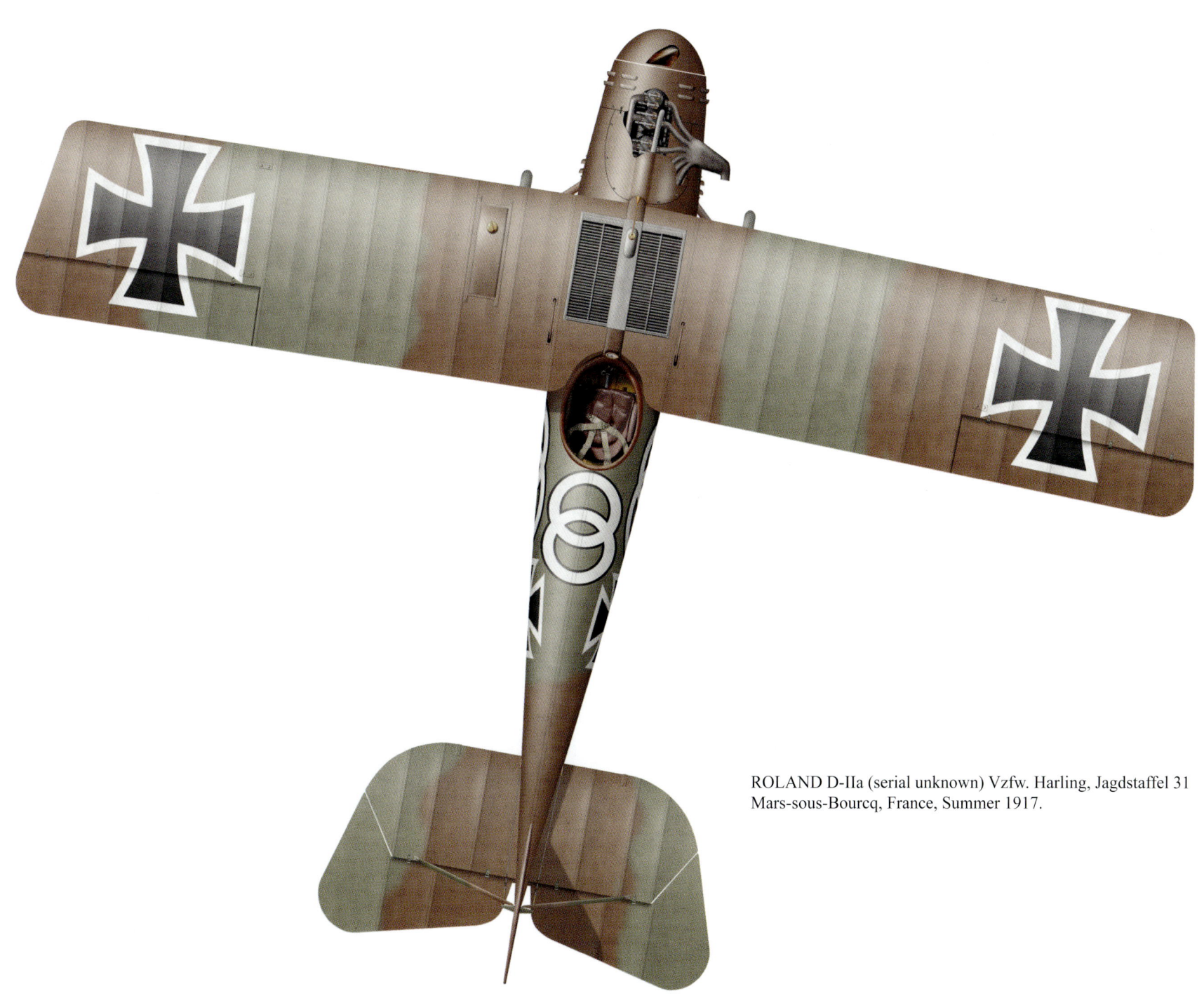

ROLAND D-IIa (serial unknown) Vzfw. Harling, Jagdstaffel 31
Mars-sous-Bourcq, France, Summer 1917.

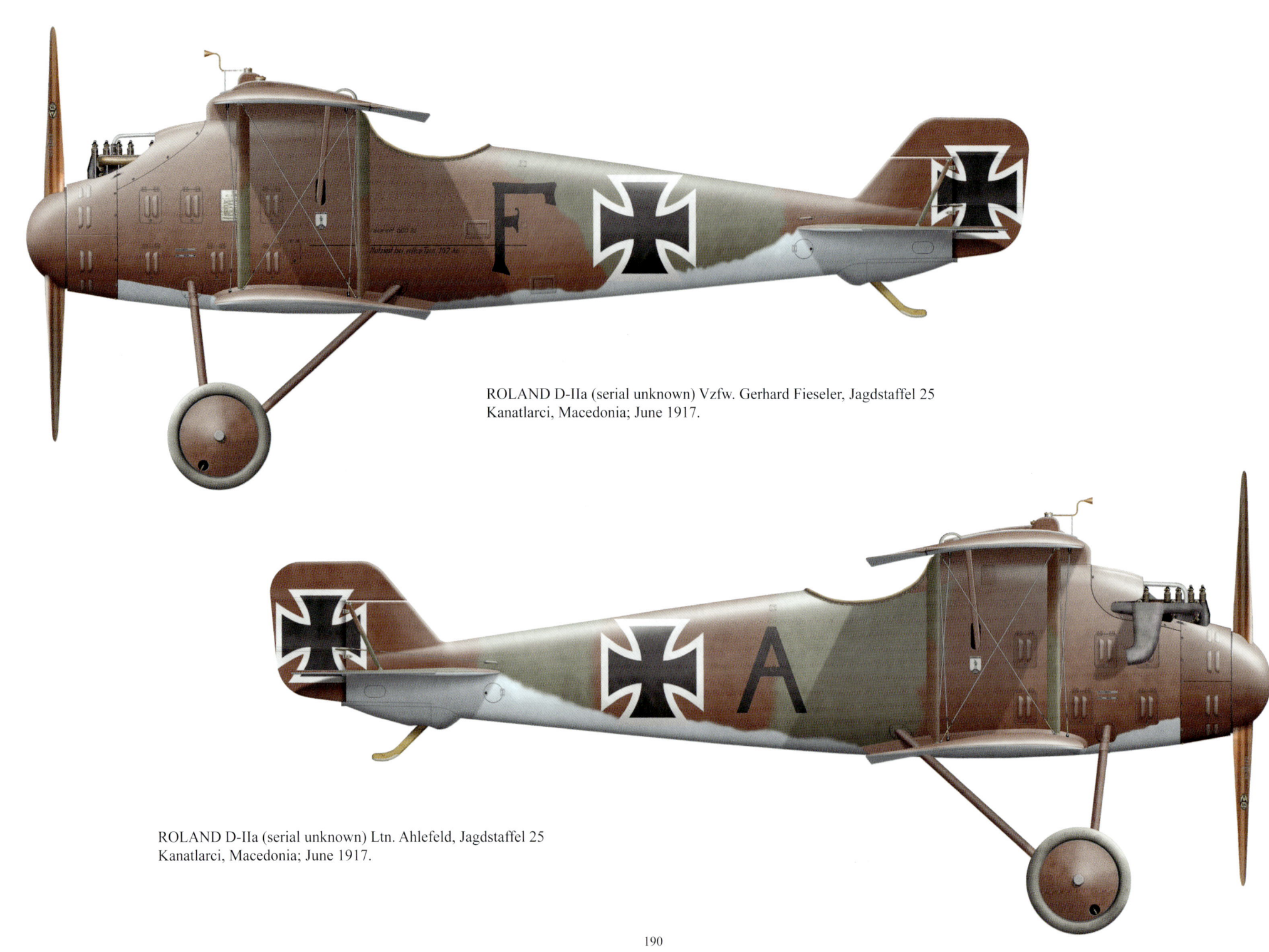

ROLAND D-IIa (serial unknown) Vzfw. Gerhard Fieseler, Jagdstaffel 25
Kanatlarci, Macedonia; June 1917.

ROLAND D-IIa (serial unknown) Ltn. Ahlefeld, Jagdstaffel 25
Kanatlarci, Macedonia; June 1917.

ROLAND D-II & D-IIa (Pfal)

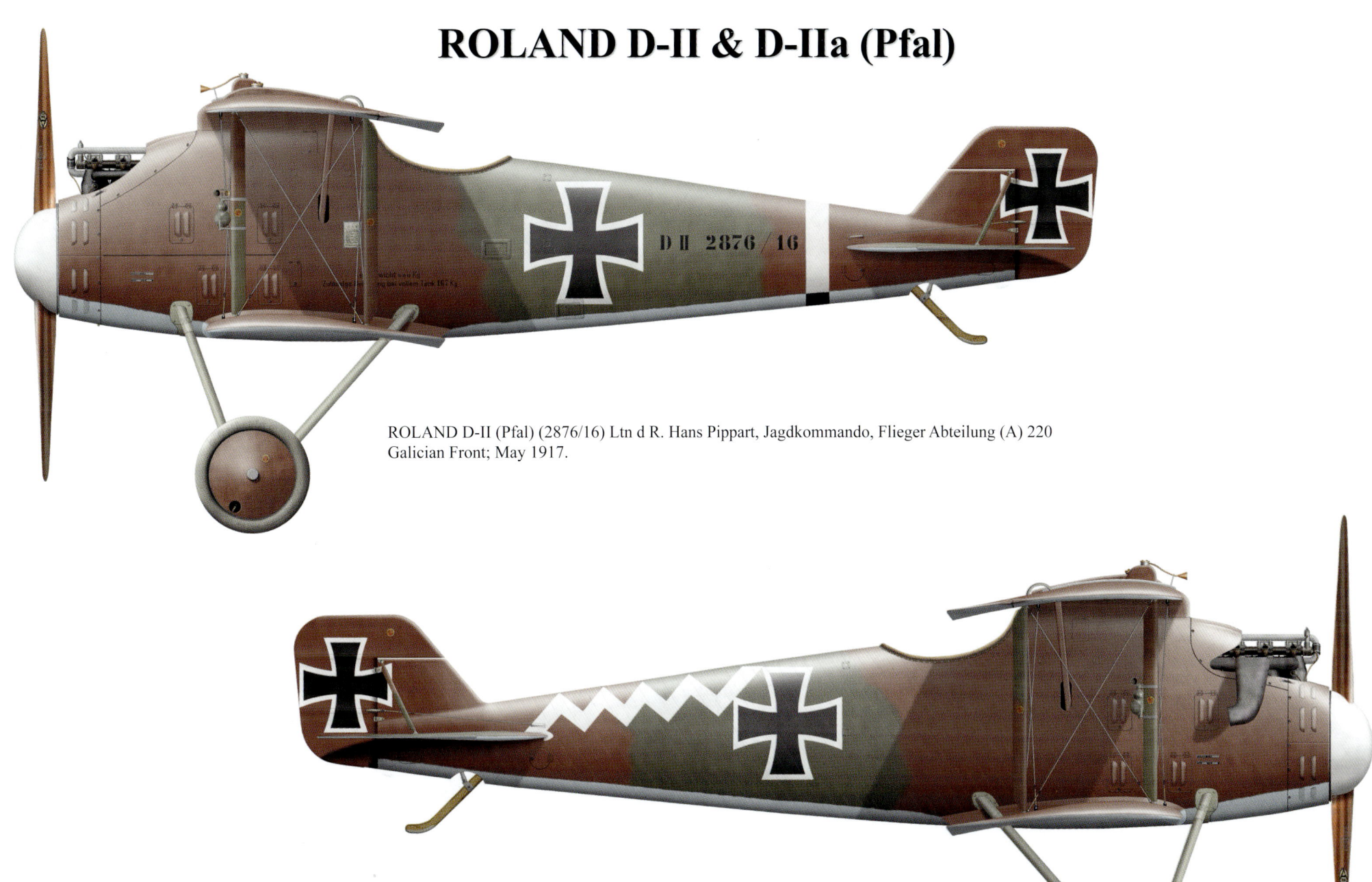

ROLAND D-II (Pfal) (2876/16) Ltn d R. Hans Pippart, Jagdkommando, Flieger Abteilung (A) 220 Galician Front; May 1917.

ROLAND D-II (Pfal) (serial unknown) Jagdkommando, Flieger Abteilung (A) 220 Galician Front, Spring 1917.

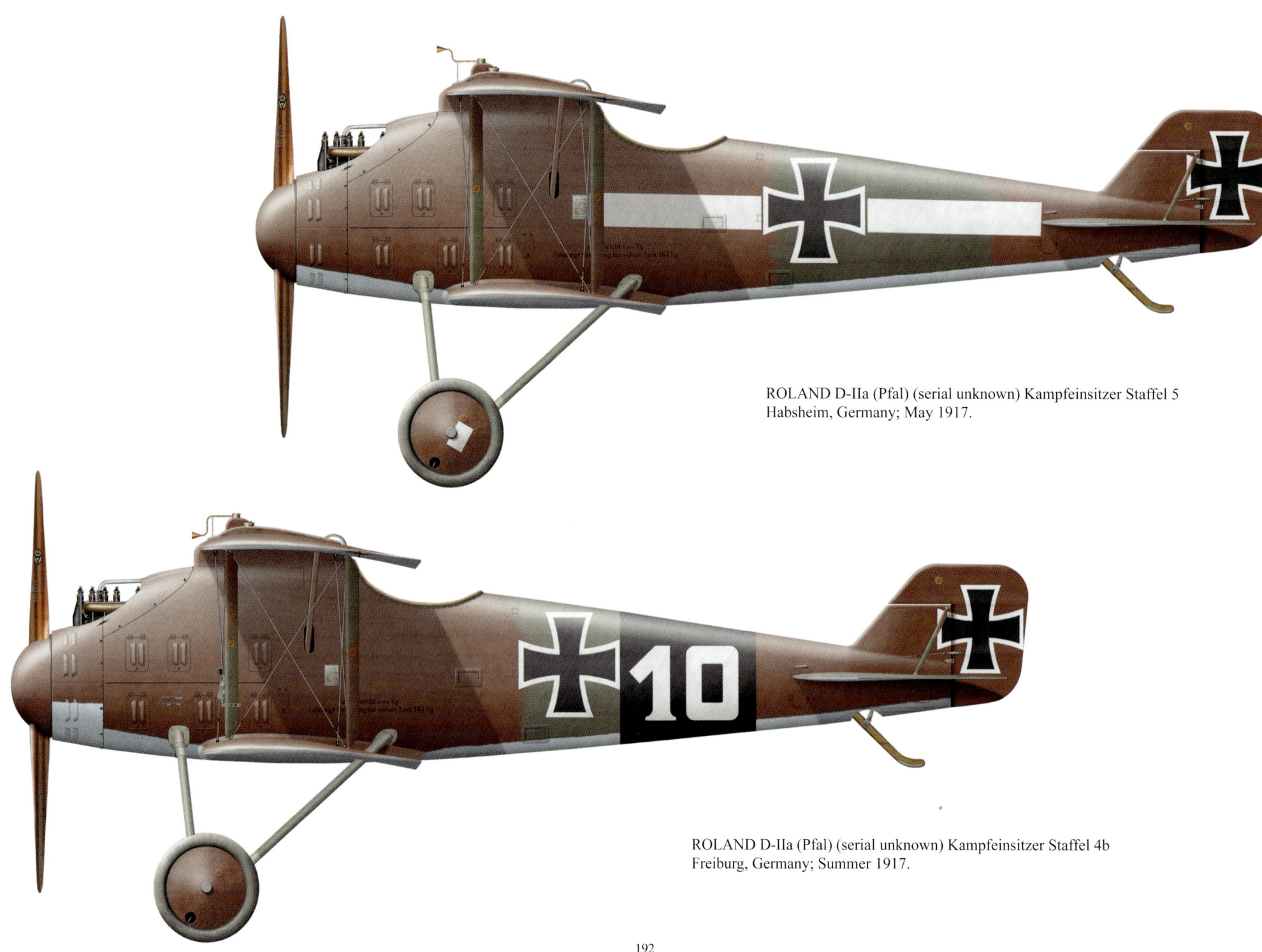

ROLAND D-IIa (Pfal) (serial unknown) Kampfeinsitzer Staffel 5
Habsheim, Germany; May 1917.

ROLAND D-IIa (Pfal) (serial unknown) Kampfeinsitzer Staffel 4b
Freiburg, Germany; Summer 1917.

SIEMENS-SCHUCKERT D-I

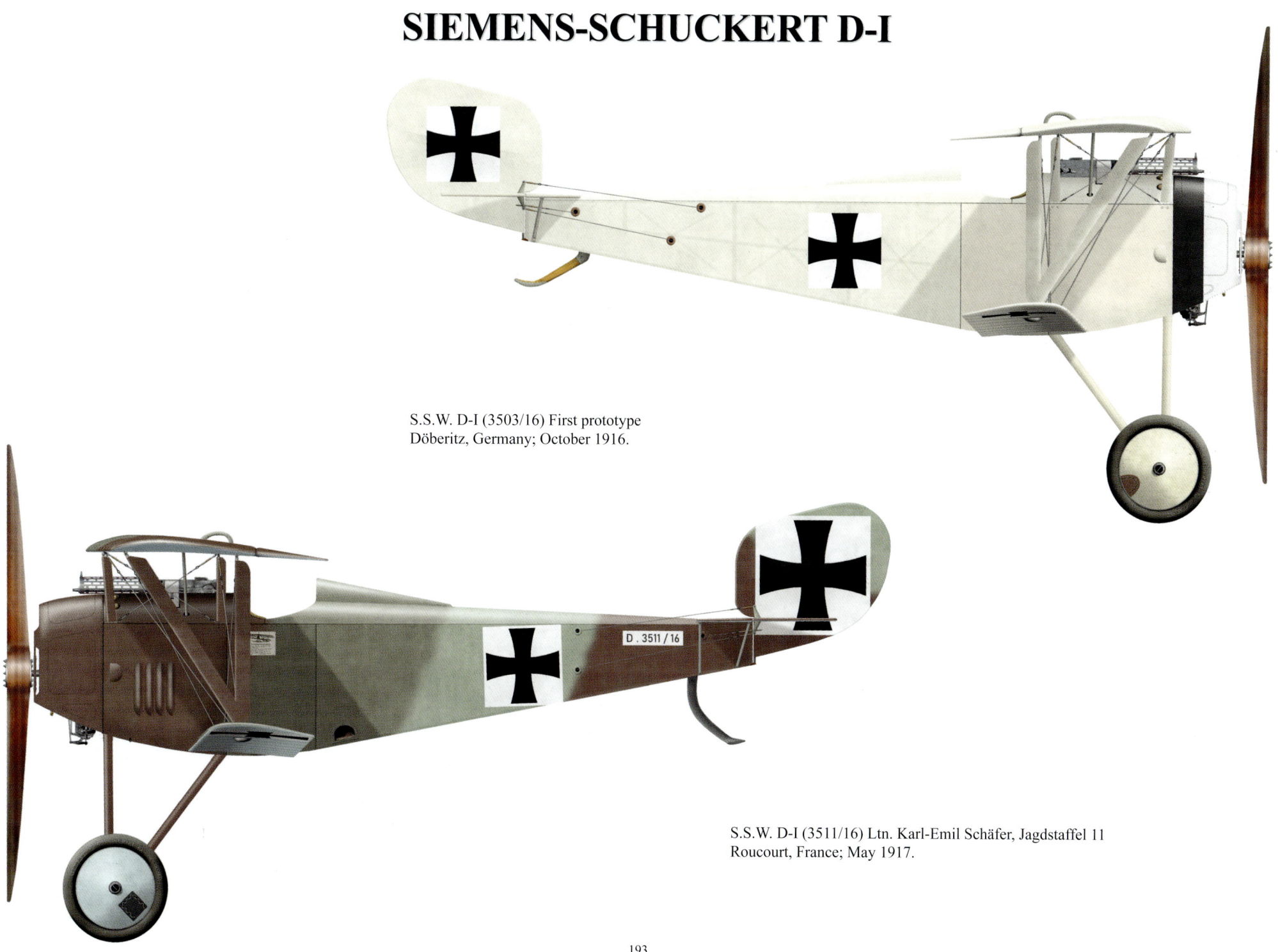

S.S.W. D-I (3503/16) First prototype
Döberitz, Germany; October 1916.

S.S.W. D-I (3511/16) Ltn. Karl-Emil Schäfer, Jagdstaffel 11
Roucourt, France; May 1917.

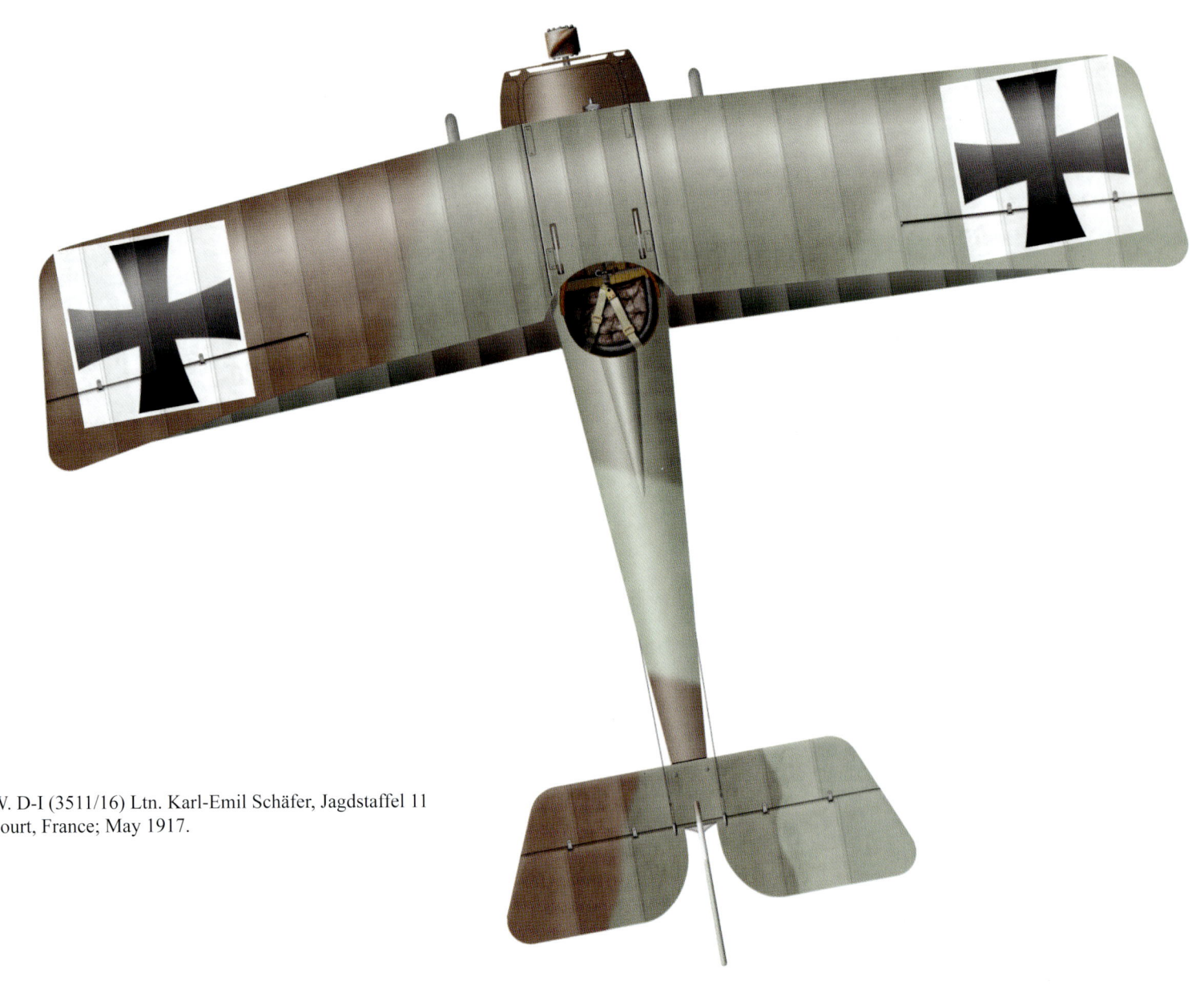

S.S.W. D-I (3511/16) Ltn. Karl-Emil Schäfer, Jagdstaffel 11
Roucourt, France; May 1917.

S.S.W. D-I (3511/16) Ltn. Karl-Emil Schäfer, Jagdstaffel 11
Roucourt, France; May 1917.

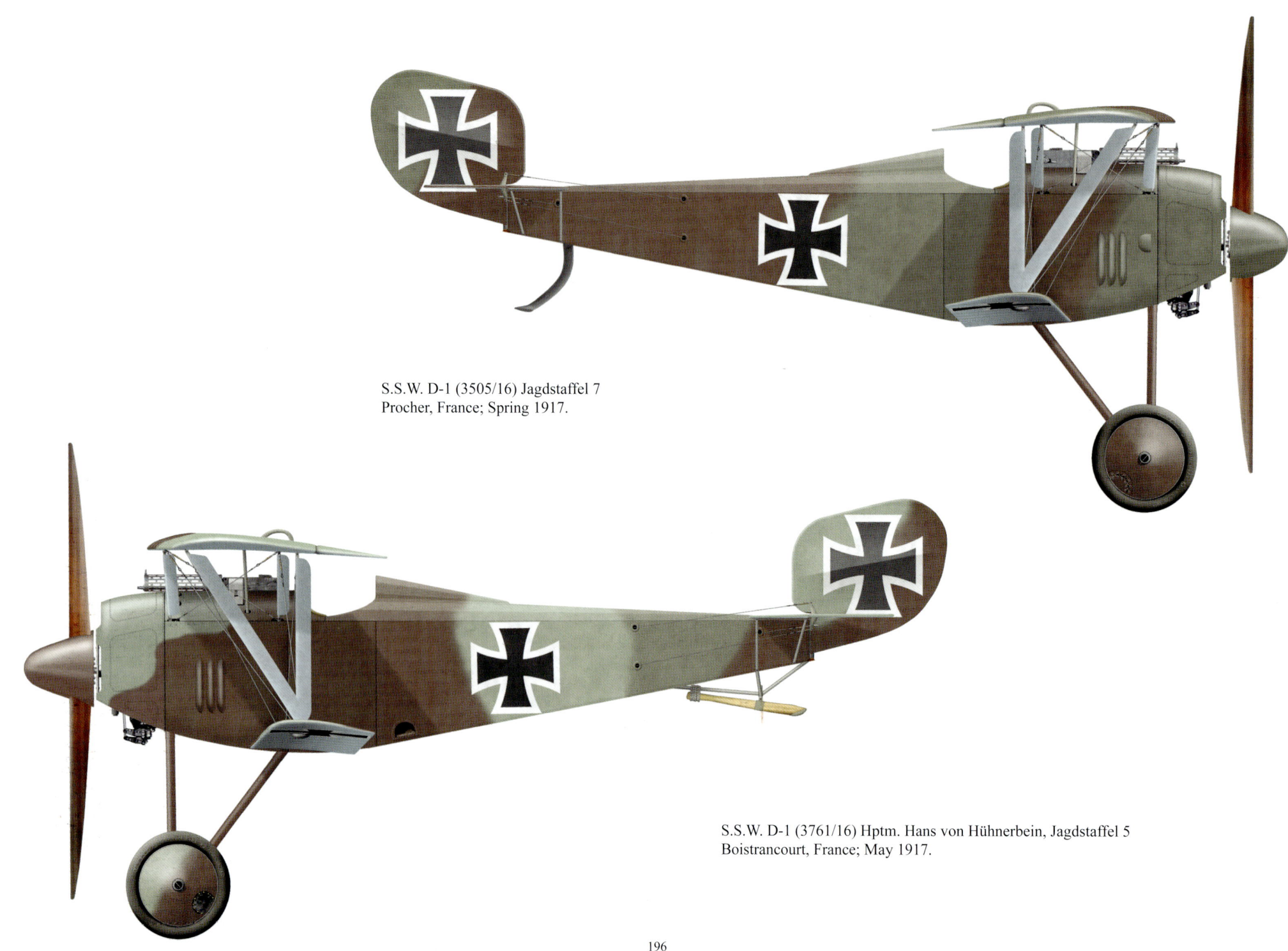

S.S.W. D-1 (3505/16) Jagdstaffel 7
Procher, France; Spring 1917.

S.S.W. D-1 (3761/16) Hptm. Hans von Hühnerbein, Jagdstaffel 5
Boistrancourt, France; May 1917.

ALBATROS D-V

ALBATROS D-V (1177/17) Rittm. Manfred von Richthofen, Jagdstaffel 11
Marckebeke, Belgium; June 1917.

ALBATROS D-V (1177/17) Rittm. Manfred von Richthofen, Jagdstaffel 11
Marckebeke, Belgium; June 1917.

ALBATROS D-V (1177/17) Rittm. Manfred von Richthofen, Jagdstaffel 11 Marckebeke, Belgium; June 1917.

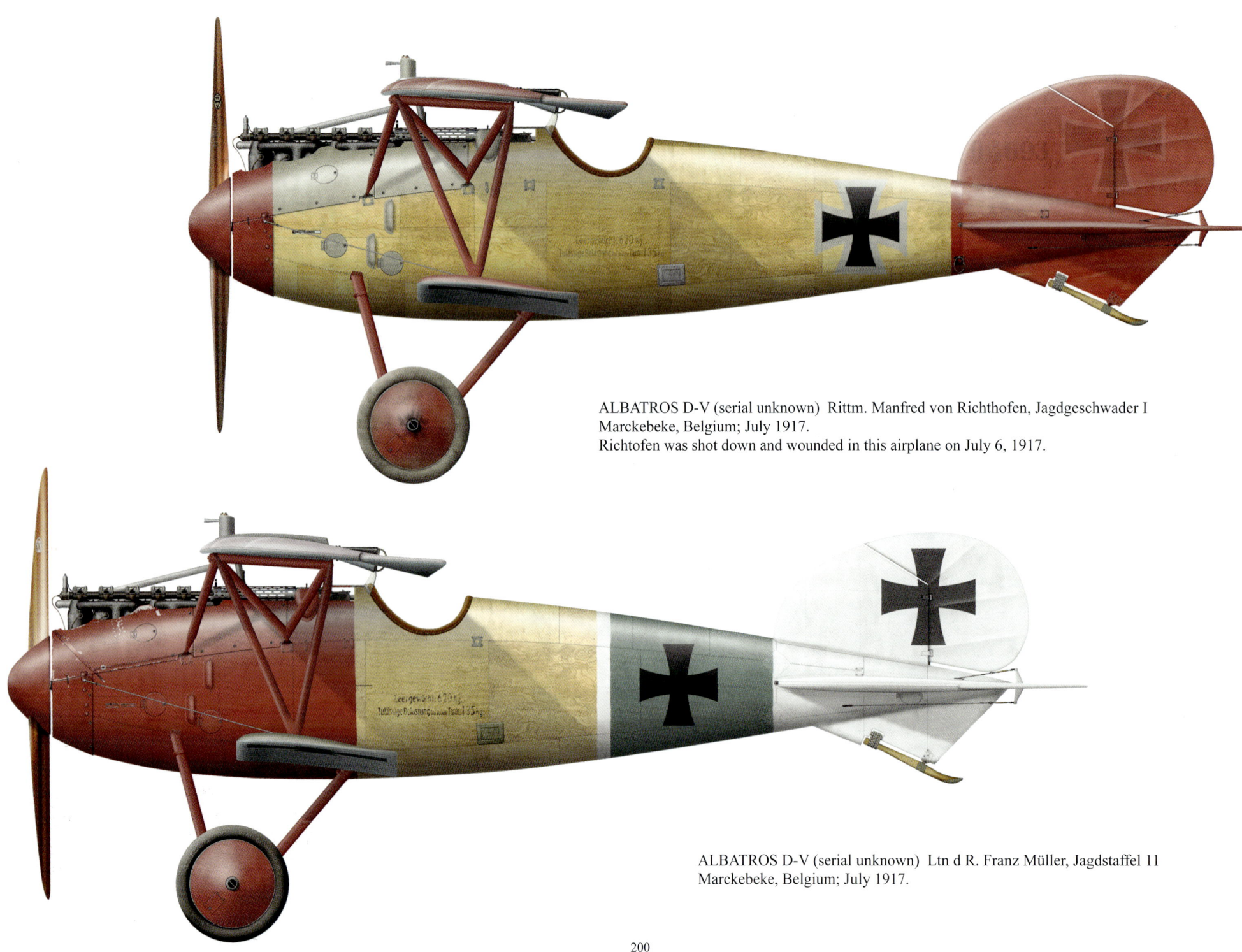

ALBATROS D-V (serial unknown) Rittm. Manfred von Richthofen, Jagdgeschwader I
Marckebeke, Belgium; July 1917.
Richtofen was shot down and wounded in this airplane on July 6, 1917.

ALBATROS D-V (serial unknown) Ltn d R. Franz Müller, Jagdstaffel 11
Marckebeke, Belgium; July 1917.

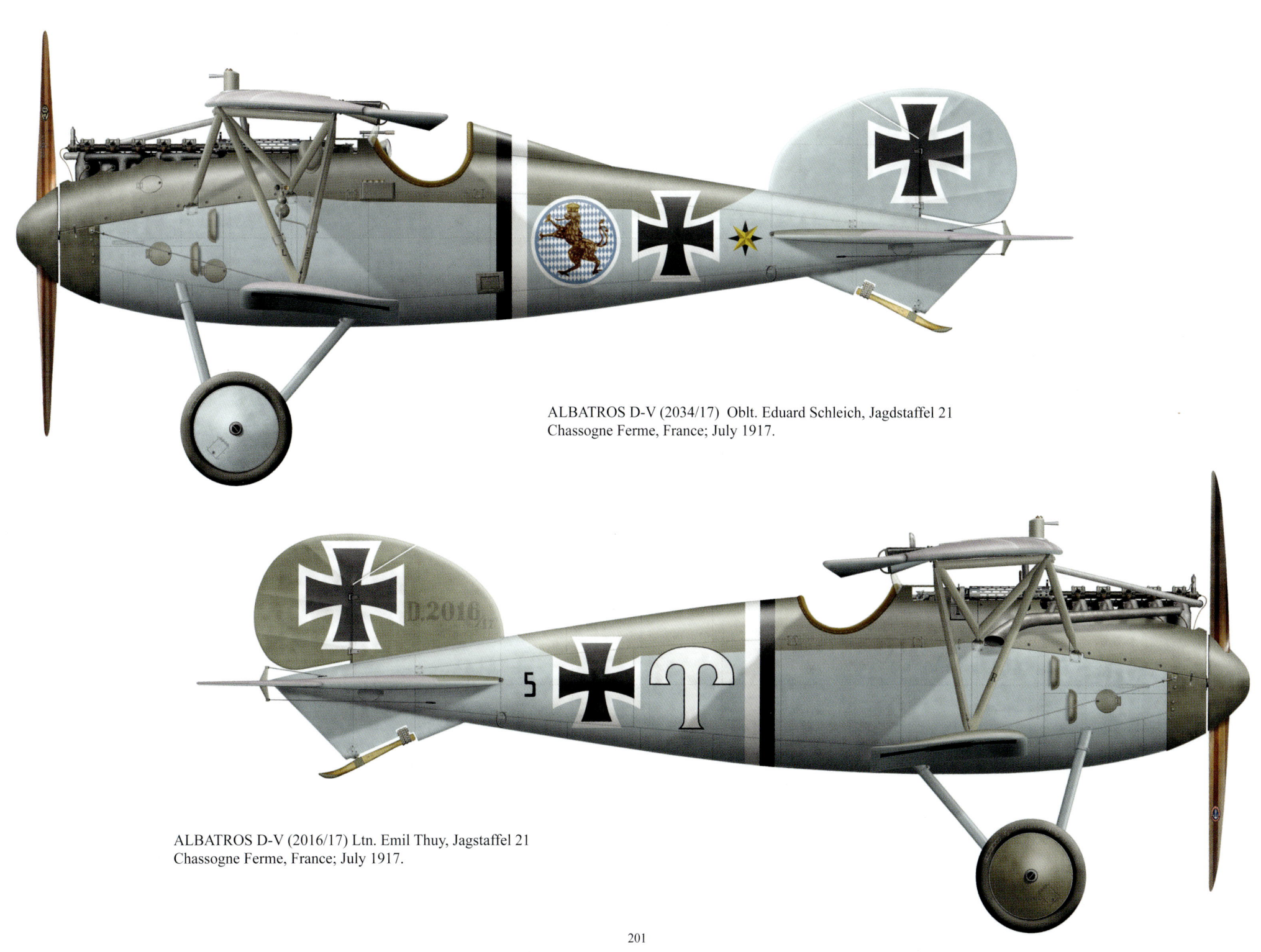

ALBATROS D-V (2034/17) Oblt. Eduard Schleich, Jagdstaffel 21
Chassogne Ferme, France; July 1917.

ALBATROS D-V (2016/17) Ltn. Emil Thuy, Jagdstaffel 21
Chassogne Ferme, France; July 1917.

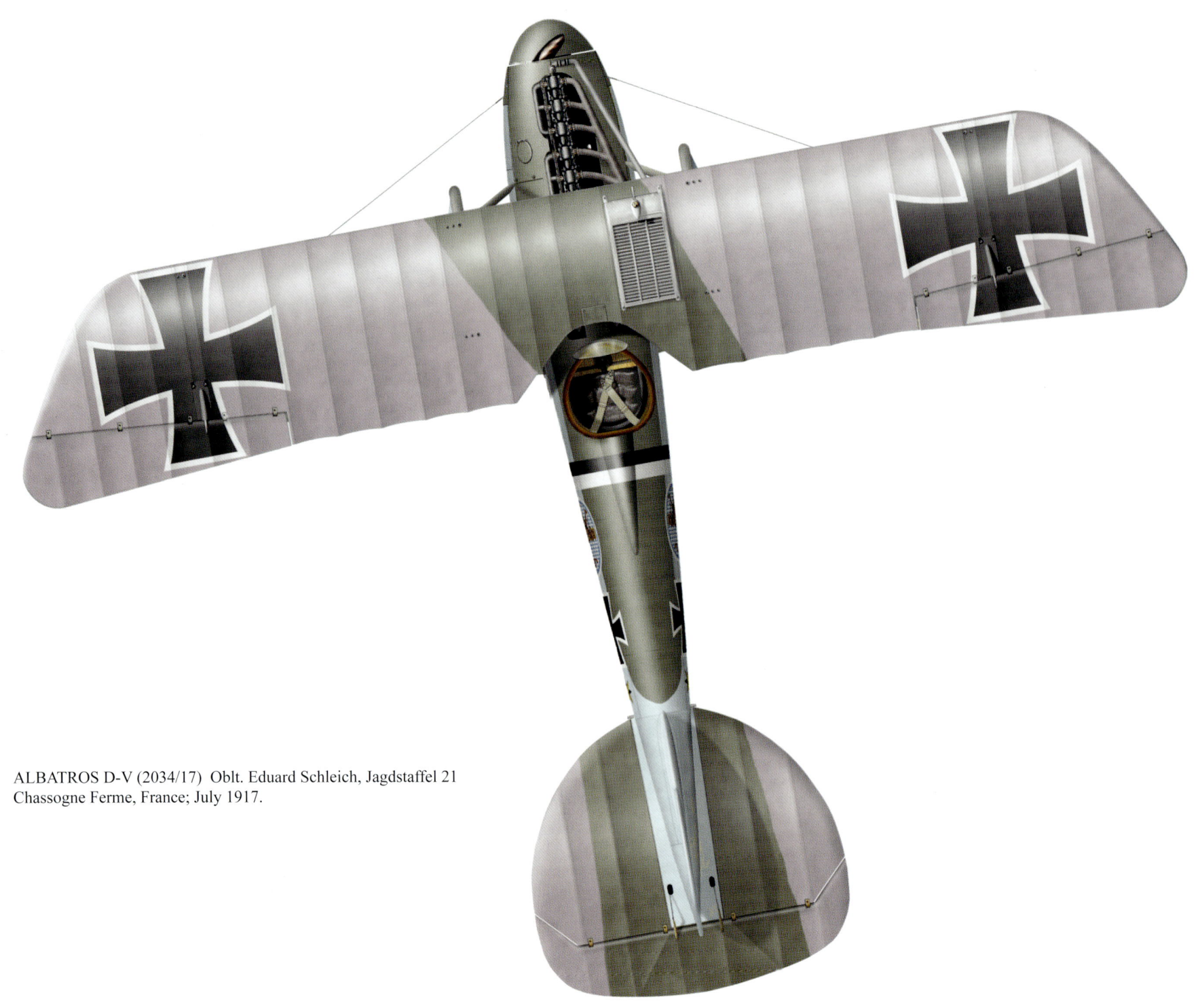

ALBATROS D-V (2034/17) Oblt. Eduard Schleich, Jagdstaffel 21
Chassogne Ferme, France; July 1917.

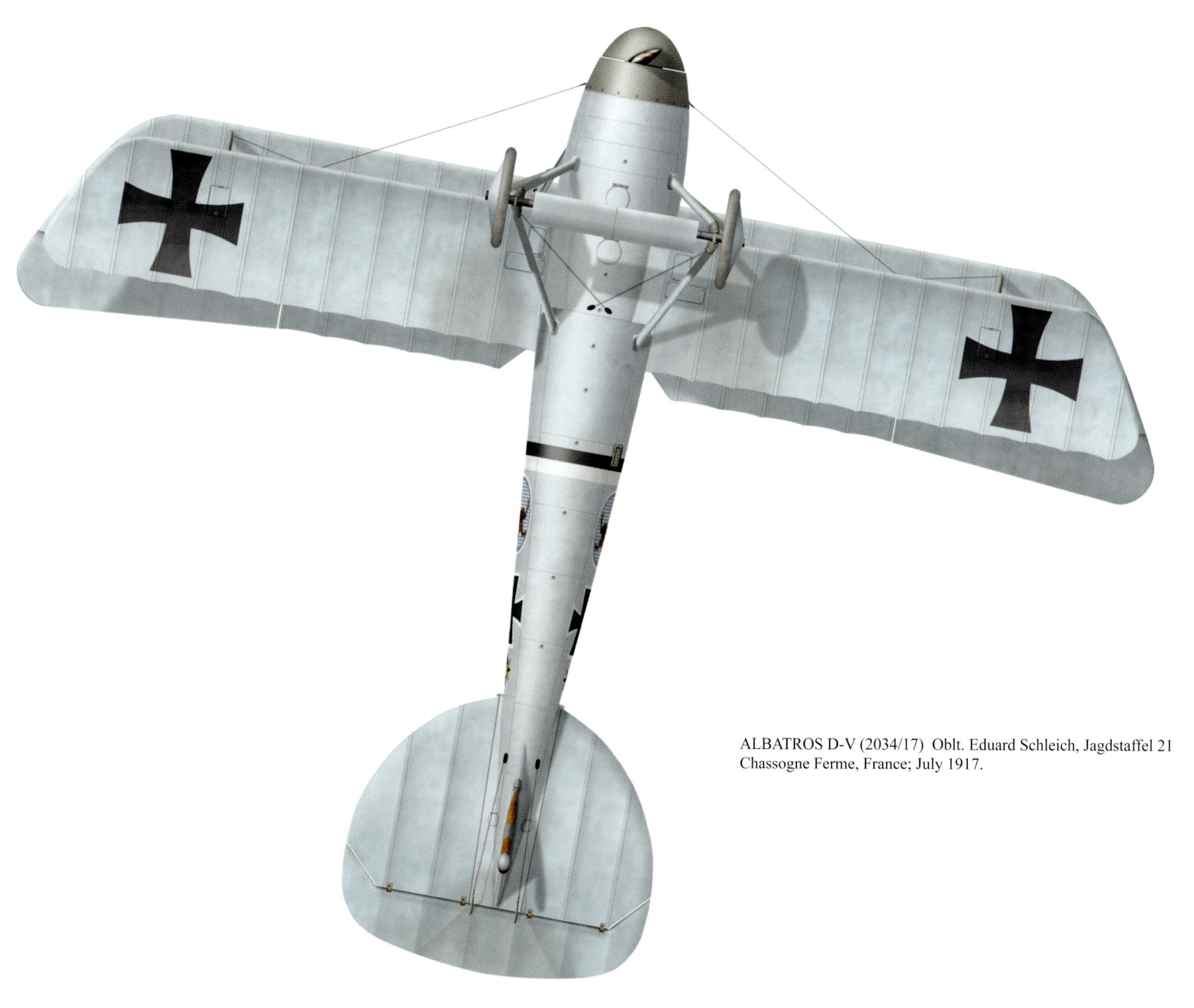

ALBATROS D-V (2034/17) Oblt. Eduard Schleich, Jagdstaffel 21
Chassogne Ferme, France; July 1917.

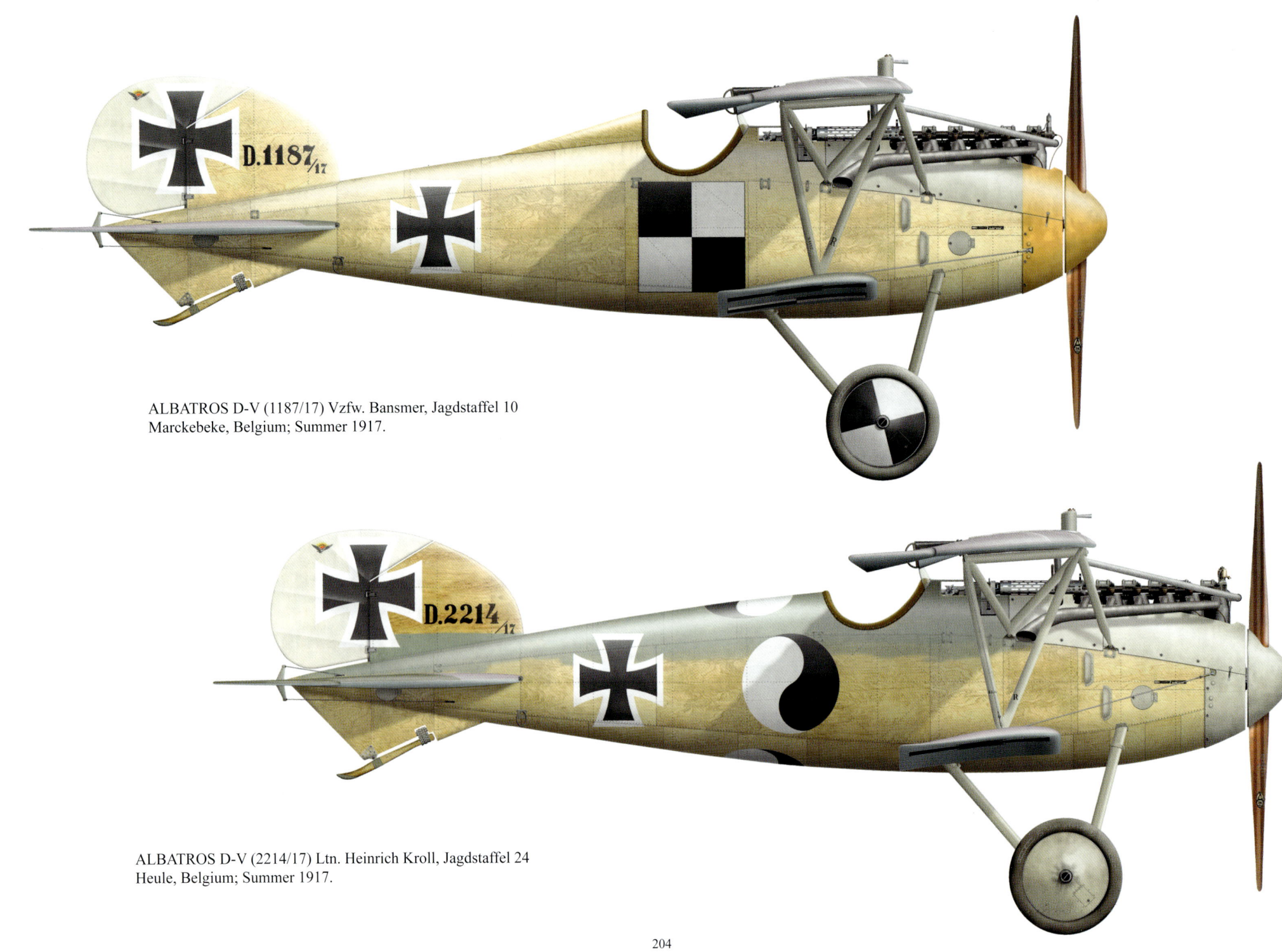

ALBATROS D-V (1187/17) Vzfw. Bansmer, Jagdstaffel 10
Marckebeke, Belgium; Summer 1917.

ALBATROS D-V (2214/17) Ltn. Heinrich Kroll, Jagdstaffel 24
Heule, Belgium; Summer 1917.

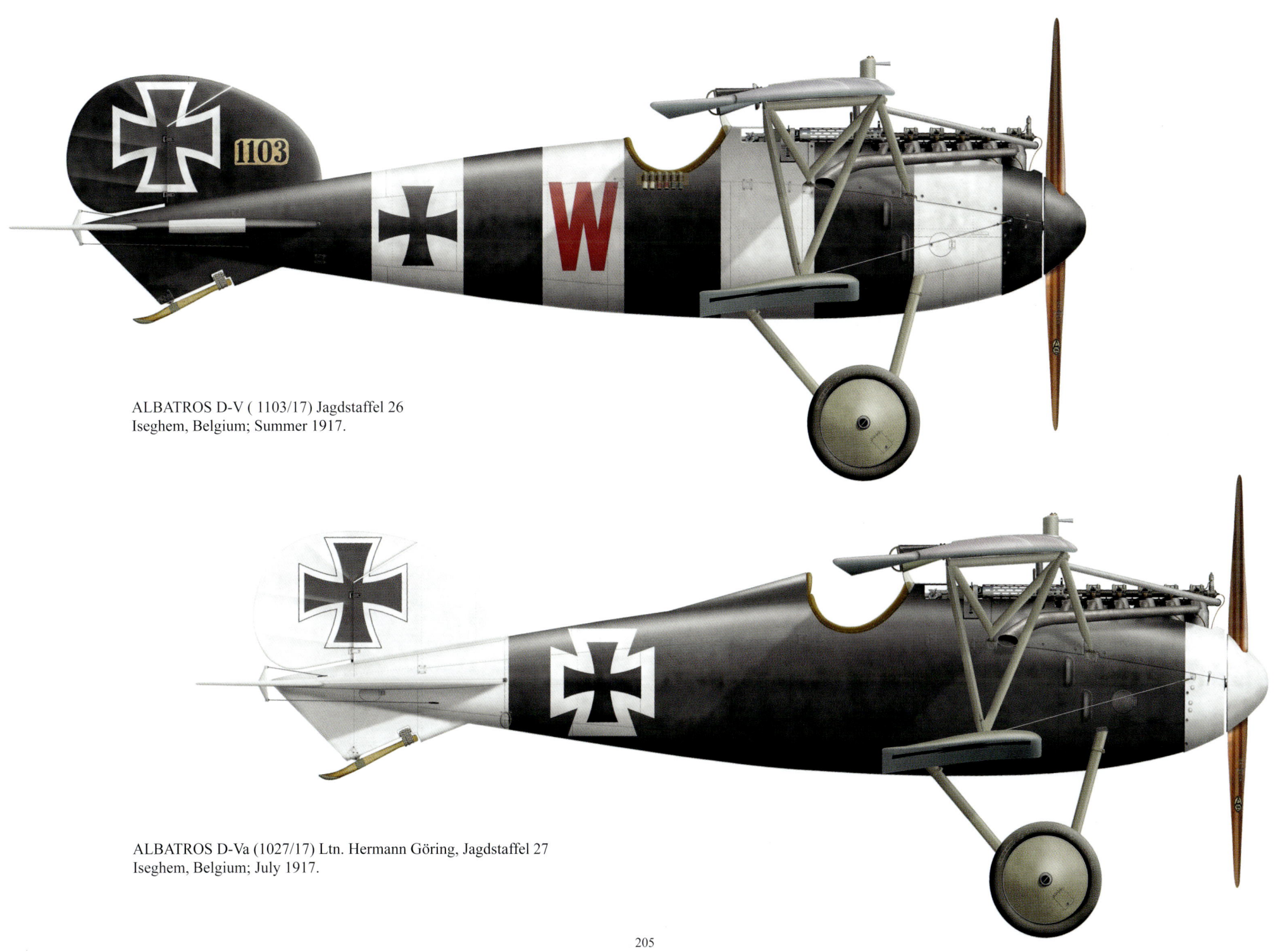

ALBATROS D-V (1103/17) Jagdstaffel 26
Iseghem, Belgium; Summer 1917.

ALBATROS D-Va (1027/17) Ltn. Hermann Göring, Jagdstaffel 27
Iseghem, Belgium; July 1917.

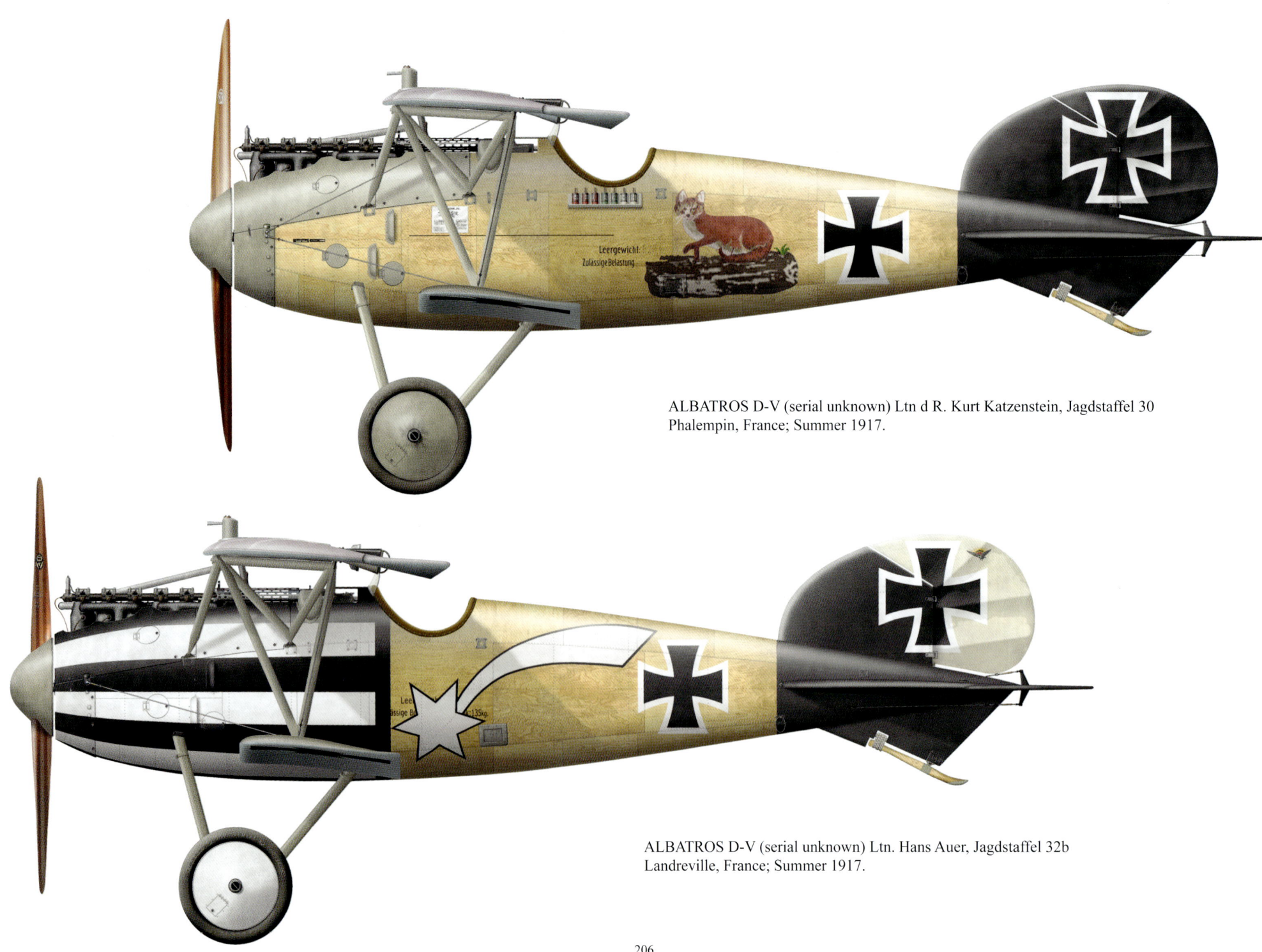

ALBATROS D-V (serial unknown) Ltn d R. Kurt Katzenstein, Jagdstaffel 30
Phalempin, France; Summer 1917.

ALBATROS D-V (serial unknown) Ltn. Hans Auer, Jagdstaffel 32b
Landreville, France; Summer 1917.

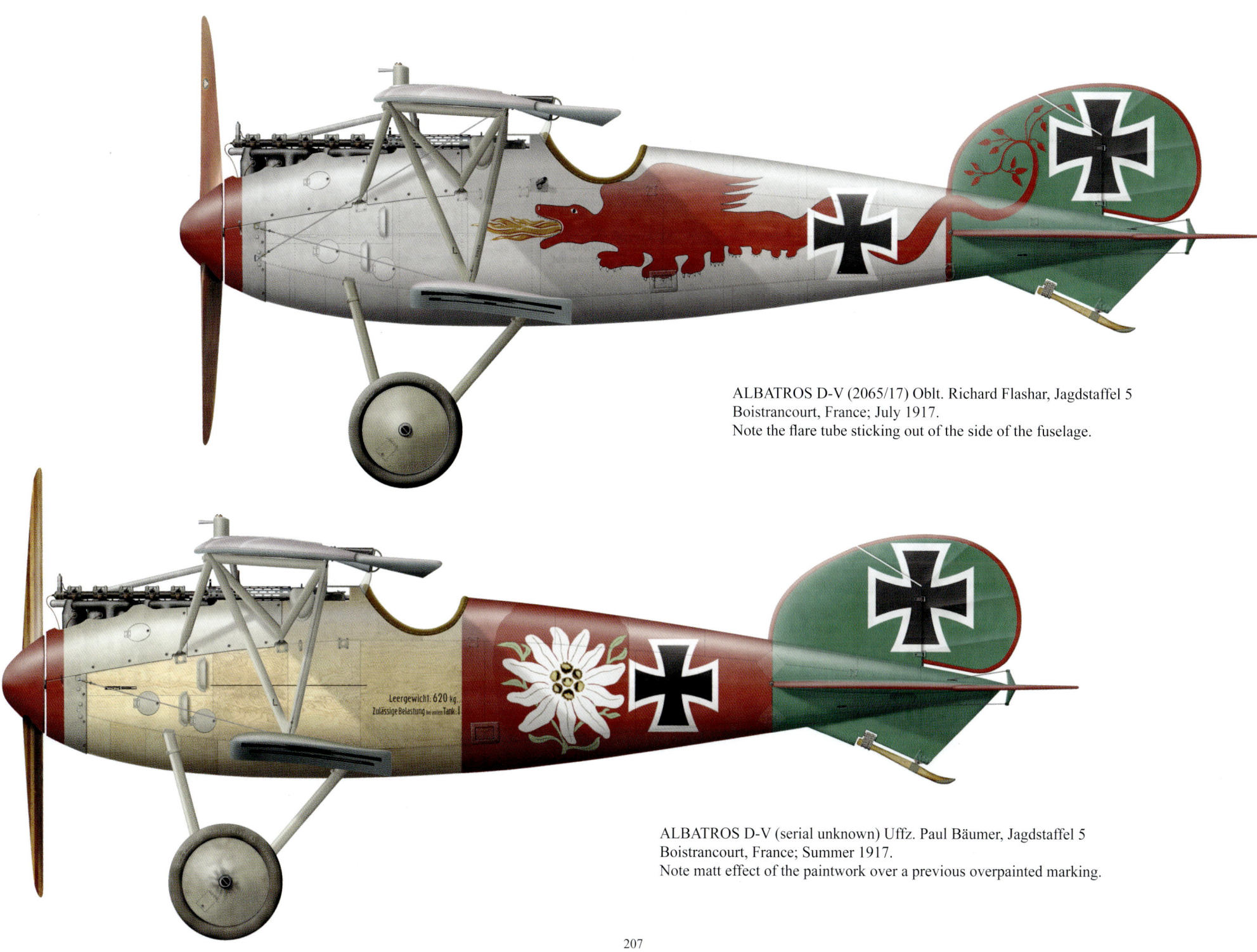

ALBATROS D-V (2065/17) Oblt. Richard Flashar, Jagdstaffel 5
Boistrancourt, France; July 1917.
Note the flare tube sticking out of the side of the fuselage.

ALBATROS D-V (serial unknown) Uffz. Paul Bäumer, Jagdstaffel 5
Boistrancourt, France; Summer 1917.
Note matt effect of the paintwork over a previous overpainted marking.

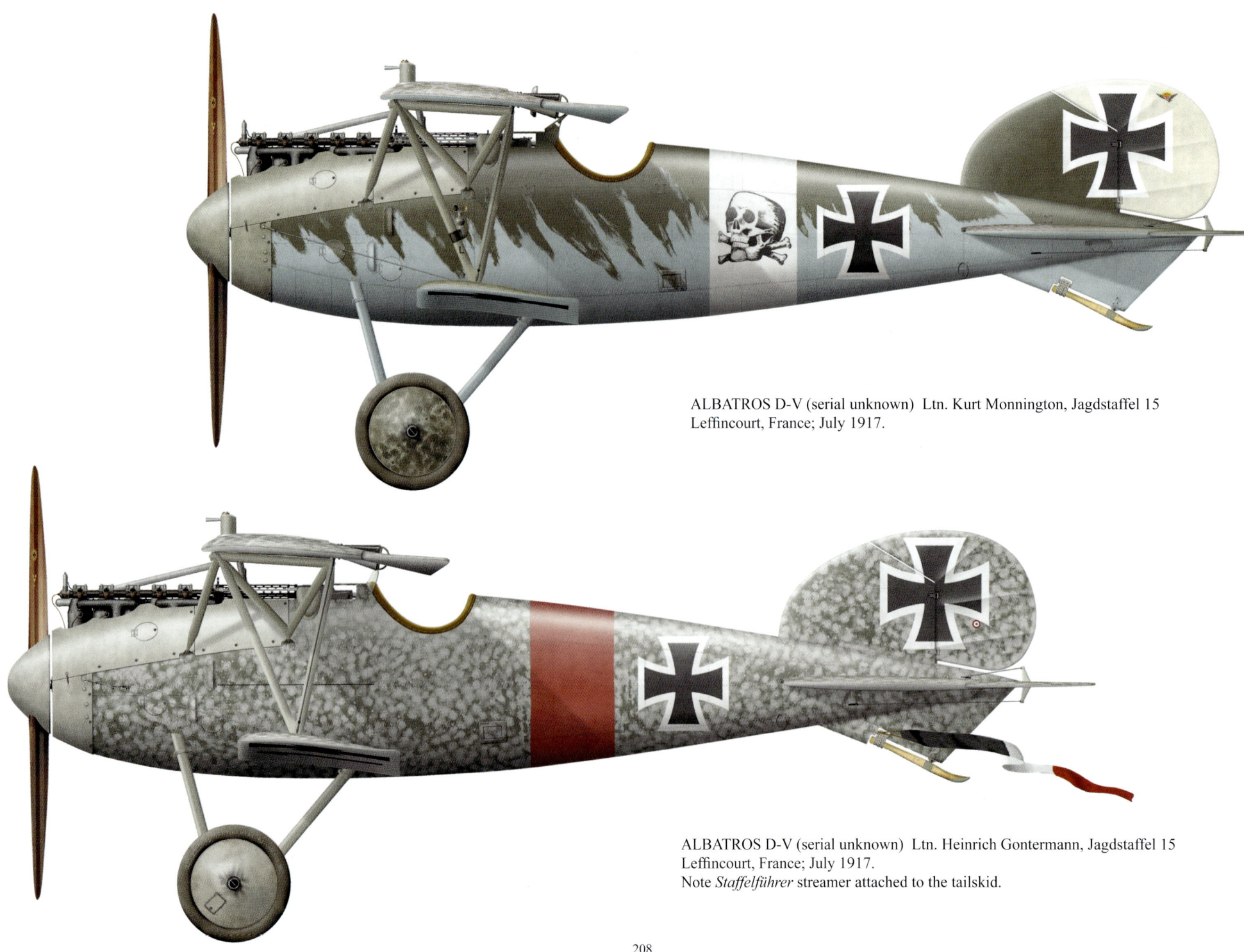

ALBATROS D-V (serial unknown) Ltn. Kurt Monnington, Jagdstaffel 15
Leffincourt, France; July 1917.

ALBATROS D-V (serial unknown) Ltn. Heinrich Gontermann, Jagdstaffel 15
Leffincourt, France; July 1917.
Note *Staffelführer* streamer attached to the tailskid.

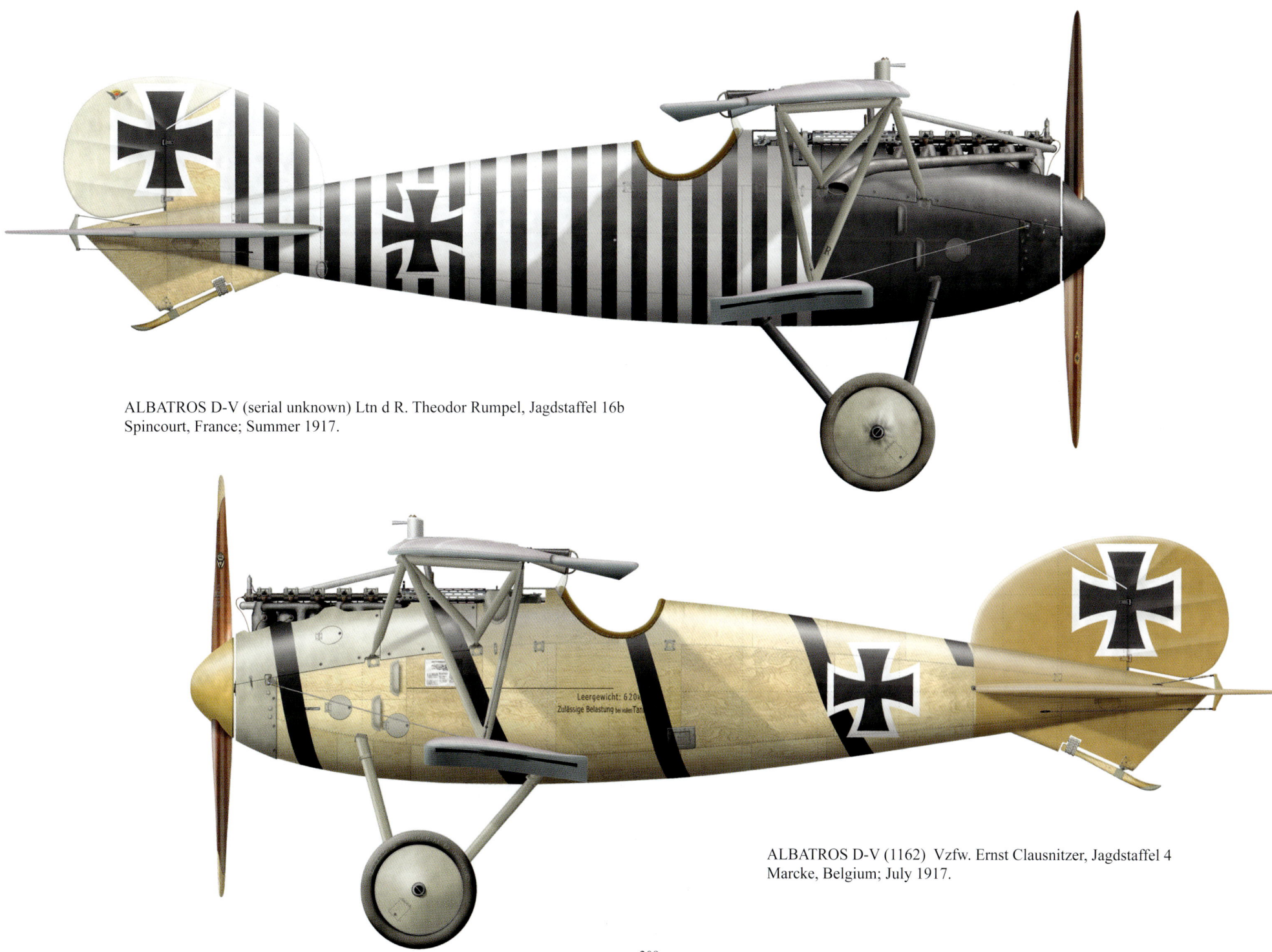

ALBATROS D-V (serial unknown) Ltn d R. Theodor Rumpel, Jagdstaffel 16b
Spincourt, France; Summer 1917.

ALBATROS D-V (1162) Vzfw. Ernst Clausnitzer, Jagdstaffel 4
Marcke, Belgium; July 1917.

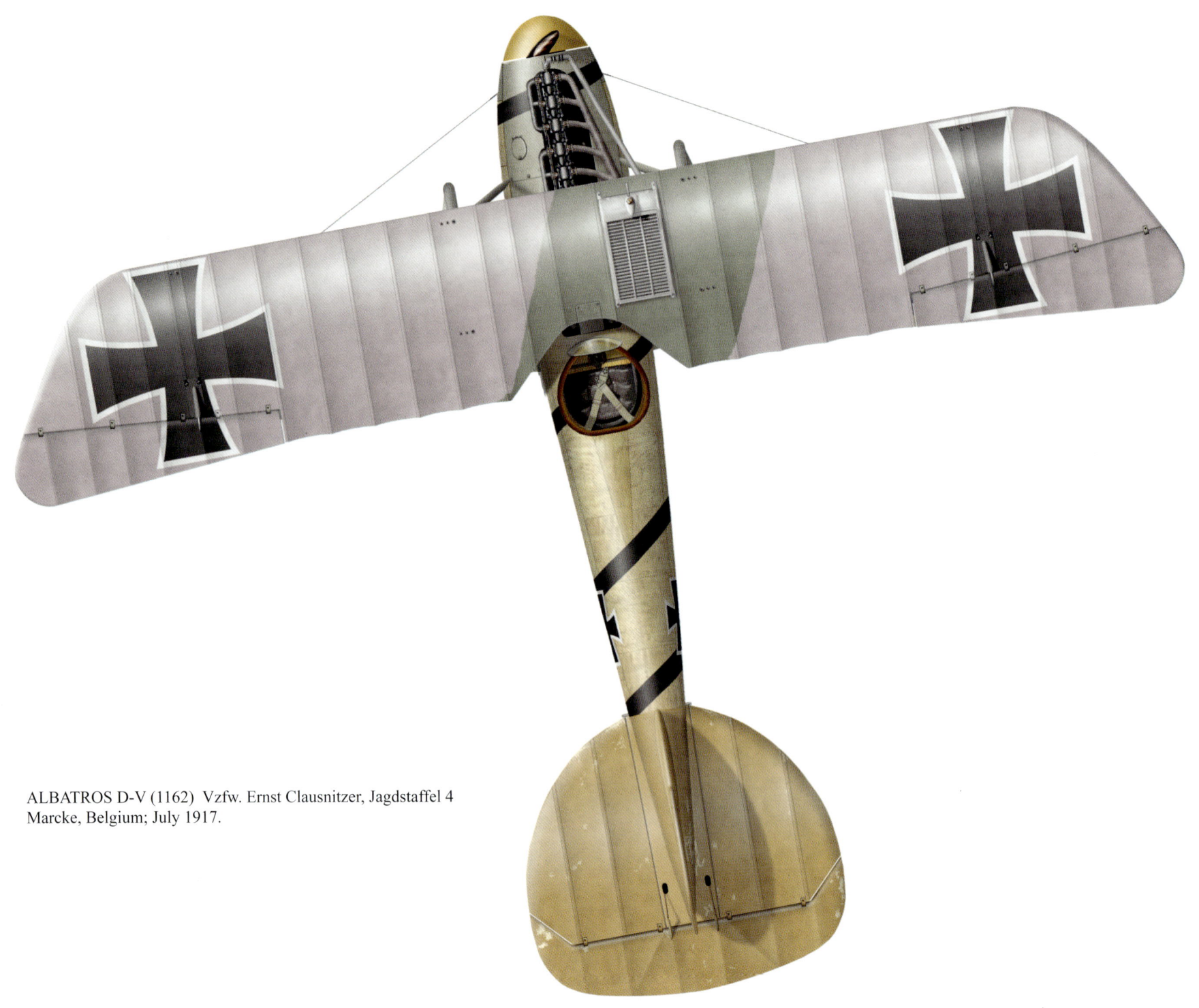

ALBATROS D-V (1162) Vzfw. Ernst Clausnitzer, Jagdstaffel 4
Marcke, Belgium; July 1917.

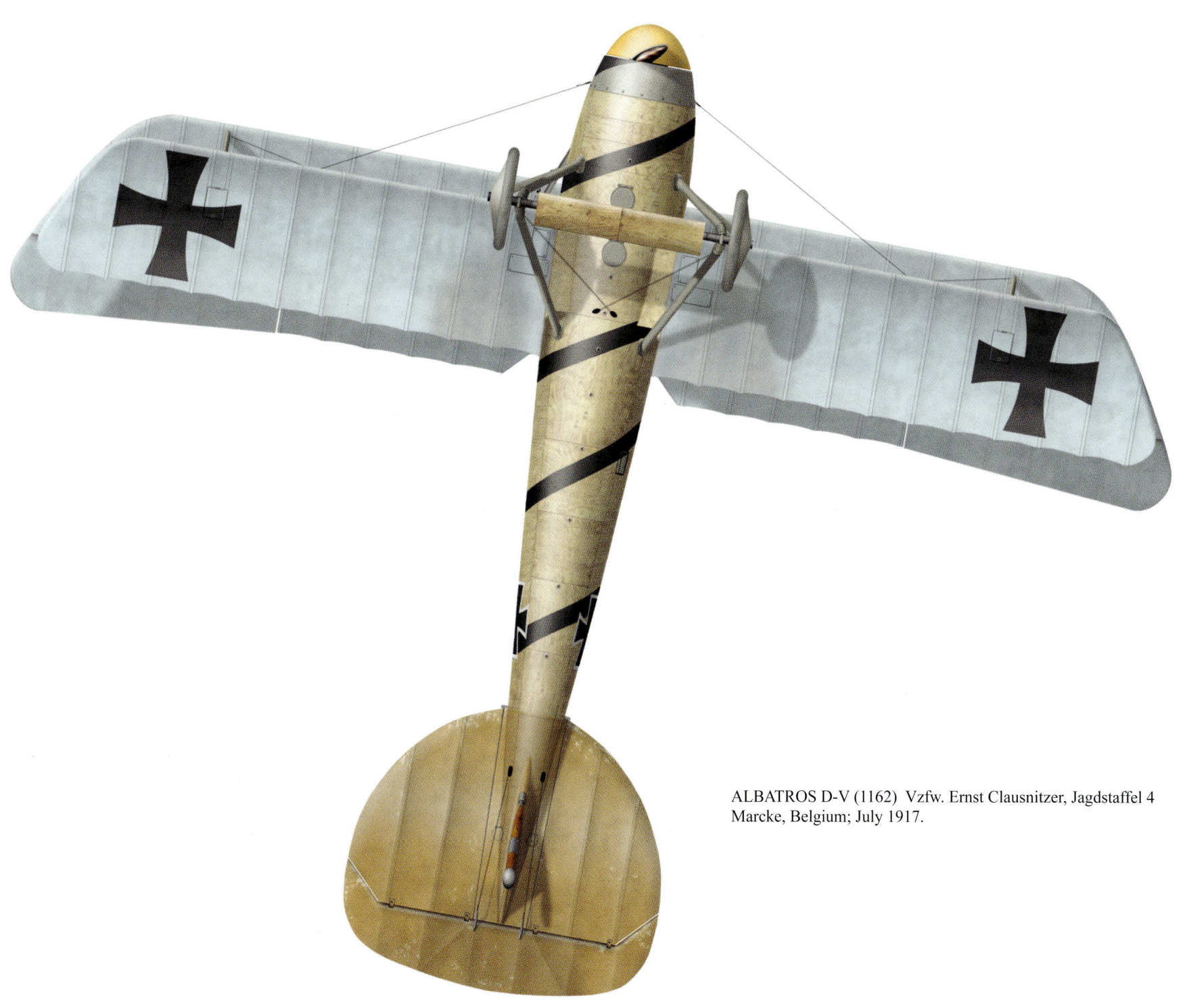

ALBATROS D-V (1162) Vzfw. Ernst Clausnitzer, Jagdstaffel 4
Marcke, Belgium; July 1917.

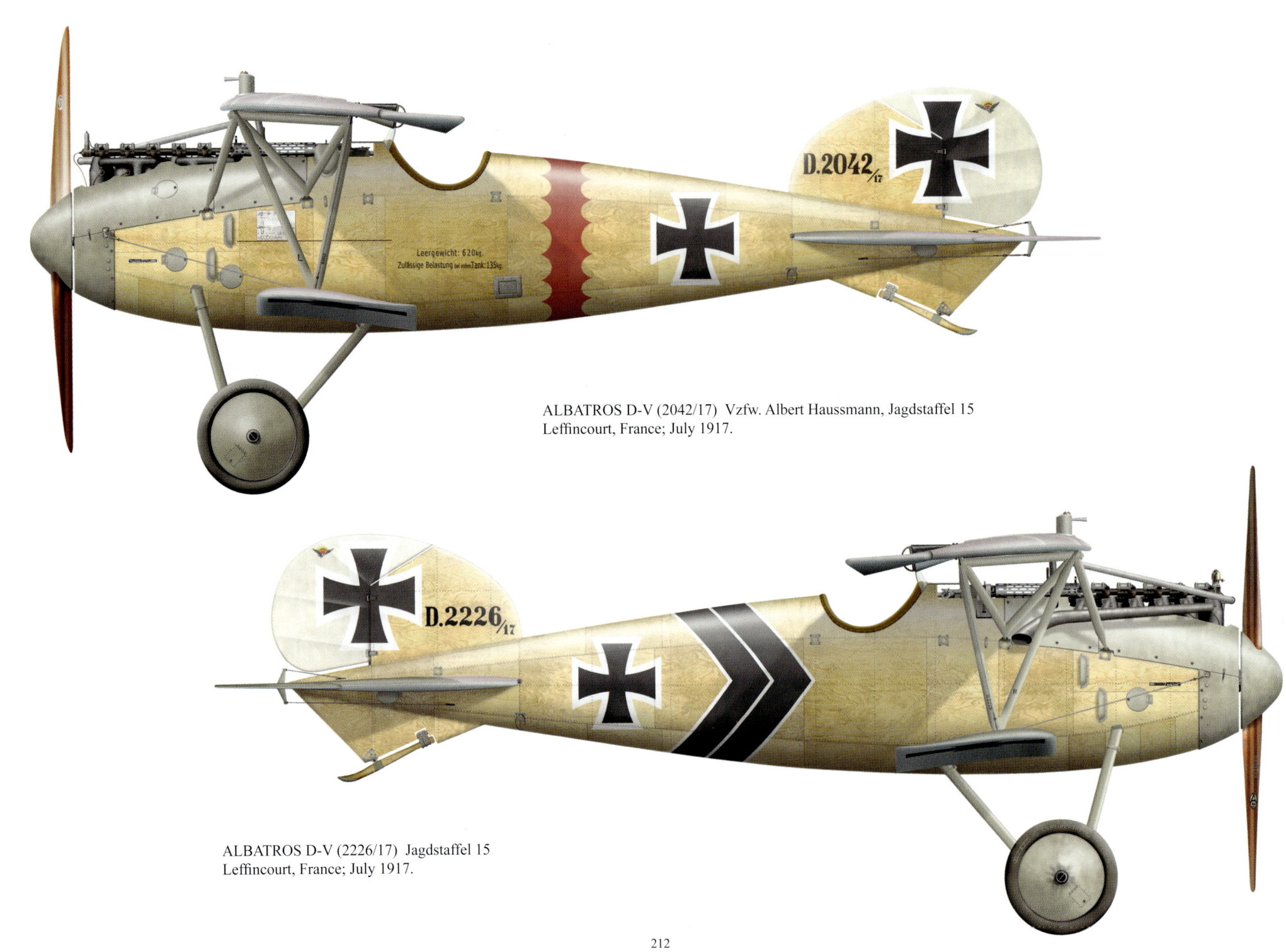

ALBATROS D-V (2042/17) Vzfw. Albert Haussmann, Jagdstaffel 15
Leffincourt, France; July 1917.

ALBATROS D-V (2226/17) Jagdstaffel 15
Leffincourt, France; July 1917.

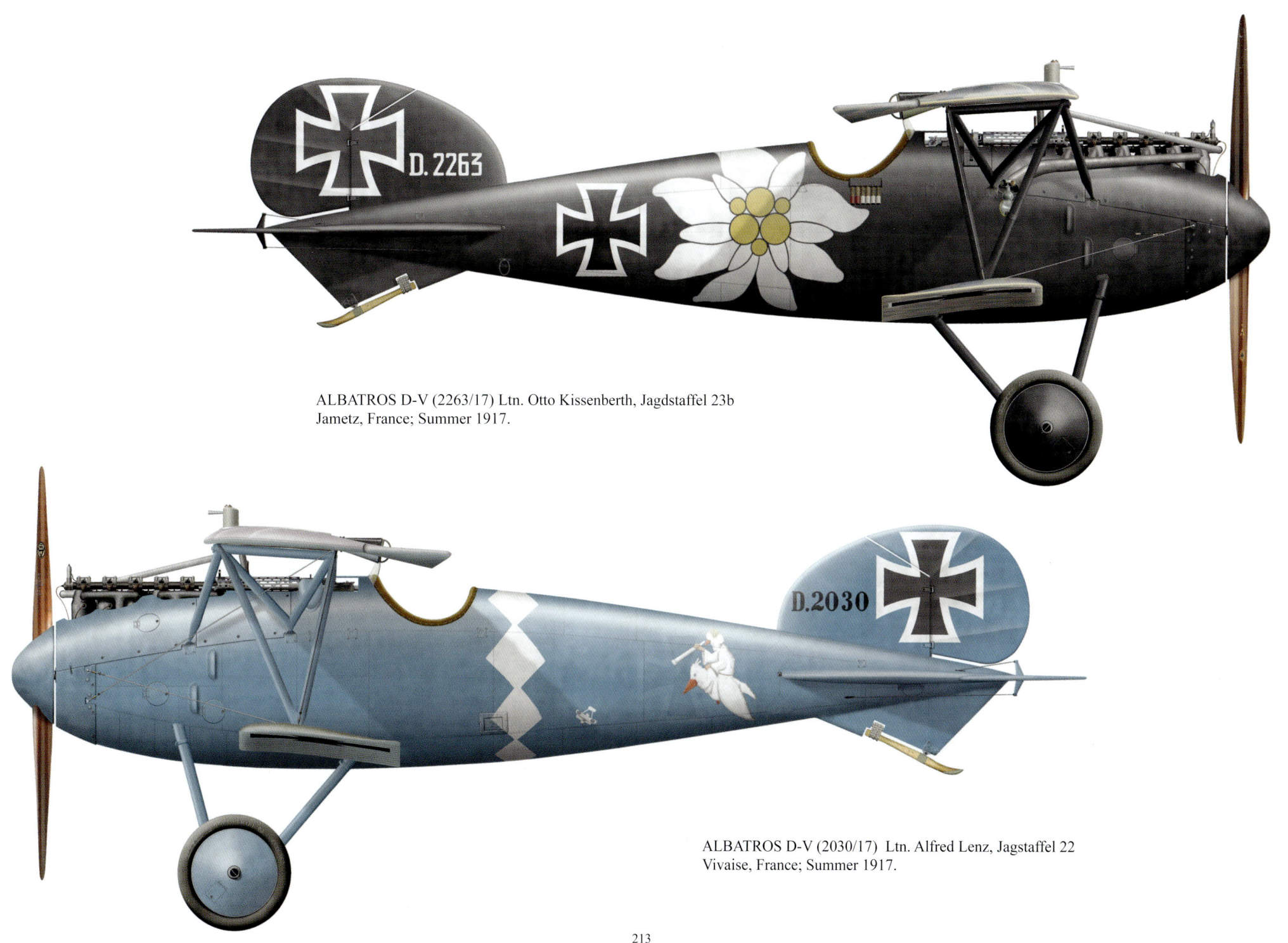

ALBATROS D-V (2263/17) Ltn. Otto Kissenberth, Jagdstaffel 23b
Jametz, France; Summer 1917.

ALBATROS D-V (2030/17) Ltn. Alfred Lenz, Jagdstaffel 22
Vivaise, France; Summer 1917.

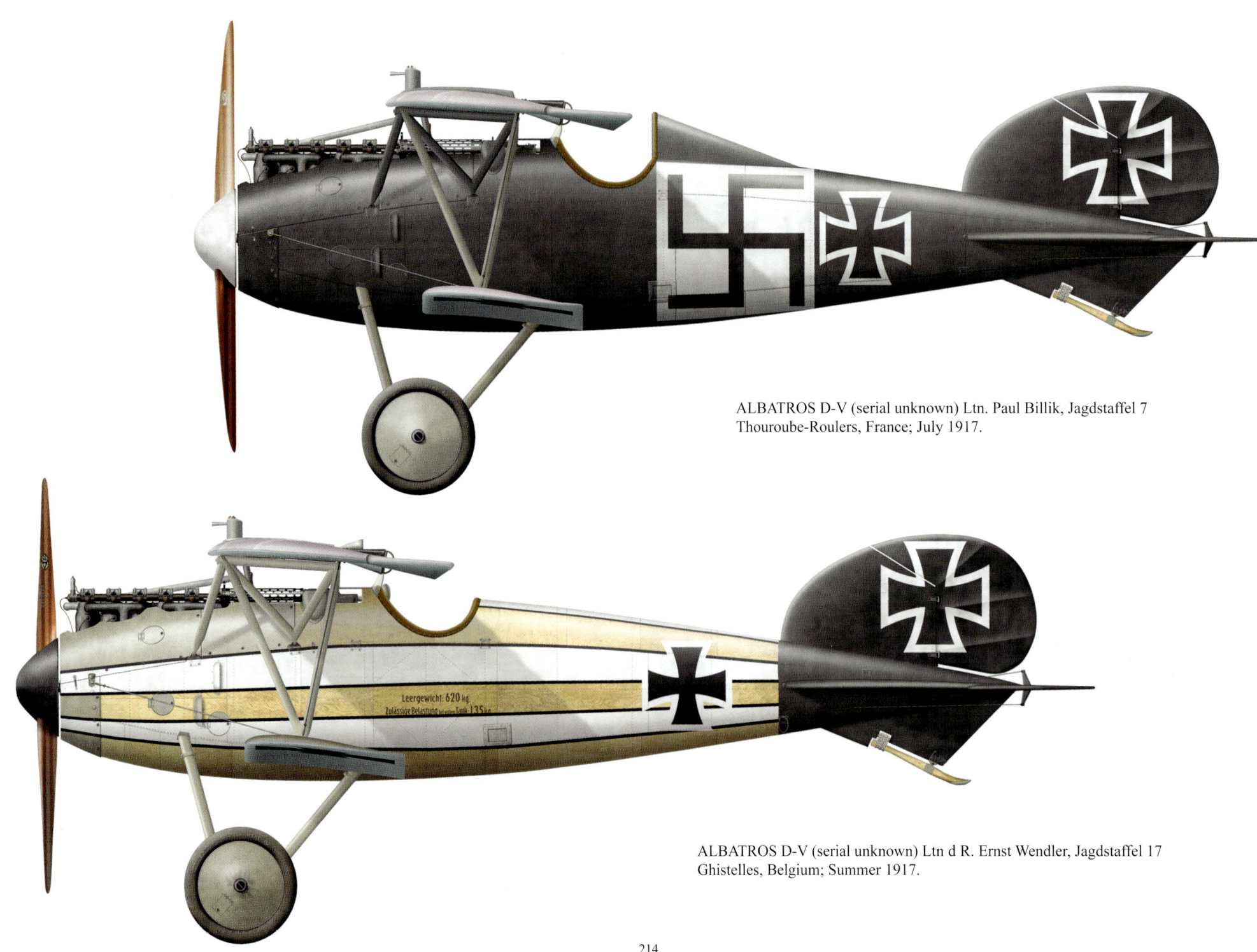

ALBATROS D-V (serial unknown) Ltn. Paul Billik, Jagdstaffel 7
Thouroube-Roulers, France; July 1917.

ALBATROS D-V (serial unknown) Ltn d R. Ernst Wendler, Jagdstaffel 17
Ghistelles, Belgium; Summer 1917.

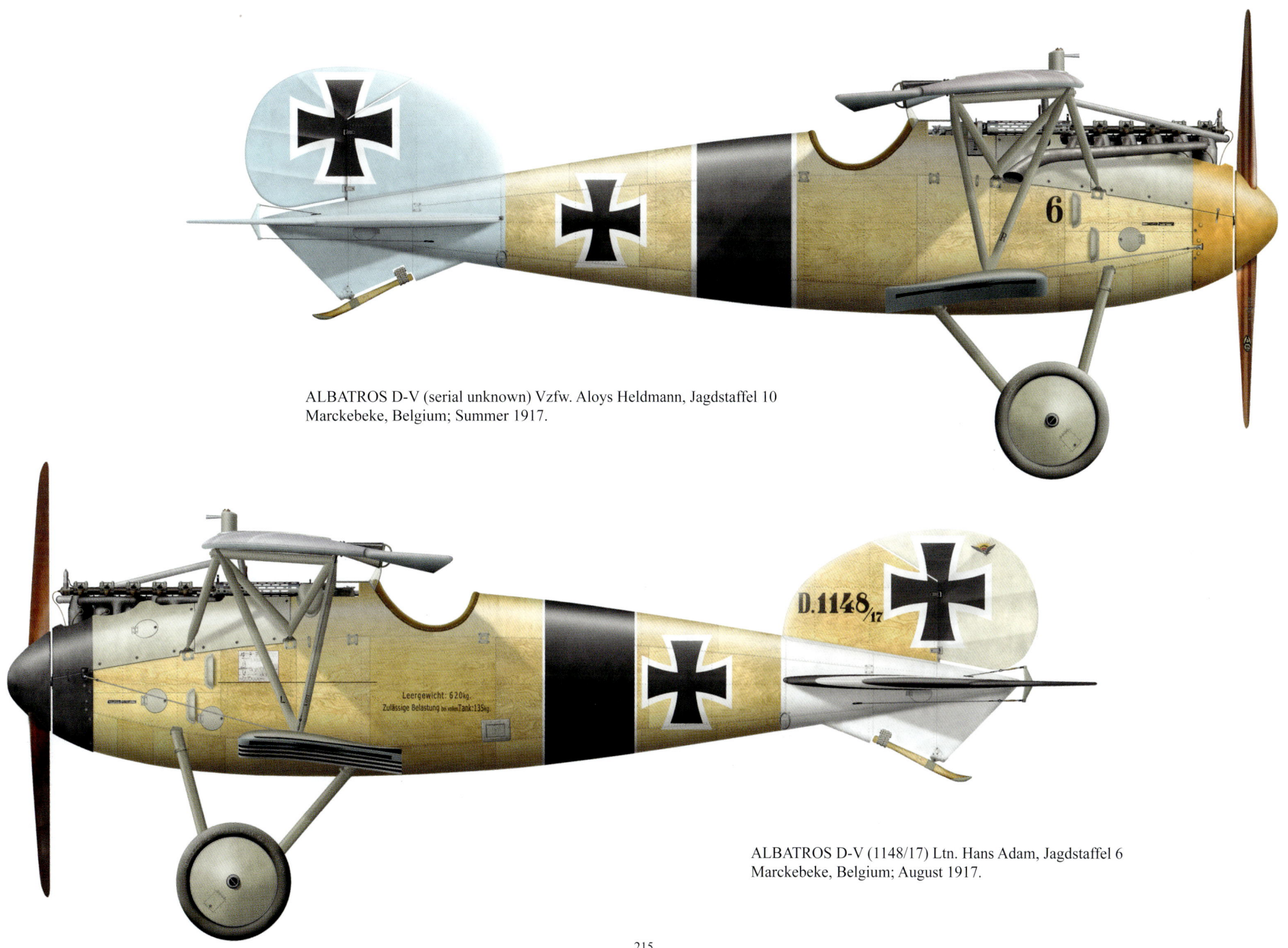

ALBATROS D-V (serial unknown) Vzfw. Aloys Heldmann, Jagdstaffel 10
Marckebeke, Belgium; Summer 1917.

ALBATROS D-V (1148/17) Ltn. Hans Adam, Jagdstaffel 6
Marckebeke, Belgium; August 1917.

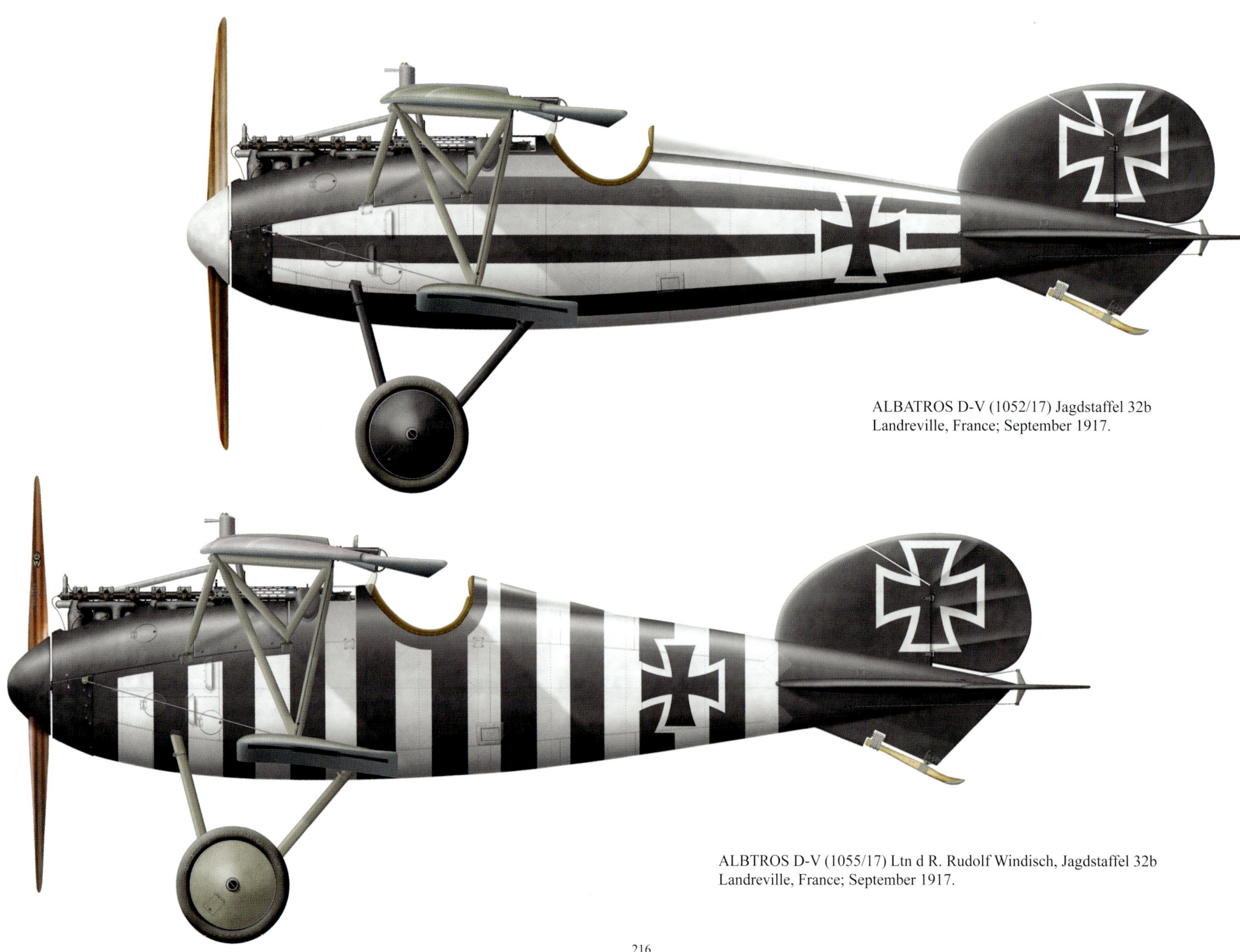

ALBATROS D-V (1052/17) Jagdstaffel 32b
Landreville, France; September 1917.

ALBTROS D-V (1055/17) Ltn d R. Rudolf Windisch, Jagdstaffel 32b
Landreville, France; September 1917.

ALBTROS D-V (1055/17) Ltn d R. Rudolf Windisch, Jagdstaffel 32b
Landreville, France; September 1917.

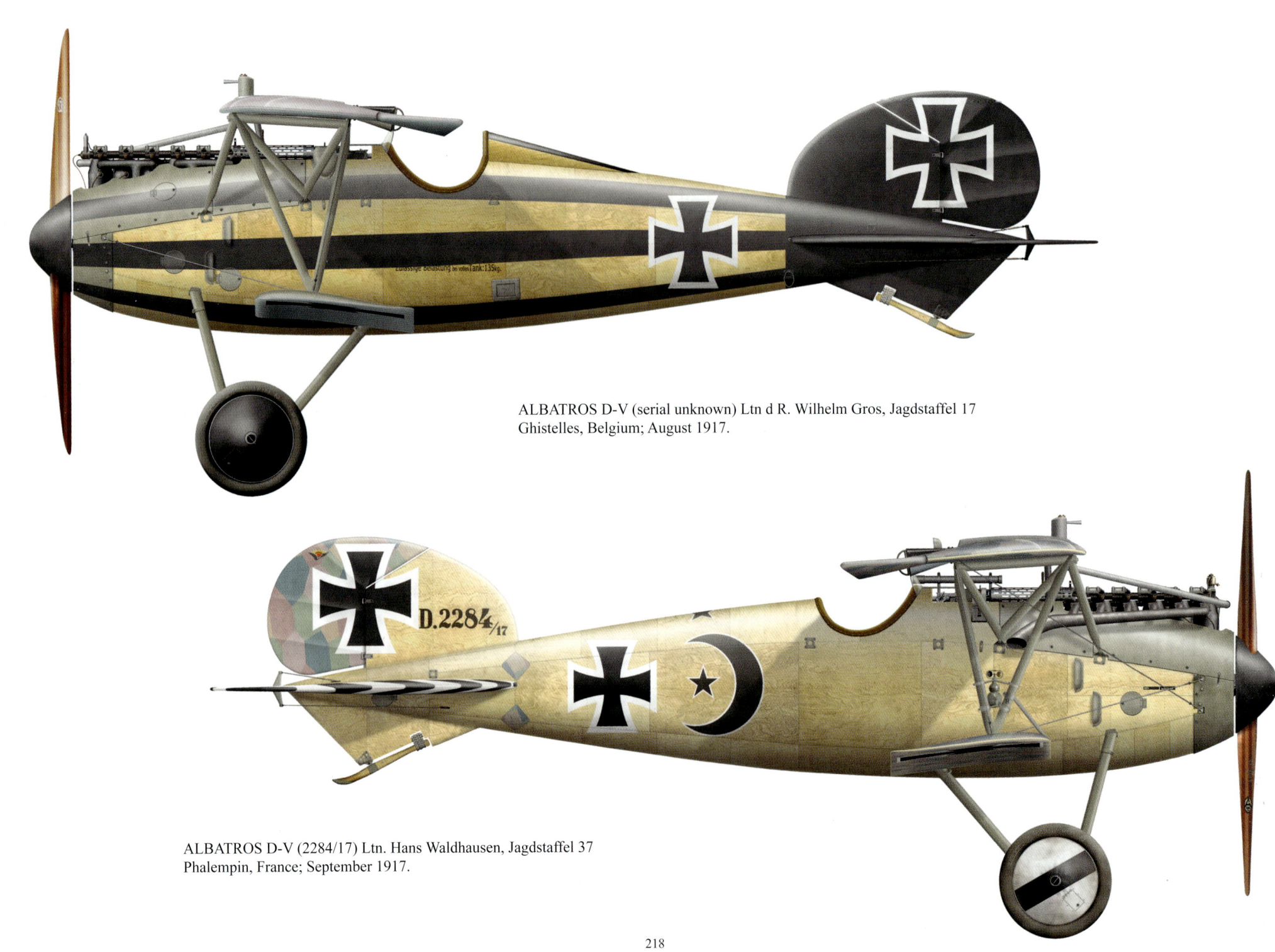

ALBATROS D-V (serial unknown) Ltn d R. Wilhelm Gros, Jagdstaffel 17
Ghistelles, Belgium; August 1917.

ALBATROS D-V (2284/17) Ltn. Hans Waldhausen, Jagdstaffel 37
Phalempin, France; September 1917.

ALBATROS D-V (2284/17) Ltn. Hans Waldhausen, Jagdstaffel 37
Phalempin, France; September 1917.

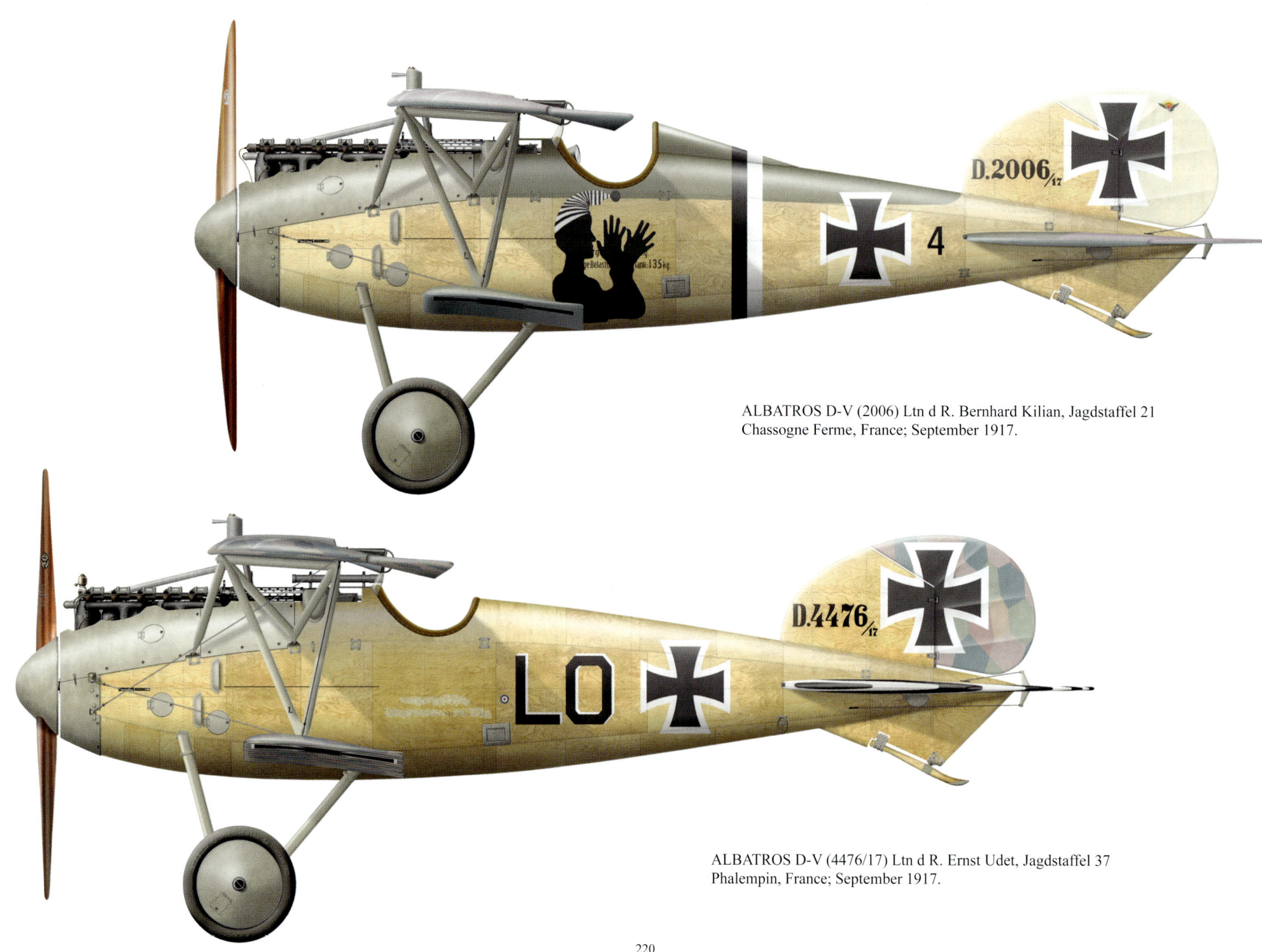

ALBATROS D-V (2006) Ltn d R. Bernhard Kilian, Jagdstaffel 21
Chassogne Ferme, France; September 1917.

ALBATROS D-V (4476/17) Ltn d R. Ernst Udet, Jagdstaffel 37
Phalempin, France; September 1917.

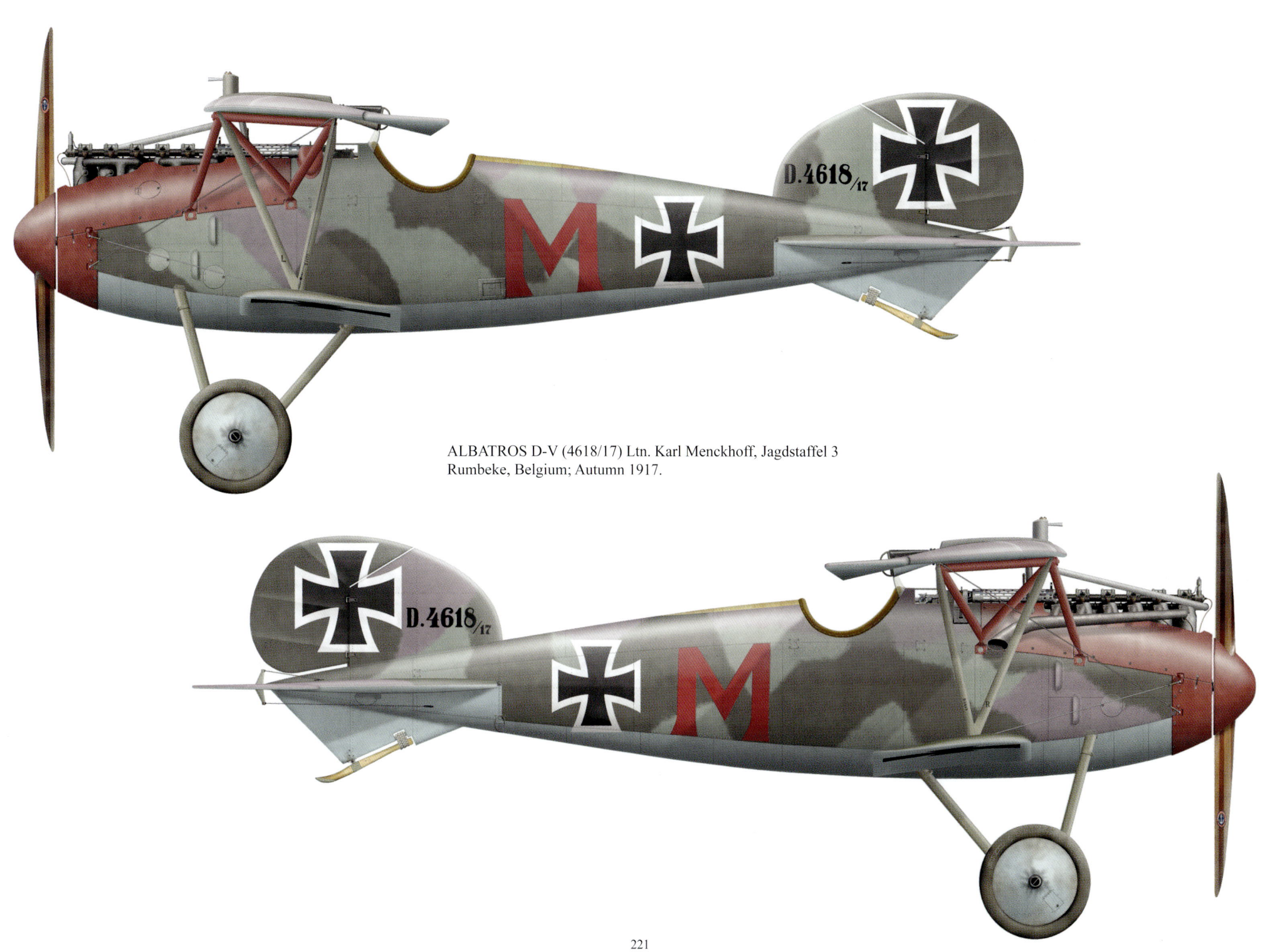

ALBATROS D-V (4618/17) Ltn. Karl Menckhoff, Jagdstaffel 3
Rumbeke, Belgium; Autumn 1917.

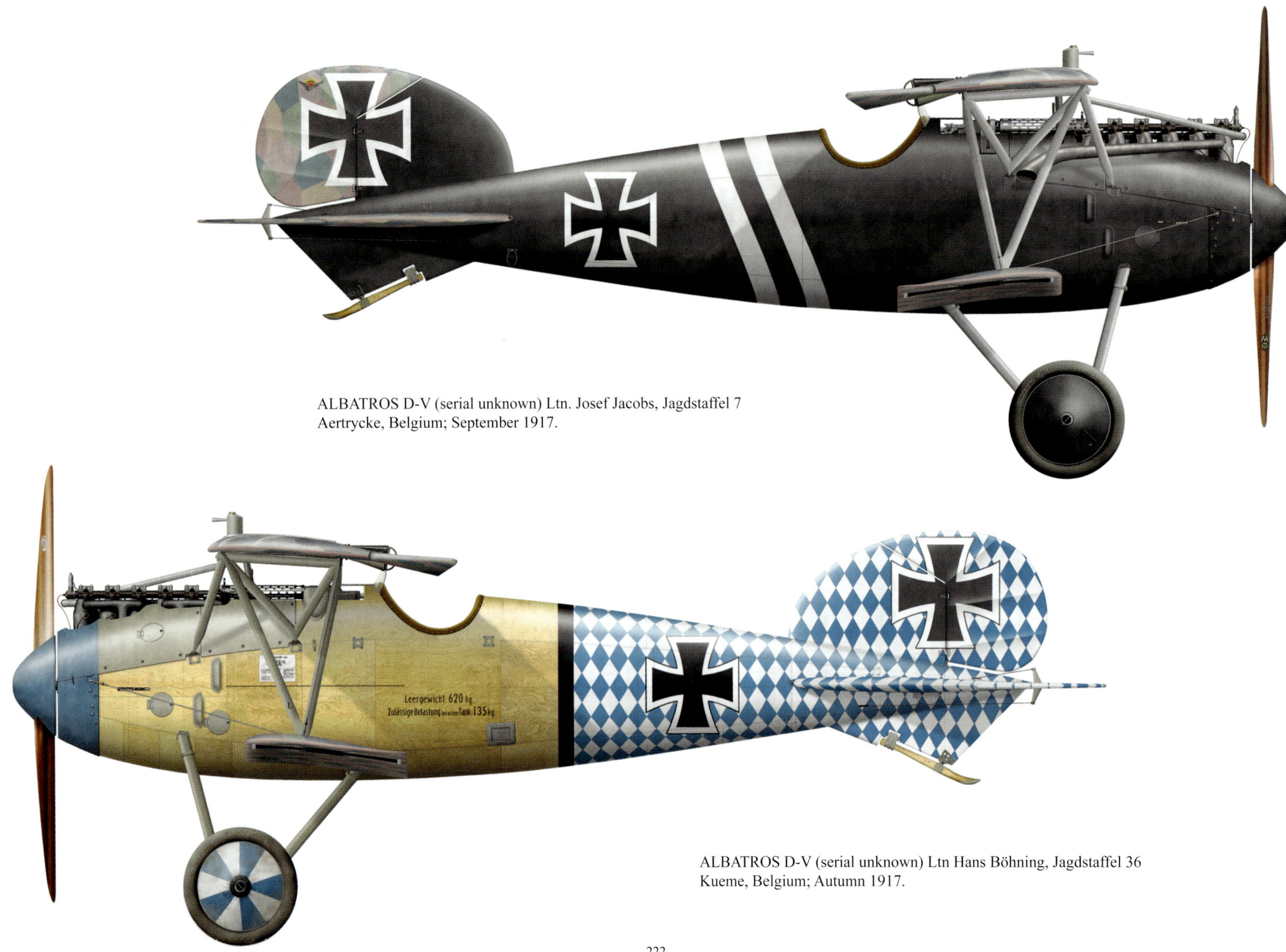

ALBATROS D-V (serial unknown) Ltn. Josef Jacobs, Jagdstaffel 7
Aertrycke, Belgium; September 1917.

ALBATROS D-V (serial unknown) Ltn Hans Böhning, Jagdstaffel 36
Kueme, Belgium; Autumn 1917.

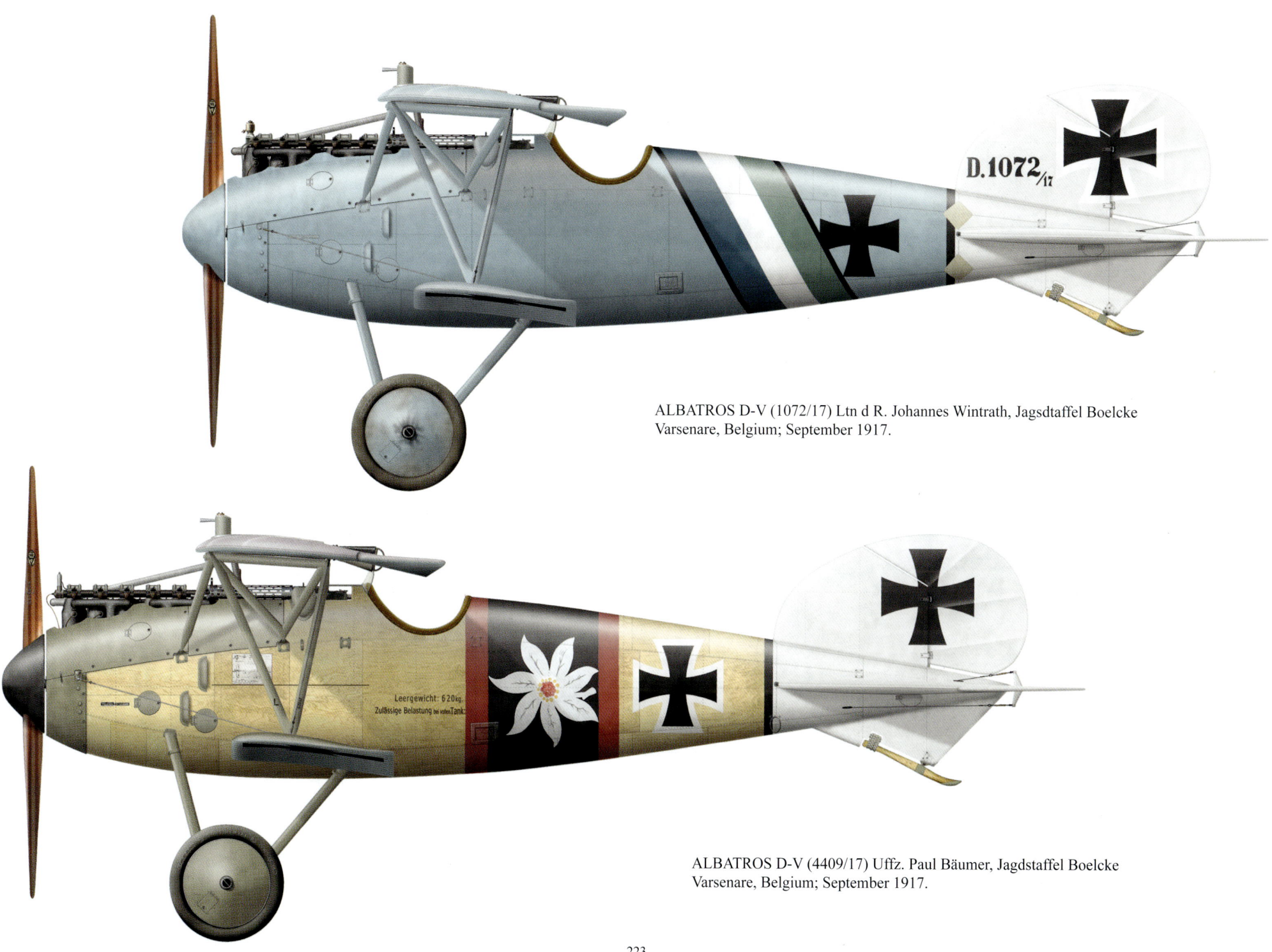

ALBATROS D-V (1072/17) Ltn d R. Johannes Wintrath, Jagsdtaffel Boelcke
Varsenare, Belgium; September 1917.

ALBATROS D-V (4409/17) Uffz. Paul Bäumer, Jagdstaffel Boelcke
Varsenare, Belgium; September 1917.

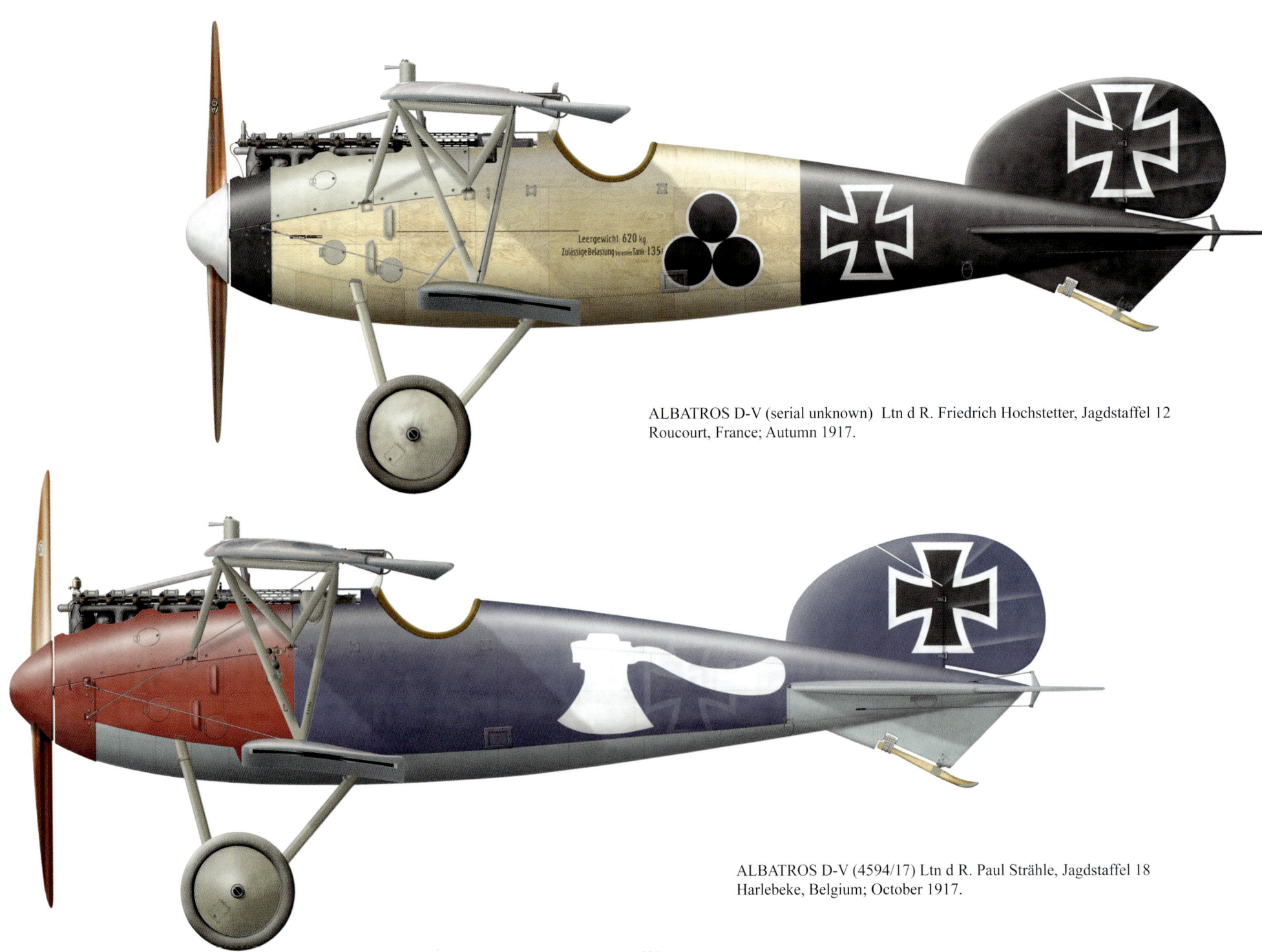

ALBATROS D-V (serial unknown) Ltn d R. Friedrich Hochstetter, Jagdstaffel 12
Roucourt, France; Autumn 1917.

ALBATROS D-V (4594/17) Ltn d R. Paul Strähle, Jagdstaffel 18
Harlebeke, Belgium; October 1917.

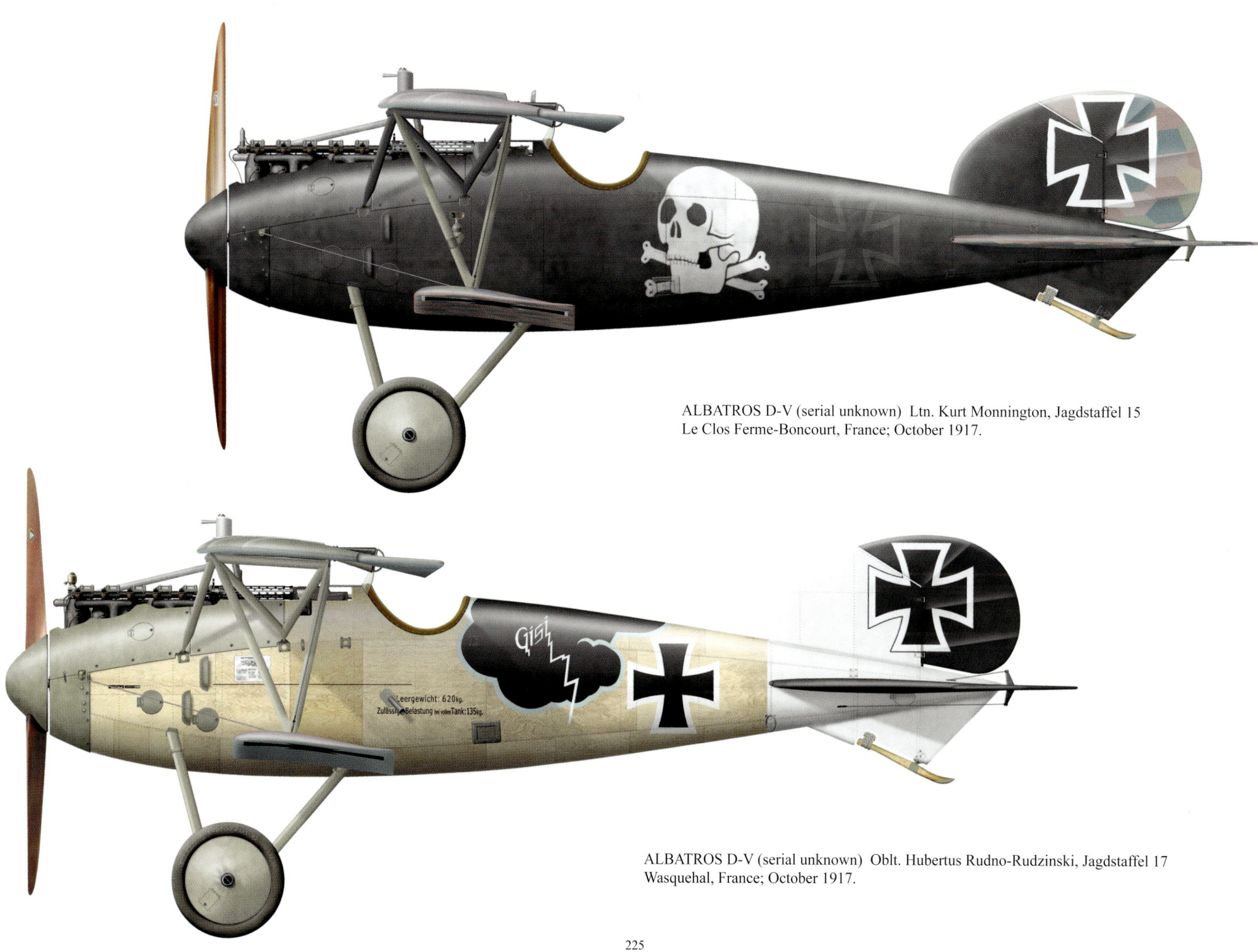

ALBATROS D-V (serial unknown) Ltn. Kurt Monnington, Jagdstaffel 15
Le Clos Ferme-Boncourt, France; October 1917.

ALBATROS D-V (serial unknown) Oblt. Hubertus Rudno-Rudzinski, Jagdstaffel 17
Wasquehal, France; October 1917.

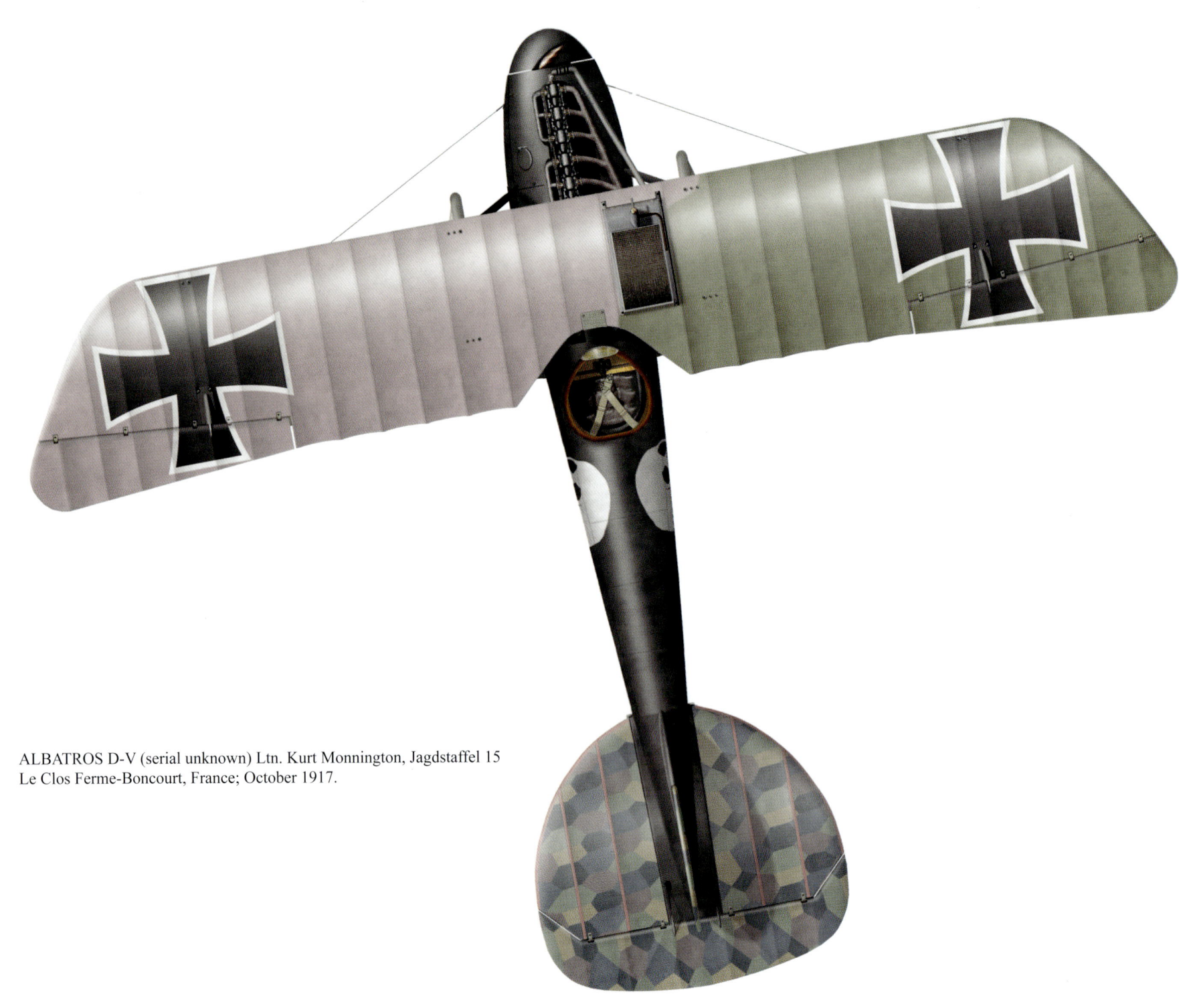

ALBATROS D-V (serial unknown) Ltn. Kurt Monnington, Jagdstaffel 15
Le Clos Ferme-Boncourt, France; October 1917.

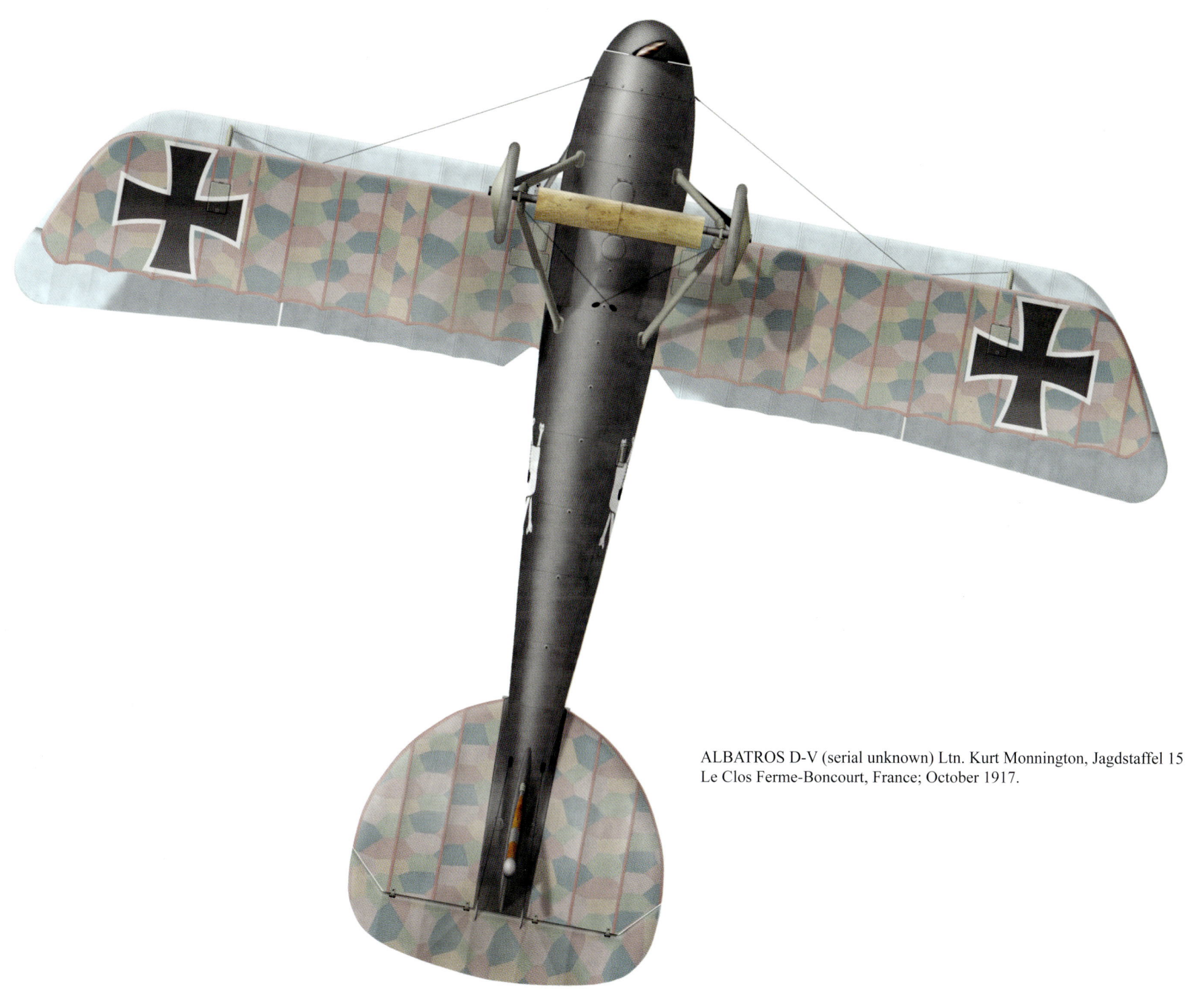

ALBATROS D-V (serial unknown) Ltn. Kurt Monnington, Jagdstaffel 15
Le Clos Ferme-Boncourt, France; October 1917.

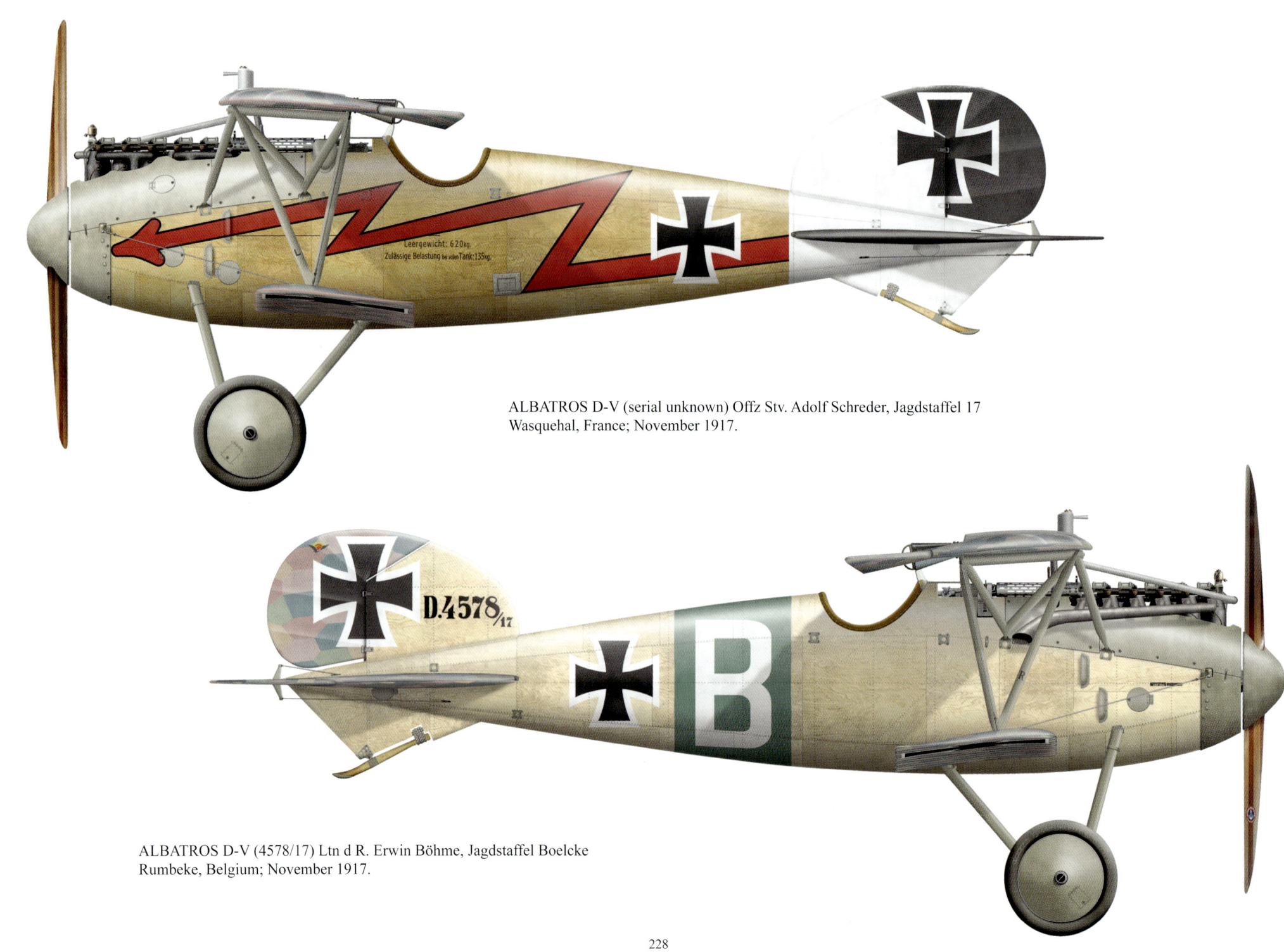

ALBATROS D-V (serial unknown) Offz Stv. Adolf Schreder, Jagdstaffel 17
Wasquehal, France; November 1917.

ALBATROS D-V (4578/17) Ltn d R. Erwin Böhme, Jagdstaffel Boelcke
Rumbeke, Belgium; November 1917.

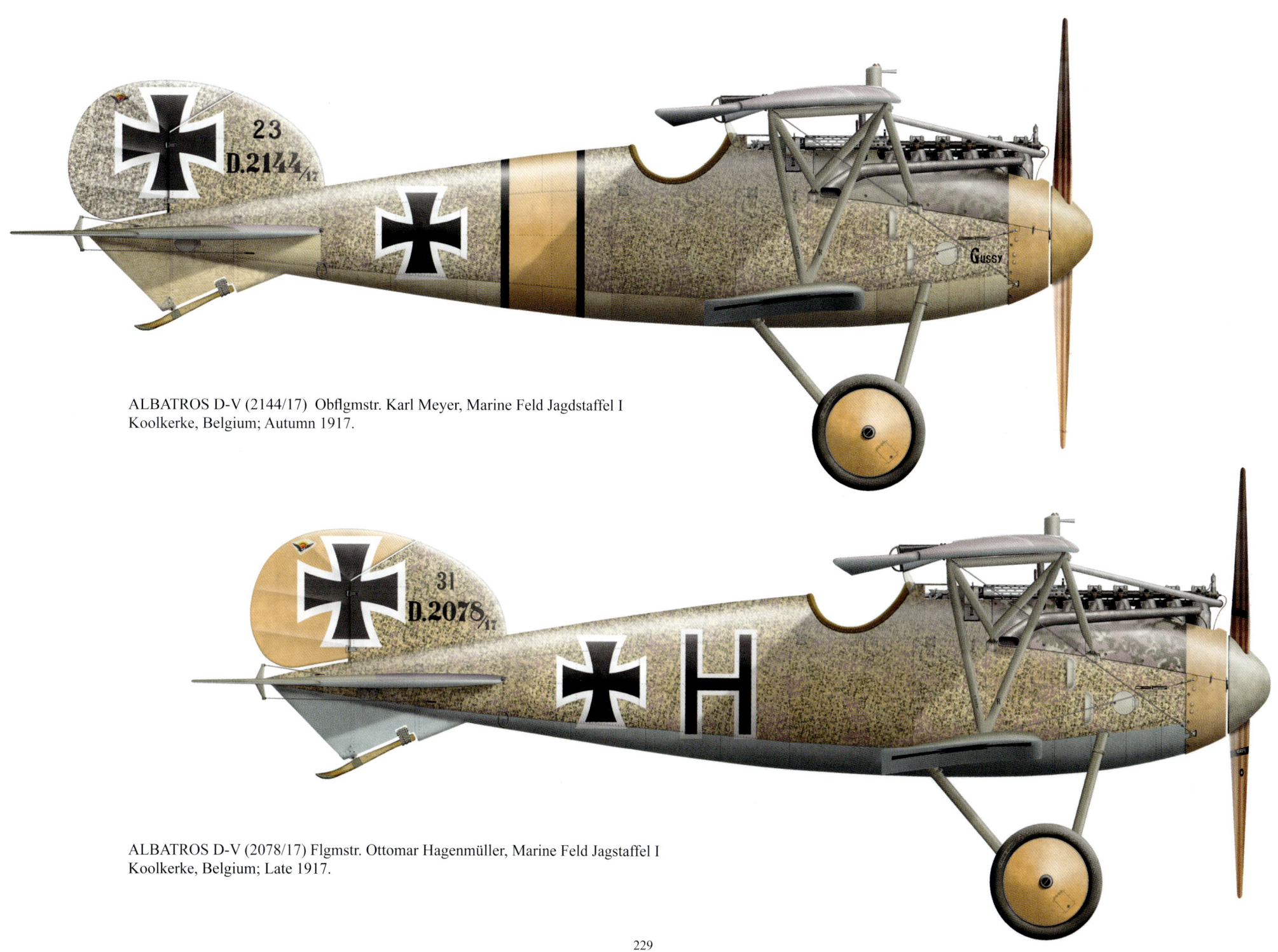

ALBATROS D-V (2144/17) Obflgmstr. Karl Meyer, Marine Feld Jagdstaffel I
Koolkerke, Belgium; Autumn 1917.

ALBATROS D-V (2078/17) Flgmstr. Ottomar Hagenmüller, Marine Feld Jagstaffel I
Koolkerke, Belgium; Late 1917.

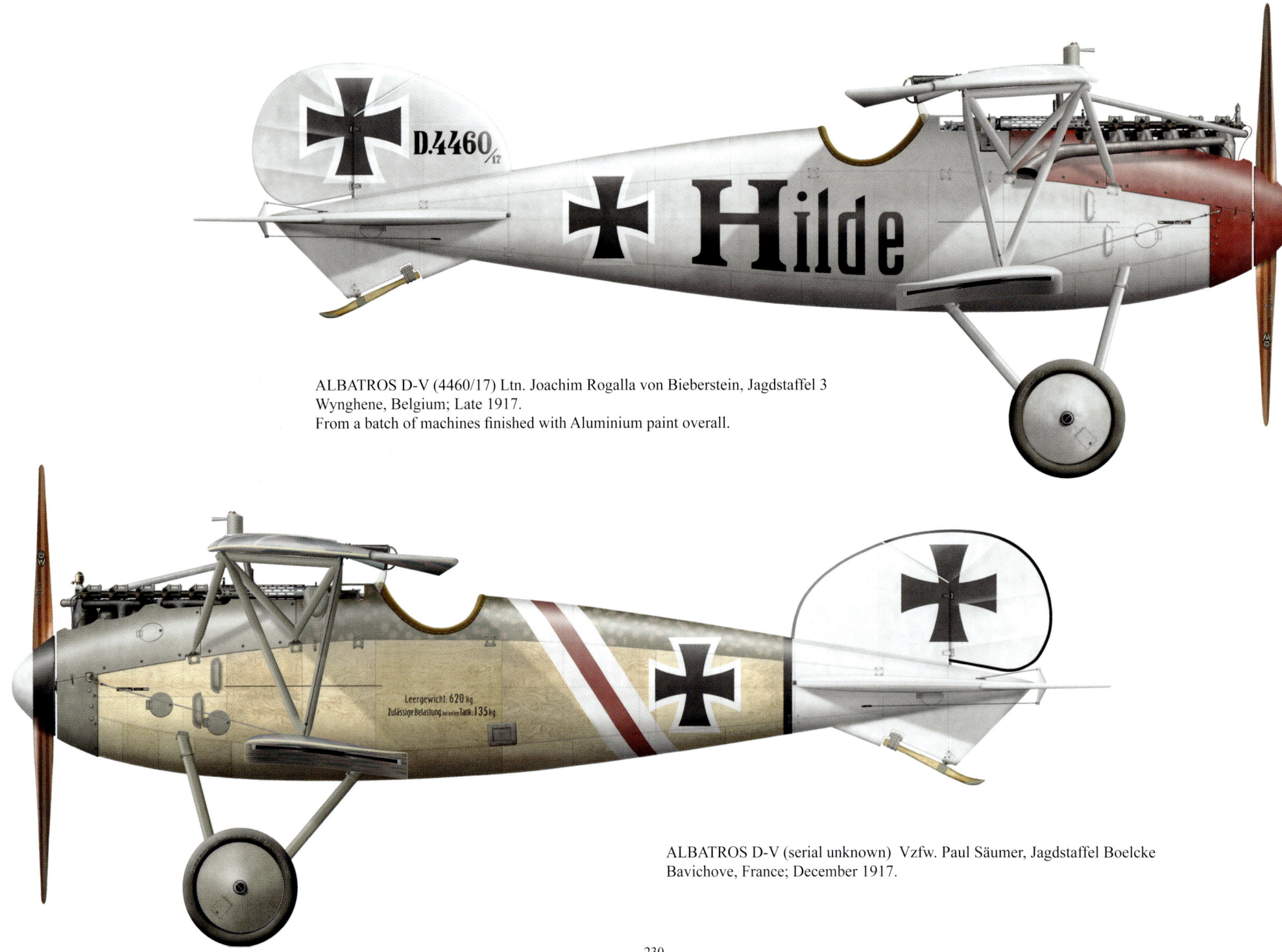

ALBATROS D-V (4460/17) Ltn. Joachim Rogalla von Bieberstein, Jagdstaffel 3
Wynghene, Belgium; Late 1917.
From a batch of machines finished with Aluminium paint overall.

ALBATROS D-V (serial unknown) Vzfw. Paul Säumer, Jagdstaffel Boelcke
Bavichove, France; December 1917.

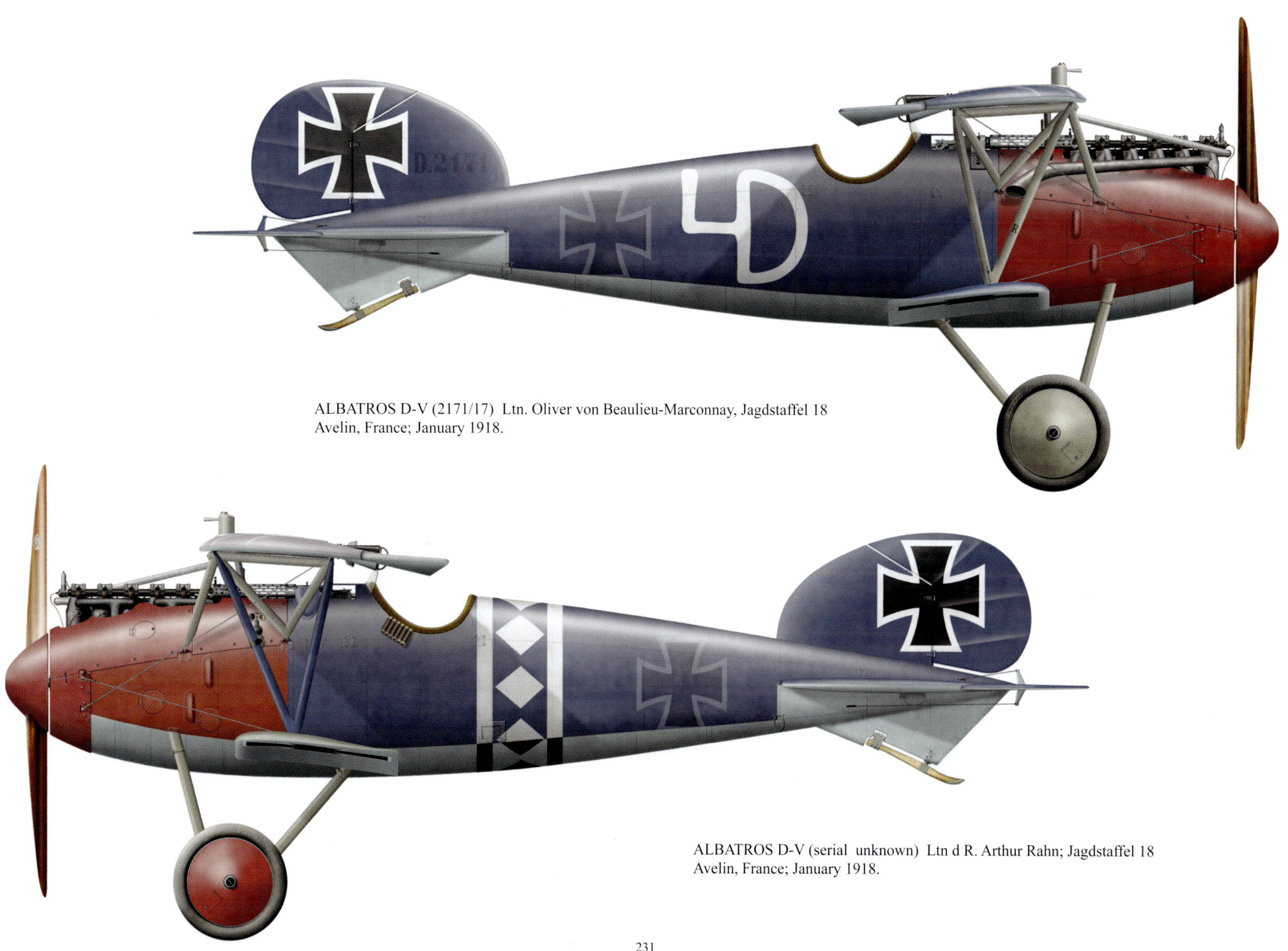

ALBATROS D-V (2171/17) Ltn. Oliver von Beaulieu-Marconnay, Jagdstaffel 18
Avelin, France; January 1918.

ALBATROS D-V (serial unknown) Ltn d R. Arthur Rahn; Jagdstaffel 18
Avelin, France; January 1918.

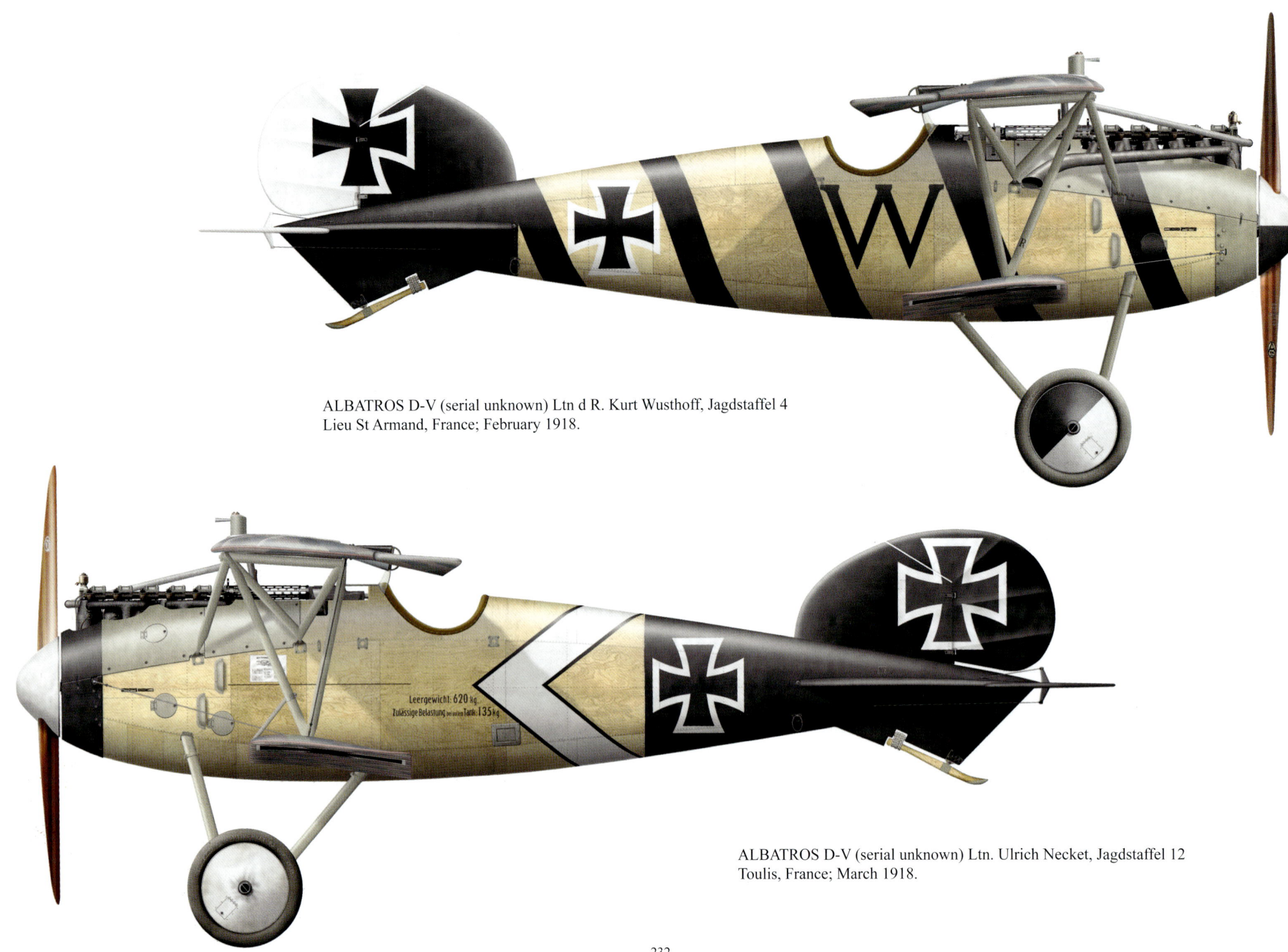

ALBATROS D-V (serial unknown) Ltn d R. Kurt Wusthoff, Jagdstaffel 4
Lieu St Armand, France; February 1918.

ALBATROS D-V (serial unknown) Ltn. Ulrich Necket, Jagdstaffel 12
Toulis, France; March 1918.

ALBATROS D-V (serial unknown) Vzfw. Fritz Rumey, Jagdstaffel 5
Boistrancourt, France; March 1918.

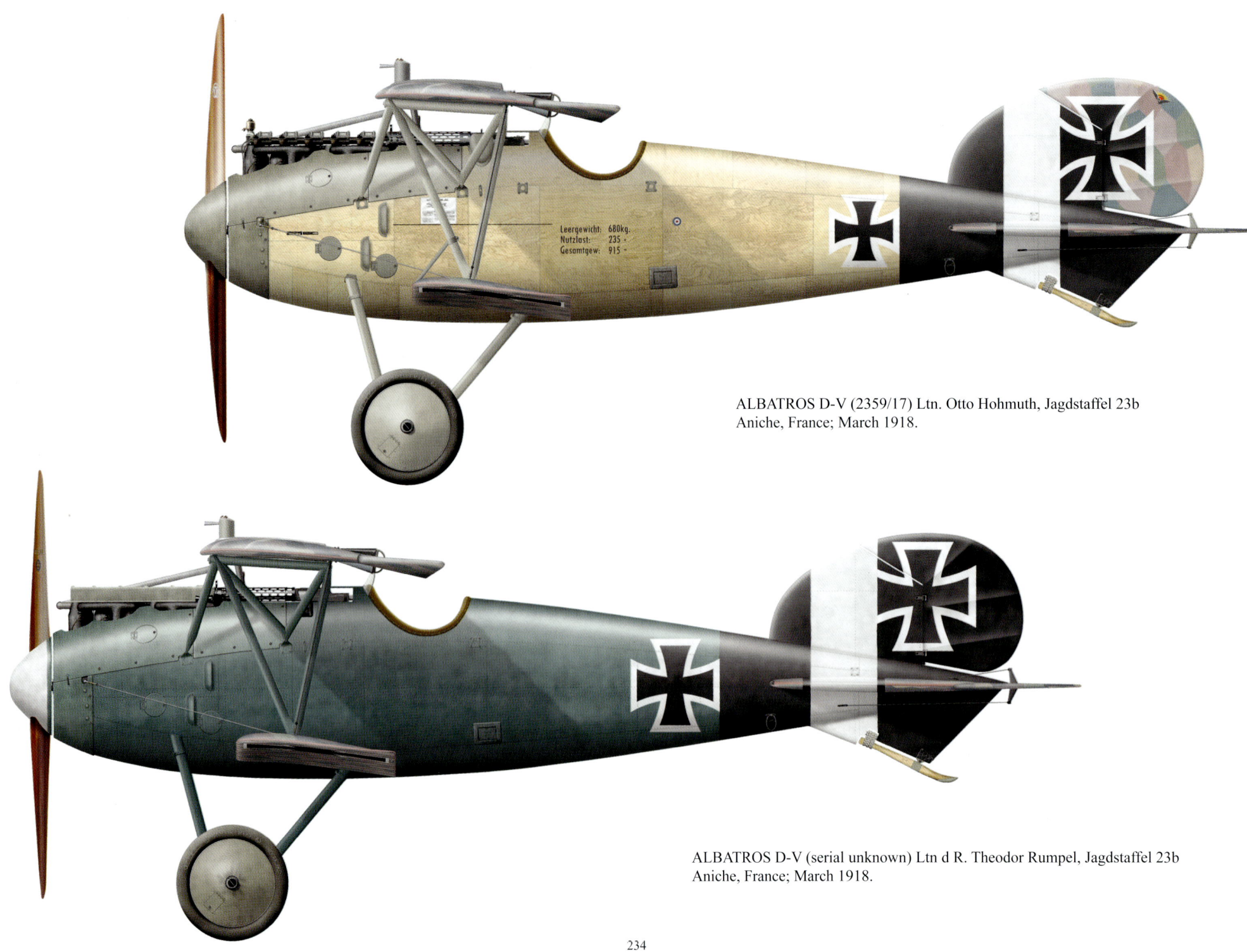

ALBATROS D-V (2359/17) Ltn. Otto Hohmuth, Jagdstaffel 23b
Aniche, France; March 1918.

ALBATROS D-V (serial unknown) Ltn d R. Theodor Rumpel, Jagdstaffel 23b
Aniche, France; March 1918.

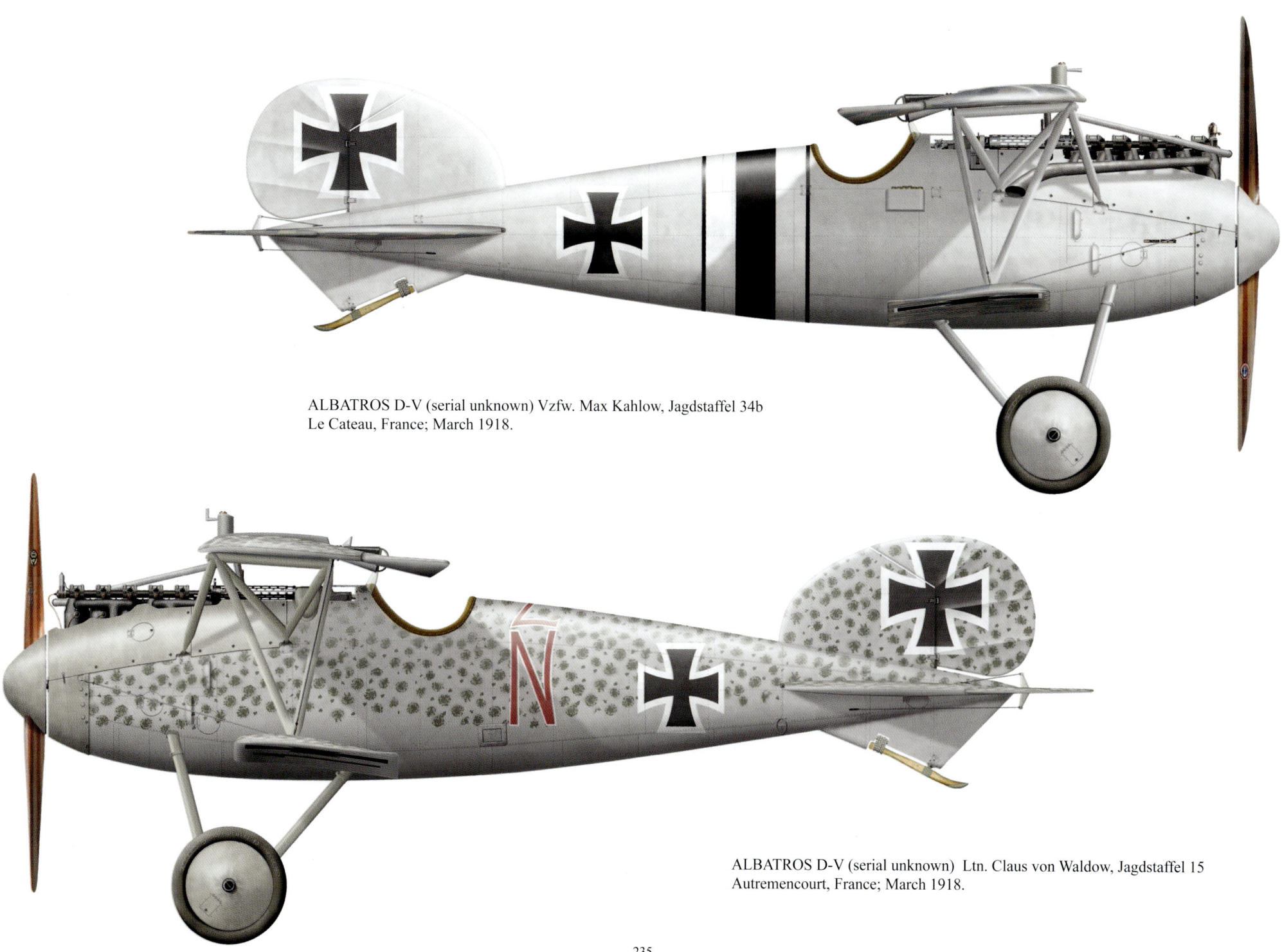

ALBATROS D-V (serial unknown) Vzfw. Max Kahlow, Jagdstaffel 34b
Le Cateau, France; March 1918.

ALBATROS D-V (serial unknown) Ltn. Claus von Waldow, Jagdstaffel 15
Autremencourt, France; March 1918.

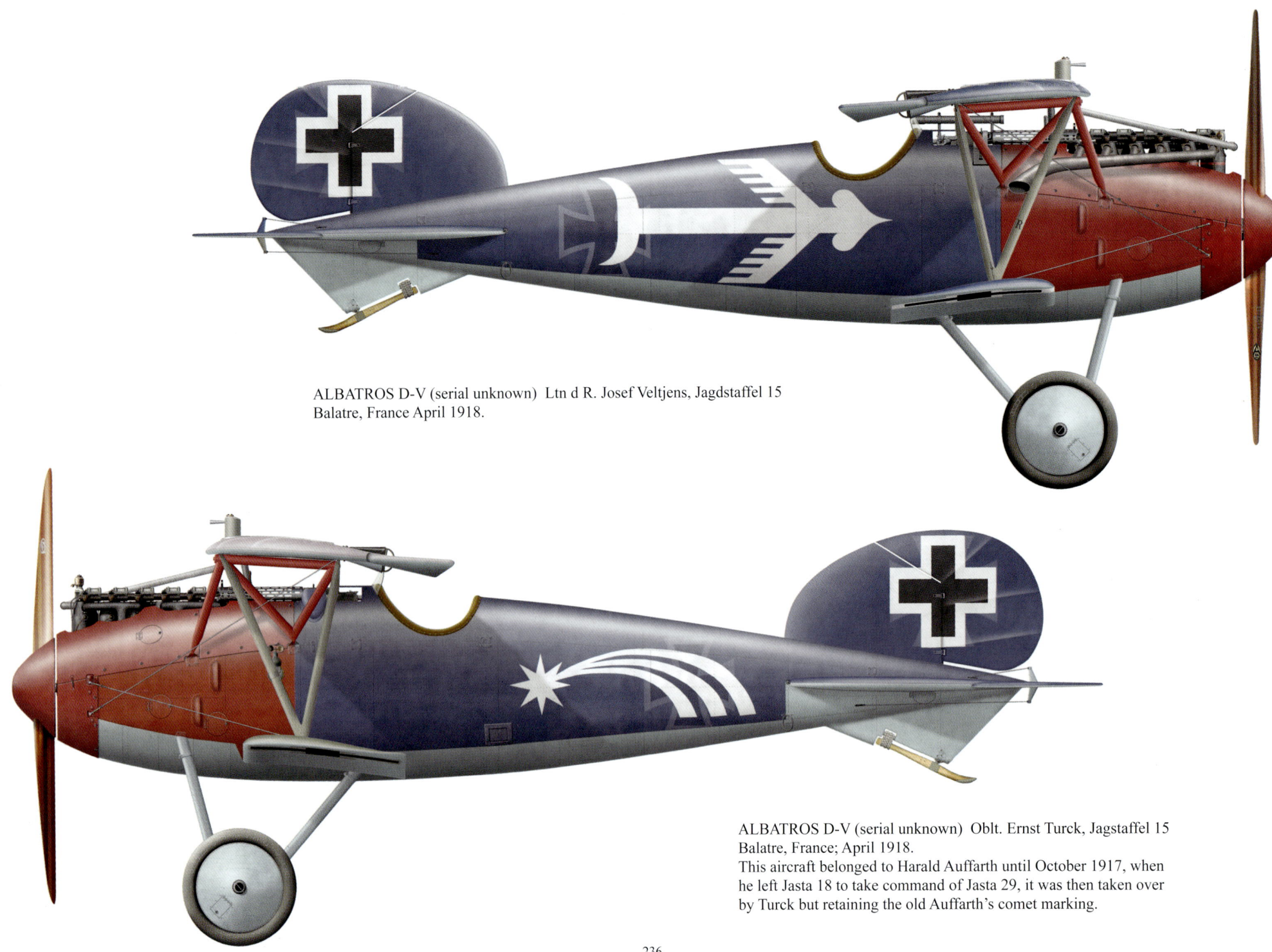

ALBATROS D-V (serial unknown) Ltn d R. Josef Veltjens, Jagdstaffel 15 Balatre, France April 1918.

ALBATROS D-V (serial unknown) Oblt. Ernst Turck, Jagdstaffel 15 Balatre, France; April 1918.
This aircraft belonged to Harald Auffarth until October 1917, when he left Jasta 18 to take command of Jasta 29, it was then taken over by Turck but retaining the old Auffarth's comet marking.

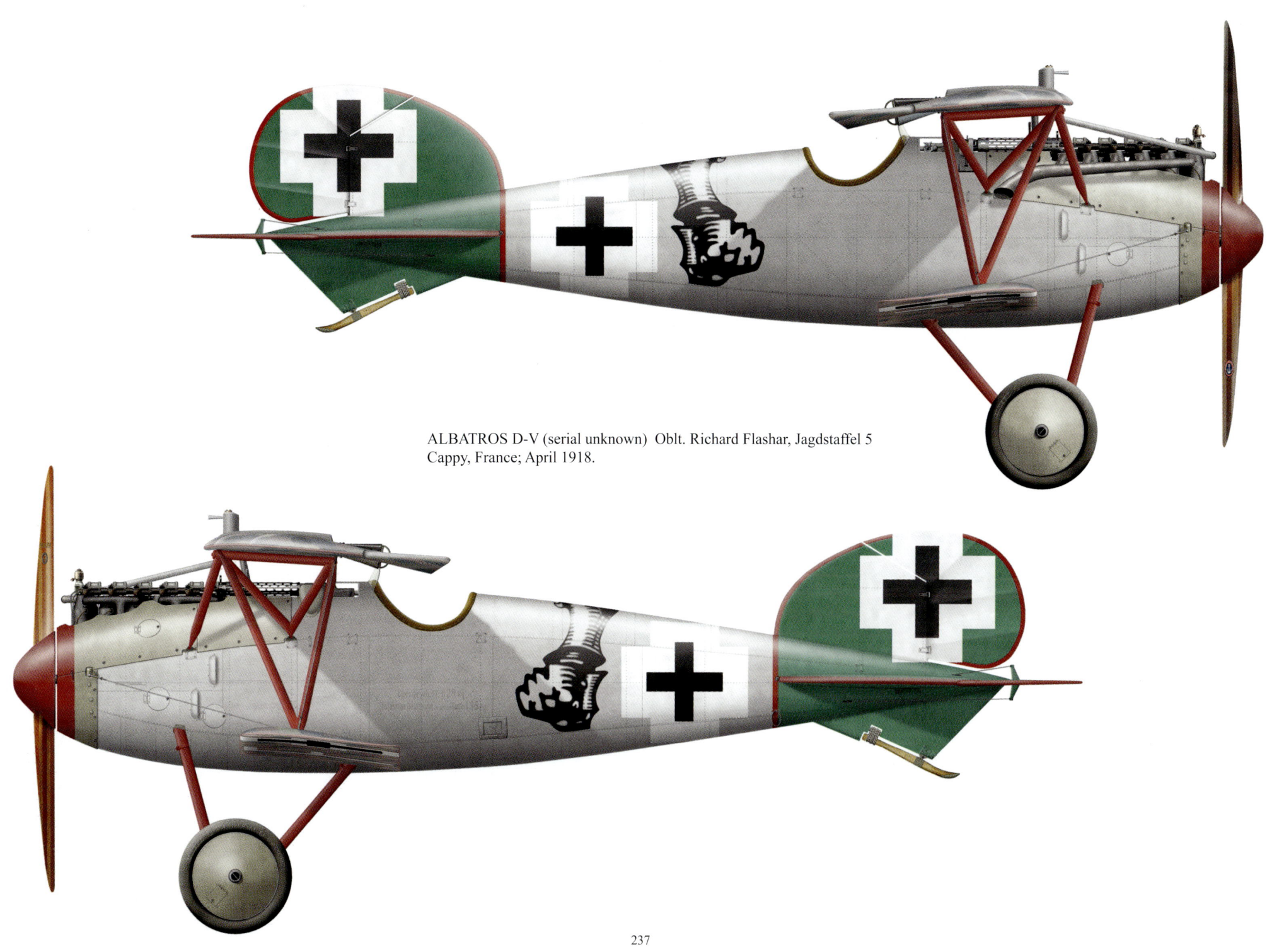

ALBATROS D-V (serial unknown) Oblt. Richard Flashar, Jagdstaffel 5
Cappy, France; April 1918.

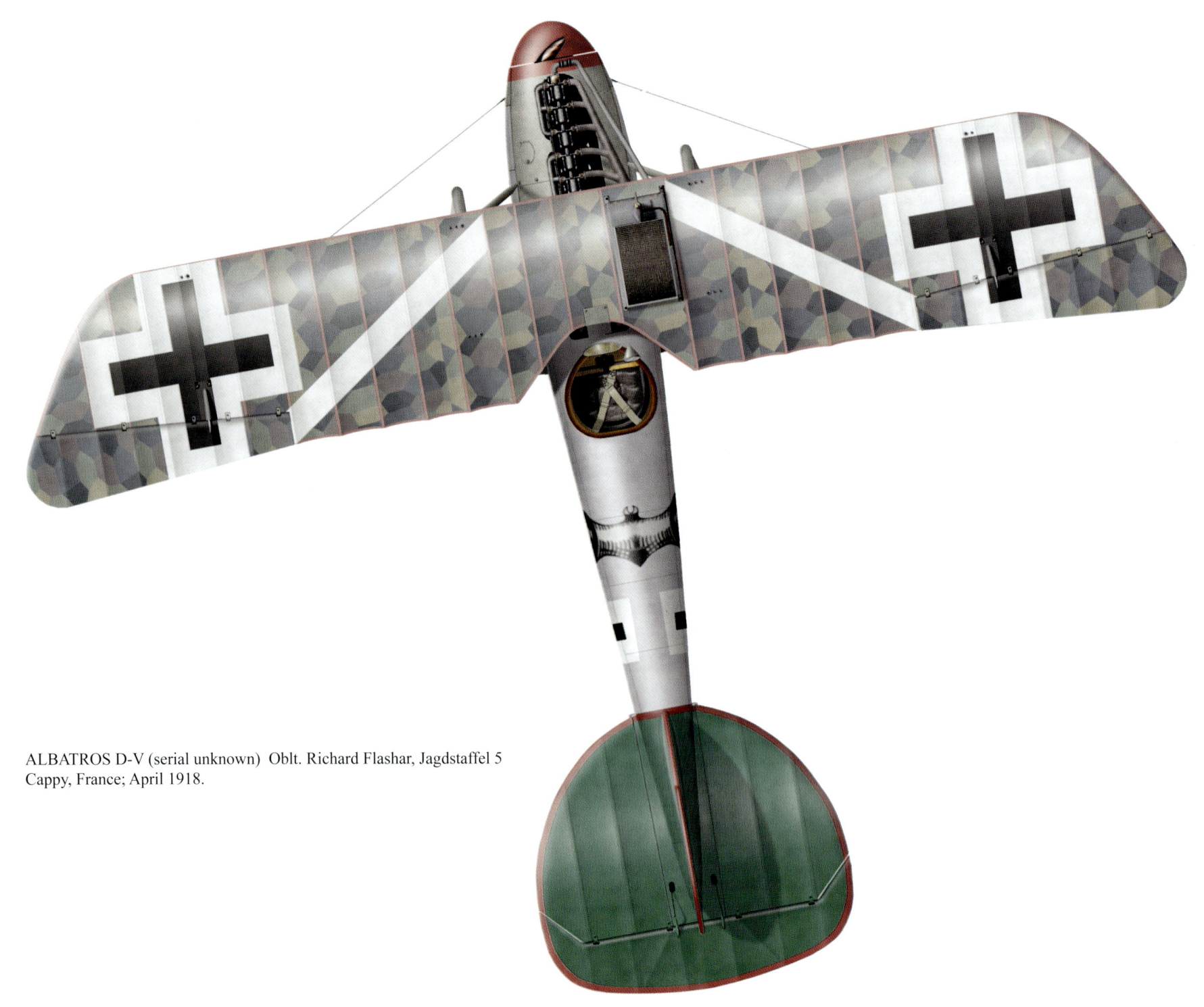

ALBATROS D-V (serial unknown) Oblt. Richard Flashar, Jagdstaffel 5
Cappy, France; April 1918.

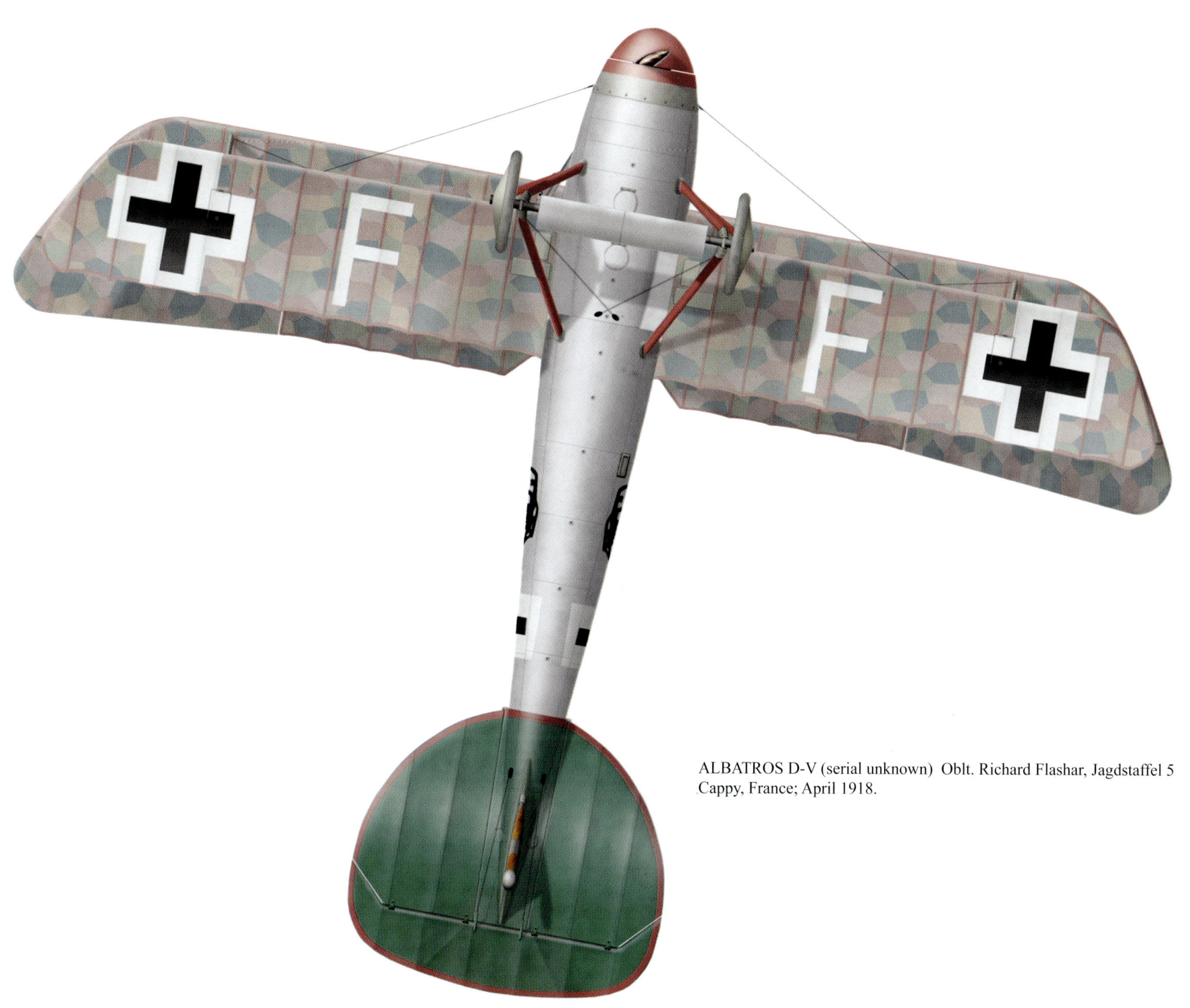

ALBATROS D-V (serial unknown) Oblt. Richard Flashar, Jagdstaffel 5
Cappy, France; April 1918.

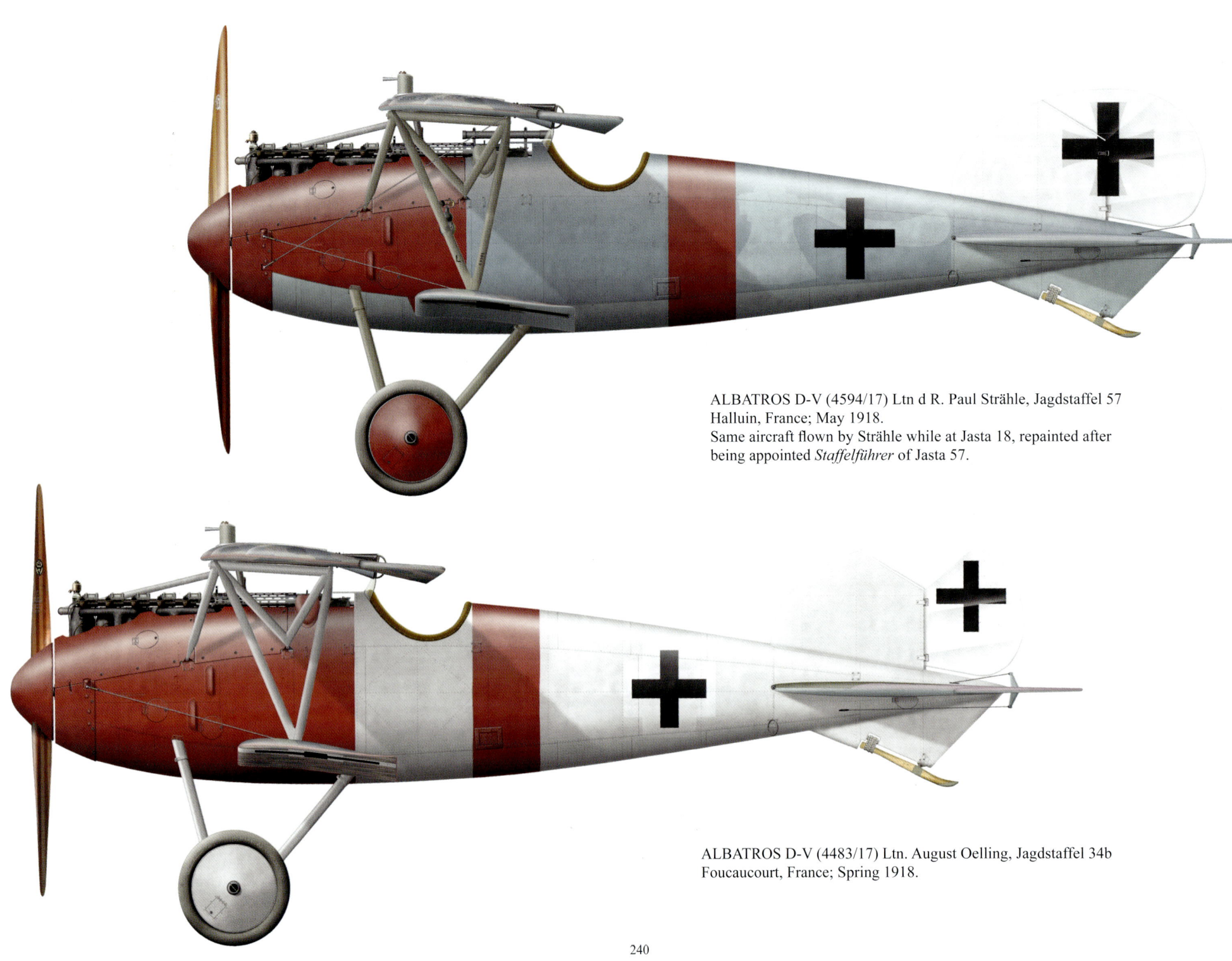

ALBATROS D-V (4594/17) Ltn d R. Paul Strähle, Jagdstaffel 57
Halluin, France; May 1918.
Same aircraft flown by Strähle while at Jasta 18, repainted after being appointed *Staffelführer* of Jasta 57.

ALBATROS D-V (4483/17) Ltn. August Oelling, Jagdstaffel 34b
Foucaucourt, France; Spring 1918.